# SHAPE THEORY

## Categorical Methods of Approximation

J.-M. Cordier

*Université de Picardie-Jules Verne*
*Faculté de Mathématiques et d'Informatique*
*Amiens, France*

T. Porter

*Emeritus Professor,*
*School of Computer Science*
*University of Bangor*
*Bangor, Wales, UK*

DOVER PUBLICATIONS, INC.
Mineola, New York

*Bibliographical Note*

This Dover edition, first published in 2008, is an unabridged republication of the work originally published in 1989 by Ellis Horwood Limited, Chichester, England.

*Library of Congress Cataloging-in-Publication Data*

Cordier, J.-M.

Shape theory : categorical methods of approximation / J.-M. Cordier, T. Porter. — Dover ed.

p. cm.

Originally published: Chichester, West Sussex, England : Ellis Horwood ; New York : Halsted Press, 1989; in series: Mathematics and its applications. Numerical analysis, statistics, and operational research.

Includes index.

ISBN-13: 978-0-486-46623-1

ISBN-10: 0-486-46623-X

1. Shape theory (Topology). I. Title.

QA474.J6 2008

512'.22—dc22

2007011655

www.doverpublications.com

# Table of Contents

# Introduction

What is shape? What is form? To say that two objects have the same shape has an intuitively obvious, but very imprecise meaning. As was pointed out by Lord and Wilson in their preface to [69], a mathematics of form description and analysis is greatly needed. This monograph summarises a theory that may help towards this goal, at least in the study of irregular shapes.

The majority of the techniques of geometric pattern description require that the objects being studied be smooth and fairly regular. Methods from differential geometry, for instance, require smoothness whereas algebraic topological methods require that the object, whether a physical or an abstract one, may be built up from cells or simplices (see Spanier [96]). Increasingly, these methods have been applied, to varying extents, to diverse practical problems of shape, and pattern description and analysis (see Faux and Pratt [39] and Gasson [46]).

Naturally occurring objects are rarely smooth and, as the work on fractals has shown, are by their very nature irregular. Within the abstract setting also, objects frequently occur that can be arbitrarily irregular, for instance closed bounded subsets of an Euclidean space. Is it possible to extend methods used in the study of smooth or regular geometric objects to such as these? With regard to the methods of algebraic topology, the answer is positive and the resulting theory is known as shape theory.

Geometric shape theory can be seen as an extension of the methods of algebraic topology to arbitrary spaces, but its general methodology is much more widely applicable and the main aim of this monograph is the study of that shape theoretic methodology. We hope to show that this methodology embodies its own form of logic, a logic that is fascinating in its own right.

Geometric shape theory as developed by Borsuk and others exemplifies a process which is common in mathematical reasoning. Typically one has a class of objects on which one has a reasonably complete set of information. In the case of Borsuk's shape theory, these objects are the finite polyhedra. This class is considered as a class of 'models' or 'prototypes' within a larger class of objects of interest; in Borsuk's shape theory, this larger class consists of the compact metric spaces. The aim of the exercise is to use approximations to the objects of interest by models to study the objects of the larger class. One may, for instance, seek to extend invariants known to give good information on the models, to be applicable to the larger class of objects. The best classical example of this is, in topology, probably the definition of Čech homology groups extending the simplicial homology groups of polyhedra.

We have used the term 'approximation' but this clearly implies some means of comparing objects. The convenient way to do this is by replacing the bare structure of the classes of interest or of models by categories, thus enabling comparison of objects and of models. Thus we suppose we have two categories, **A**, of models and **B** of objects of interest. Often **A** is a full subcategory of **B** but it is useful for the development of the theory to free oneself from this restriction to specify a functor $\mathbf{K}: \mathbf{A} \to \mathbf{B}$. In Borsuk's geometric shape theory, **A** is the homotopy category of finite polyhedra, and **B** is the homotopy category of compact metric spaces.

Borsuk's theory works well because of the classical result that tells one that any compact metric space can be embedded in the Hilbert cube

$$I^\infty = \prod_{n \in \mathbb{N}} \left[0, \frac{1}{n}\right].$$

Such a compact metric space is thus an intersection of polyhedral subspaces of $I^\infty$, i.e. one can give a reasonable geometric sense to the notion of approximating an object by models in this case.

A somewhat similar situation occurs in the theory of topological groups. It was known in the 1920s from the work of von Neumann and Weyl that every compact group could be 'approximated' by compact linear groups and therefore by compact Lie groups. The philosophy, used by Gleason, Montgomery, Iwasawa and others, was thus to extend invariants of Lie groups to locally compact groups (see the article by Hofmann [57] for a categorical treatment of this area of topological algebra).

A similar but dual philosophy was used in Galois theory. There, the study of field extensions could be, to some extent, reduced to that of studying all the finite field extensions within the given one. Thus the Galois group that resulted was a limit of the Galois groups of the finite intermediate algebraic extensions. One arrived at the category of profinite groups, and, for instance, Serre's famous work on Galois cohomology in part consisted of trying to extend the invariants defined on discrete finite groups to ones on this useful larger category of profinite groups.

Leaving pure mathematics aside for the moment, there are strong indications that the same ideas provide the mathematical basis for a methatheory of pattern recognition. Such a metatheory also provides a qualitative model for certain aspects of perception. At present the theory directly derivable from the material presented here is too simple to give an adequate description of the deeper phenomena of perception, but even at this simple level it does give a useful insight into classification and recognition problems. The problem with this theory is that its deterministic nature does not allow one to describe the stochastic behaviour observable in naturally occurring systems. Similarly for automatic pattern recognition, it would be necessary to consider optimal strategies and stochastic searches for the models (systems of approximations etc.) to be of significant practical use. However, notwithstanding these points, it would seem that many of the ideas from this approach have considerable relevance to perception theory and pattern recognition. It is worth noting that certain workers in pattern recognition have expressed the opinion that there may be a useful link between the ideas and methods of geometric shape theory and the complex problems of automatic pattern recognition (see Pavel [83]).

Thus the basic idea of approximation by models and extension of invariants is a recurrent one in mathematics and its applications. How can one create a formal categorical theory of such situations? To do so would allow one to formalise the various processes involved and hence hopefully to see the extent to which these processes might be applied in new situations. Of course, it would also show the limitations on the use of such methods. In this work we set out the basic category theory of such 'shape theoretic' situations. We cannot be exhaustive since although interest in geometric shape theory has perhaps waned since the peak of activity in the mid-1970s, the categorical aspect is still producing a fairly constant stream of results on extensions of the theory to wider contexts.

Before we give a chapter-by-chapter breakdown of the book, let us first give some brief idea of how the theory goes and how we have structured the description of that theory.

One thus starts with a functor

$$\mathbf{K}: \mathbf{A} \to \mathbf{B}.$$

If $B$ is an object of $\mathbf{B}$, one can form the comma category $B\downarrow\mathbf{K}$ whose objects are pairs $(B \xrightarrow{f} \mathbf{K}A, A)$ consisting of a morphism with codomain in the image of $\mathbf{K}$ and a corresponding object of $\mathbf{A}$. A morphism from $(f, A)$ *to* $(g, A')$ is a morphism $a: A \to A'$ such that $\mathbf{K}a \circ f = g$. If $B \xrightarrow{h} B'$ is a morphism in $\mathbf{B}$, there is an induced functor, $h^*: B'\downarrow\mathbf{K} \to B\downarrow\mathbf{K}$ obtained by composition in an obvious way. This functor preserves the codomain, $h^*(f, A) = (fh, A)$. Taking up this property one introduces new morphisms between objects in $\mathbf{B}$, namely functors preserving codomains between the corresponding comma categories. This gives a shape category, $\mathbf{S_K}$. One of the main problems is to link up properties of $\mathbf{K}$ with manageable descriptions of $\mathbf{S_K}$.

To illustrate this, consider two examples. In the first, let $\mathbf{A}$ be the category of groups and $\mathbf{B}$ the category of sets with $\mathbf{K}$ the forgetful functor between them. In this case $\mathbf{S_K}$ is isomorphic to the category of free groups. This behaviour is typical; if $\mathbf{K}$ has a left adjoint, $\mathbf{S_K}$ is a category of free algebras of some sort. The second case is when $\mathbf{A}$ is the category of finite groups and $\mathbf{B}$ that of all groups with $\mathbf{K}$ the inclusion. Now $\mathbf{K}$ does not have an adjoint, but one can still describe $\mathbf{S_K}$ quite easily, as it is a subcategory of the category of profinite groups. This appearance of projective systems, or as we shall call them 'pro-objects', in a category is typical and we will spend a considerable time exploring categories of projective systems thoroughly.

We have assumed that the reader has a reasonable knowledge of category theory. In general the material from, say, the first six chapters of MacLane's book [71] (or any equivalent text) should suffice. Although we have mentioned other examples, we have concentrated our attention on geometric shape theory as a case study, for motivation and illustration. This is usually because the ideas have developed or have been abstracted from that theory, but it is also relevant to note that both authors of this book come to the categorical form of shape theory from the geometric one. Apart from these considerations, it should also be mentioned that the geometric form of shape does have a considerable geometric and intuitive impact even if we have chosen not to emphasize this to any great extent, feeling that it is already well represented in the shape theoretic literature.

We now turn to a chapter-by-chapter description of the book. To understand the categorisation of shape theory, it is necessary to have at least some slight knowledge of the geometric form. We give a brief introduction to this in Chapter 1. The important process to watch for in this theory is that of the passage to a more abstract categorical approach via ANR systems. Unless the reader wishes to study the techniques used in detail, this section can be perused lightly at a first reading; however, the detailed constructions will be needed later on when we check that the categorisation process has encapsulated the original geometric information.

In Chapter 2, we introduce the details of the construction of $\mathbf{S_K}$ sketched out earlier in this introduction, and relate it to an abstract definition of a shape theory. After a set of examples of various shape theories, there is a long section giving the detailed theory of procategories. Although this theory is essentially well known, it is difficult to find a readily accessible treatment of it. This is followed by a section linking up procategories with shape theory. We then turn to the problem of extending invariants from the models to the shape category. It should come as no surprise that the process involves Kan extensions. Kan extensions again are important in the last section of this chapter where codensity monads are used to give a neat categorical description of $\mathbf{S_K}$ in many cases.

Chapter 3 returns to the geometric base treating simplicial complexes and numerable covers before turning to Morita's form of shape theory and making a comparison between the categorical form and the various geometric versions of shape theory.

We mentioned earlier that if $\mathbf{K}$ had a left adjoint then $\mathbf{S_K}$ was a category of free algebras in a certain sense. This same idea plays a rôle in the discussion involving codensity monads in Chapter 2 and is taken up in Chapter 4 using Bénabou's theory of distributors. This theory allows one to add to the category of small categories formal adjoints to all functors; hence the basic ideas coming from codensity monads etc. can be applied in a completely general context. This allows one to analyse completely the conditions on $\mathbf{K}$ which imply that $\mathbf{S_K}$ is part of an abstract shape theory in the sense of Chapter 2.

An obvious question to ask is: '*How does the shape category change if one changes the models?*' This is investigated in Chapter 5 using the theory of exact squares. This latter theory is closely linked with Kan extensions.

In the final chapter, we introduce the notion of a stable object. An object $B$ is stable if $B\downarrow\mathbf{K}$ has an initial object. Stability leads, in the topological case, to some very useful properties involving Čech homology sequences.

Working in categories of pro-objects in an Abelian category, we then use results of Verdier to analyse stability in this context where additivity and the existence of kernels and cokernels make life easier. After a brief look at the derived functors of the limit functor, we turn to the study of movability, a weaker condition than stability that resulted from an analysis by Borsuk of the systems of neighbourhoods of certain spaces that resembled polyhedra in some of their aspects. In the final section we introduce a stronger version of movability and prove that in many cases it gives an 'internal' characterisation of stability.

As mentioned earlier, we have not attempted to be exhaustive even on the categorical side and certainly not on the geometric side. We recommend that any reader who wishes to pursue the geometric form of shape theory consult the books by Borsuk [13] and Mardešić and Segal [78] and for a shorter study, the lecture notes of Dydak and Segal [32].

We would like to thank the British Council and our universities for the help that enabled this synthesis to be made; also, of course, the several typists at Bangor, who had the ungrateful task of typing the manuscript, and finally Professors Andrée Charles Ehresmann and Ronald Brown for their encouragement.

# Advice to the Reader

This monograph can, of course, be used in various ways, depending on the demands placed on it by the reader.

First, for the reader who has a reasonable knowledge of (algebraic) topology, the monograph aims to provide the necessary categorical machinery for a thorough treatment of the categorical foundations of shape theory. This approach would be invaluable for researchers intending to work in the newly developing branch of strong shape theory, since although it is only slowly emerging what the problems of that new subject will be, it is increasingly clear that a very categorical approach will be necessary for any real progress to be made. For such a reader, the best course through the book will be a direct one. The main topological emphasis is to be found in Chapters 1, 3 and 6, but particularly useful material (i.e. the detailed study of procategories etc.) is found in Chapter 2.

The second type of reader is someone with a good categorical background, for instance a worker in topos theory who wishes to see how algebraic topological ideas might be extended from locally nice spaces to all spaces and thence to locales and toposes. Such a reader will need to use the topological Chapters 1 and 3 somewhat differently. The type of topological argument involved can seem very dry to a non-topologist so such a reader should try to read Chapter 1, in particular 'once over lightly' and then use it as a reference when later in the monograph topological ideas are needed to check that the categorical formulation has faithfully encoded the topological problems. Remember the topological examples are intended as a case study of the more widely applicable ideas within the categorical framework.

A third class of reader will be someone who has used the approximation theoretic approach in another part of mathematics, e.g. topological algebras. To such a reader the general points of the above advice to the categorically minded apply, with the rider that a knowledge of basic category theory (equivalent to the first five or six chapters of MacLane's book [71]) is assumed.

Finally we could hope and expect there to be readers from a less pure mathematical background. The implications of this general categorical approach for the theory of pattern recognition and perception are as yet largely unexplored. It is hoped that given some knowledge of categorical language the main ideas of relevance to these potential applications can be extracted from Chapters 2, 4 and 5. Although we cannot attempt to provide an introduction to category theory in this monograph we have included at the end of the monograph a brief account of a possible interpretation of the abstract theory in terms of a non-technical language more suited to a description of perception or of pattern or symbol recognition.

# 1

# Borsuk's Shape Theory for Compact Metric Spaces

Before we start on a detailed introduction to Borsuk's shape theory, it will pay to give a brief description of what it aimed to do, why, and how, thus indicating the main ideas to be developed later on in this chapter. We cannot hope to give here more than an idea of the overall theory, but this should not matter as the main point of this chapter is to lay down some of the foundations for our use of the geometric side of shape theory as a 'case-study' in the general ideas behind the use of approximations, and in the general philosophy of shape theory. It should also be pointed out that there are several good introductory articles on geometric shape theory plus at least three books giving fully detailed accounts (namely Borsuk [13], Dydak and Segal [32] and Mardešić and Segal [78]).

In 1978, Borsuk introduced the theory of shape in an article [10]. This aimed to give a classification of compact metric spaces that was coarser than homotopy but which would coincide with homotopy theory on spaces that had reasonable local properties, namely the ANRs (absolute neighbourhood retracts—see section 1.1). These latter spaces have some of the same homotopy theoretic properties as polyhedra, and one can usually replace general ANRs with polyhedra in their application within shape theory.

Shape is often stated to be a sort of Čech homotopy theory. Using the results of Alexandroff, Čech had managed to extend homological and cohomological invariants (in particular the Betti numbers) from polyhedra to arbitrary compact metric spaces. (In so doing he incidentally laid down some of the ideas that were later to lead to the development of category theory.) Borsuk's work led to an extension of many of the ideas of homotopy theory in a somewhat similar way. To illustrate this, consider the following example.

Let $X$ be the circle $S^1$ and $Y$ the space defined as the union of the closure of the graph of $y = \sin(1/x)$, $-1/\pi \leqslant x \leqslant 1/\pi$, $x \neq 0$ and an arc joining $(-1/\pi, 0)$ and $(1/\pi, 0)$ which is disjoint from the graph except at its endpoints.

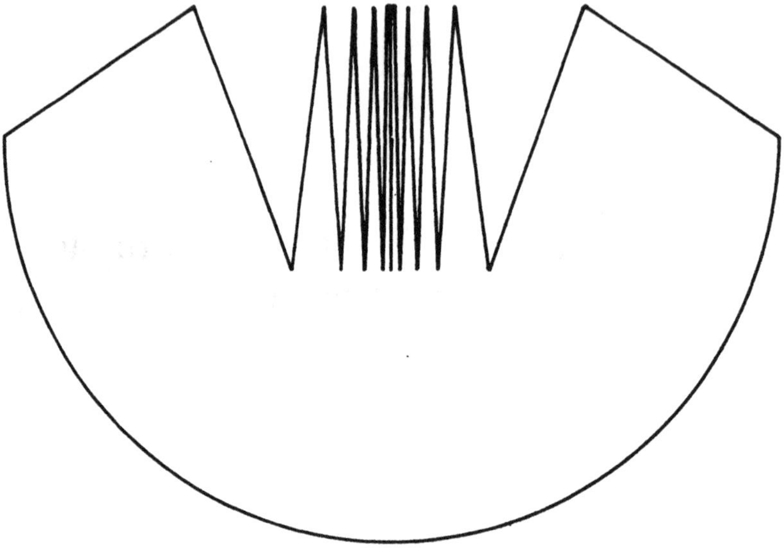

Stylised drawing of $Y$.

Then $X$ and $Y$ considered as subspaces of $\mathbb{R}^2$ have the properties that they both divide $\mathbb{R}^2$ into two components, and in fact any open neighbourhood of $X$ in $\mathbb{R}^2$ is of the same homotopy type as an open neighbourhood of $Y$ in $\mathbb{R}^2$ and vice versa. However, these spaces fail to have the same homotopy type since there are not enough continuous maps from $X$ to $Y$ to make this possible, $Y$ having local singularities. We shall see (in Theorem 1 of section 1.2) that they have the same shape and that this corresponds loosely to the comment just made about their open neighbourhoods in $\mathbb{R}^2$.

The key development of Borsuk was to realise the importance of systems of interrelated mappings between neighbourhoods of spaces as a means of studying the spaces, especially when, as in the above example, because of the presence of local singularities, there are not enough maps for a direct comparison. These systems of mappings were called fundamental sequences by Borsuk (see section 1.2).

The next major development from our point of view was due to Mardešić and Segal [76] (although their ideas like Borsuk's, had first been put forward over 20 years earlier by Christie [17]). Their papers mark the first treatment where categorical notions start playing an important role. They translated Borsuk's treatment into one in terms of inverse sequences of polyhedra. Borsuk's theory strongly depends on the result of Kuratowski and Wojdysławski that any compact metric space can be embedded in the Hilbert cube $I^\infty$ (see the Introduction and section 1.1). By their replacement of fundamental sequences by maps of ANR-sequences, Mardešić and Segal were able to extend Borsuk's theory to apply to general compact Hausdorff spaces. We introduce their ideas in section 1.3.

We would suggest that at a first reading, this chapter is only read 'once over lightly'. To someone without a good basis of topology, the details will no doubt be somewhat heavy going. It would seem advisable therefore to use this chapter as a reference section, referring to it as needed in those later sections which depend on it. It is the basic ideas of approximation to a space by polyhedra and of using the approximating systems as a means of defining generalisations of mappings between the spaces that are the important ones to retain.

## 1.1 ABSOLUTE RETRACTS AND ABSOLUTE NEIGHBOURHOOD RETRACTS

In this section all spaces being considered will be metrisable.

**Definition**

A space $X$ is said to be an *absolute retract* ($X$ is an AR) if for any embedding of $X$ as a closed subspace of a space $Y$, $X$ is a retract of $Y$, that is, there is a continuous mapping, $r: Y \to X$ with $r|X = \mathrm{Id}_X$, the identity on $X$.

**Example**

Any convex subspace of a Banach space is an AR. In particular the Hilbert cube $I^\infty = \prod_{n=1}^{\infty} [0, 1/n]$ is a compact AR.

Any Euclidean space is an AR.

Absolute retracts are characterised as being homeomorphic images of retracts of convex subsets of Banach spaces [9]. One can also characterise them in the following way.

**Proposition 1**

*A space $X$ is an AR if and only if for $Y'$, a closed subset of a space, $Y$, and $f: Y' \to X$ a continuous mapping, there is an extension $f^*$ of $f$ to $Y$.*

**Definition**

A space $X$ is said to be an *absolute neighbourhood retract* ($X$ is an ANR) if for any embedding of $X$ as a closed subspace of a space $Y$, there is a neighbourhood $V$ of $X$ in $Y$ such that $X$ is a retract of $V$.

Clearly any AR is an ANR, and any open subspace of an ANR is an ANR. ANRs can be characterised in two ways as follows· ANRs are exactly the $r$-images of open convex subsets of Banach spaces, that is, if $Y$ is an ANR, there is a convex subset, $C$, of some Banach space, $U$ an open subset of $C$, and a continuous mapping $f: U \to Y$, which is right invertible [9].

**Proposition 2**

*A space $X$ is an ANR if and only if for every closed subspace $Y'$ of a space $Y$ and $f: Y' \to X$ continuous, $f$ has an extension $f^*$ over some neighbourhood of $Y'$ in $Y$.*

One of the most important properties of ANRs that we will need is the homotopy extension property.

**Theorem 1**

*Let $Y'$ be a closed subspace of a space $Y$, and let $H: Y' \times [0, 1] \to X$ be a homotopy, $X$ being an ANR, such that $H|Y' \times \{0\} = f_0$, $H|Y' \times \{1\} = f_1$. If $f_0^*: Y \to X$ is an* *extension of $f$, then $H$ has an extension*

$$H^*: Y \times [0, 1] \to X$$

*joining $f_0^*$ and an extension, $f_1^*: Y \to X$, of $f_1$.*

We will also need two lemmas.

**Lemma 1**

*Let $X$ be an AR, then there is a continuous $\lambda: X \times X \times [0, 1] \to X$ such that*

(1) $\lambda(x, y, 0) = x, \quad \lambda(x, y, 1) = y$ *for all* $(x, y) \in X \times X$

(2) $\lambda(y, y, t) = y$ *for all* $(y, t) \in X \times [0, 1]$.

*Proof*

As $X$ is an AR, we may consider $X$ to be a closed subset of some convex set $C$ in a Banach space. Let $s: C \to X$ be a retraction, then $\lambda(x, y, t) = s((1-t)x + ty)$ satisfies (1) and (2) and is continuous.

**Lemma 2**

*Let $X$ be an AR and $Y$ a closed subspace of $X$ with $Y$ an ANR. If $W$ is a neighbourhood of $Y$ in $X$ and $r: W \to Y$ a retraction, then for every neighbourhood $V$ of $Y$ in $X$, there is a neighbourhood $V'$ of $Y$ in $X$ and $H$ a homotopy from $V' \times [0, 1]$ to $V$ such that*

$$H(y, 0) = y \quad \text{and} \quad H(y, 1) = r(y) \quad \text{for all } y \in V'.$$

*Proof*

Let $\lambda: X \times X \times [0, 1] \to X$ be as in Lemma 1, defined using the retraction $s: C \to X$. Let $H: W \times [0, 1] \to X$ be defined by

$$H(x, t) = \lambda(x, r(x), t),$$

and let $V$ be any neighbourhood of $Y$ in $X$; then there is a neighbourhood $\hat{V}$ of $Y$ in $C$ such that $s(\hat{V}) \subset V \cap W$. For each $y \in Y$, let $B_y$ be an open ball containing $y$, with $B_y \cap C \subset \hat{V}$ and set

$$V' = \bigcup_{y \in Y} (r^{-1}(B_y \cap Y) \cap (B_y \cap X));$$

$V'$ is a neighbourhood of $Y$ in $X$ such that, if $(y', t) \in V' \times [0, 1]$, $H(y', t) \in V$.

**Proposition 3**

*Let $X'$ be a closed subspace of $X$, $Y'$ a closed subspace of $Y$ with $Y'$ an AR; then, if $f_0, f_1$ are two maps from $X'$ to $Y'$ homotopic in some open neighbourhood $V$ of $Y'$ in $Y$ and if $f_0^*, f_1^*$ are two extensions of $f_0$ and $f_1$ from $X$ to $Y$, there exists a neighbourhood $U$ of $X'$ in $X$ such that $f_0^*|U \simeq f_1^*|U$ in $V$ ($f_0^*|U$ is homotopic to $f_1^*|U$ in $V$).*

*Proof*

Let $H$ be the homotopy $H: X' \times [0, 1] \to V$, with $H(x, 0) = f_0(x)$, $H(x, 1) = f_1(x)$ and let $W$ be a closed non-trivial neighbourhood of $X'$ in $X$ such that $f_0^*(W) \cup f_1^*(W) \subset V$.

As $A = (W \times \{0\}) \cup (W \times \{1\}) \cup (X' \times [0, 1])$ is a closed subset of $W \times [0, 1]$ and $\bar{H}: A \to V$ defined by

$$\bar{H}(x, 0) = f_0^*(x), \qquad \bar{H}(x, 1) = f_1^*(x)$$
$$\bar{H}(x, t) = H(x, t) \quad \text{for } (x, t) \in X' \times [0, 1]$$

is continuous, we can find, using the fact that $V$ is an ANR (as an open subset of an AR), an open $U' \subset W \times [0, 1]$ and a continuous extension $H^*$ of $\bar{H}$, $H^*: U' \to V$.

As $X' \times [0, 1] \subset A \subset U' \subset W \times [0, 1]$, there is an open neighbourhood $U$ of $X'$ in $X$ with $U \times [0, 1] \subset U'$. $H^*$ then restricts to a homotopy joining $f_0^*|U$ with $f_1^*|U$ as required.

## 1.2 BORSUK'S SHAPE THEORY

Let $X$ be a metric space, then by a theorem of Kuratowski and Wojdysławski, $X$ is homeomorphic to a closed subspace of a convex subset of a Banach space. As each convex subset is an AR, we can choose for each space $X$, an AR containing $X$ as a closed subset. If $X$ is embedded in an AR, $M$, we will denote by $M(X)$ the set of all neighbourhoods of $X$ in $M$.

**Definition**
Let $X$ and $Y$ be two metric spaces, $X \to M$, $Y \to N$ embeddings with $M$, $N$ absolute retracts. A *fundamental sequence* $\underline{f} = (f_n)_{n \in \mathbb{N}}$ from $X$ to $Y$ is a sequence of continuous mappings from $M$ to $N$ such that given any $V \in N(Y)$, there is a $U \in M(X)$ and $n_0 \in \mathbb{N}$ such that for all $n, n' \geqslant n_0$,

$$f_n|U \simeq f_{n'}|U \text{ within } V.$$

We shall say that two fundamental sequences, $\underline{f} = (f_n)_{n \in \mathbb{N}}$, $\underline{f}' = (f'_n)_{n \in \mathbb{N}}$, are *homotopic* if given any $V \in N(Y)$, there is a $U \in M(X)$ and $n_0 \in \mathbb{N}$ such that for all $n \geqslant n_0$, $f_n|U \simeq f'_n|U$ within $V$.

The homotopy of fundamental sequences is an equivalence relation. We denote by $[\underline{f}]$ the homotopy class of the fundamental sequence, $\underline{f}$.

**Definition**
A fundamental sequence $\underline{f} = (f_n)_{n \in \mathbb{N}}$ is said to be *generated* by $f: X \to Y$ if for all $n \in \mathbb{N}$ and $x \in X$, $f_n(x) = f(x)$.

If $f$ is a continuous map from $X$ to $Y$, then $f$ generates a fundamental sequence: as $N$ is an AR, let $f^*$ be the extension of $f$, $f^*: M \to N$, and let $\underline{f}^* = (f_n)_{n \in \mathbb{N}}$ where $f_n = f^*$ for each $n$; $\underline{f}^*$ is a fundamental sequence generated by $f$.

**Definition**
Let $\underline{f} = (f_n)_{n \in \mathbb{N}}$ and $\underline{f}' = (f'_n)_{n \in \mathbb{N}}$ be two fundamental sequences from $X$ to $Y$, then $\underline{f}$ and $\underline{f}'$ are said to be *associated* if for all $n \in \mathbb{N}$ and $x \in X$,

$$f_n(x) = f'_n(x).$$

**Proposition 1**
*Let $\underline{f}$ and $\underline{f}'$ be two associated sequences, then* $[\underline{f}] = [\underline{f}']$.

*Proof*
Let $V \in N(Y)$; then there is some $U \in M(X)$ and $n_0 \in \mathbb{N}$, such that if $n \geqslant n_0$, $f_{n_0}|U \simeq f_n|U$ and $f'_n|U \simeq f'_{n_0}|U$ within $V$.

As $N$ is an AR, it follows from Lemma 1 of section 1.1 that there is some $\lambda: N \times N \times [0, 1] \to N$, such that for all $(y, t) \in N \times [0, 1]$, $\lambda(y, y, t) = y$. As $f_{n_0}(x) =$

$f'_{n0}(x)$ for all $x \in X$, there is some $U_0 \subset U$ such that for all $(x, t) \in U_0 \times [0, 1]$, $\lambda(f_n(x), f'_{n0}(x), t) \in V$. Let $H$ be the homotopy from $U \times [0, 1]$ to $V$ defined by

$$H(x, t) = \lambda(f_{n0}(x), f'_{n0}(x), t).$$

$H$ is then a homotopy joining $f_{n0}|U_0$ to $f'_{n0}|U_0$; one thus has for all $n \geqslant n_0$,

$$f_n|U_0 \simeq f_{n0}|U_0 \simeq f'_{n0}|U_0 \simeq f'_n|U_0 \text{ within } V.$$

**Corollary**

*If $\underline{f}$ and $\underline{f}'$ are two fundamental sequences generated by a mapping $f: X \to Y$ then $[\underline{f}] = [\underline{f}']$.*

**Proposition 2**

*Let $f, f': X \to Y$ be homotopic. If $f^*$ and $f'^*$ are two extensions from $M$ to $N$ of $f$ and $f'$ respectively, then $[f^*] = [f'^*]$.*

*Proof*

As $f \simeq f': X \to Y$, for any open set $V \in N(Y)$, $f \simeq f': X \to V$ (with the usual abuse of notation). From Proposition 3 of section 1.1, there is a $U \in M(X)$ such that $f^*|U \simeq f'^*|U$ within $V$.

**Proposition 3**

*Let $\underline{f} = (f_n)$ be a fundamental sequence from $X$ to $Y$, where $Y$ is an ANR. Then there is a unique homotopy class, $[f]$, with $f$ a mapping from $X$ to $Y$ such that $[\underline{f}] = [\underline{f}^*]$.*

*Proof*

Let $X$ be embedded in $M$, and $Y$ in $N$. As $Y$ is an ANR, there is a closed neighbourhood $W$ of $Y$ in $N$ and a retraction of $W$ onto $Y$. Let $r^*$ be an extension of $r$, $r^*: N \to N$, and let $U \in M(X)$, $n_0 \in \mathbb{N}$ be such that for all $n \geqslant n_0$, $f_n|U \simeq f_{n0}|U$.

Consider $f: X \to Y$ defined by $f(z) = rf_{n0}(x)$ for all $x \in X$ and $f^*: M \to N$ defined by $f^* = r^*f_{n0}$. $f^*$ is an extension of $f$. Let $\underline{f}^*$ be the fundamental sequence $(f'_n)_{n \in \mathbb{N}}$ with $f'_n = f^*$ for each $n$. We then have $[\underline{f}^*] = [\underline{f}]$.

To prove this, let $V \in N(Y)$, $V \subset W$; it follows from lemma 2 of section 1.1, and the fact that $Y$ is an ANR, that there is a $V' \in N(Y)$ and a homotopy $H$ from $V' \times [0, 1]$ to $V$ such that for all $y \in V'$, $H(y, 0) = y$ and $H(y, 1) = r(y)$.

As $V' \in N(Y)$, there is a $U' \subset U$, $U' \in M(X)$ and $n'_0$ such that for all $n \geqslant n'_0$, $f_n|U' \simeq f'_{n0}|U'$ within $V'$. Then for all $n \geqslant n'_0$, we have $f_n(U') \subset V'$ and $f_n|U' \simeq rf_n|U'$ within $V$ by considering the homotopy $H'(x, t) = H(f_n(x), t)$. Now as for all $n \geqslant n_0$, $f_n|U \simeq f_{n0}|U$ within $W$, $r^*f_n|U \simeq r^*f_{n0}|U$ within $r^*(W) = Y \subset U$, also for all $n \geqslant n_0$, $f_n(U) \subset W$ so $r^*f_n|U = rf_n|U$, thus we have for all $n \geqslant \max(n_0, n'_0)$,

$$f_n|U' \simeq rf_n|U' = r^*f_n|U' \simeq r^*f_{n0}|U',$$

where $r^*f_{n0} = f'_n$ for all $n$. This proves what we claimed earlier.

If $f'$ is another mapping from $X$ to $Y$ so that $f'$ generates $[\underline{f}']$, it should be clear that $f \simeq f'$.

Let **B** be the category having as objects compact metric spaces, $X$ (together with an embedding of $X$ in an AR, $M$) and having, as morphisms, homotopy classes of fundamental sequences, to each object $X$ corresponding the class of its identity mapping $[\mathrm{Id}_X^*]$.

Let **MC** be the category of compact metric spaces and **HMC** be the homotopy category corresponding to **MC**, $[\ ]: \mathbf{MC} \to \mathbf{HMC}$ being the homotopy functor.

**Definition**

The category **B**, defined above, is called the *Borsuk shape category* of compact metric spaces; the functor $\mathbf{F}: \mathbf{MC} \to \mathbf{B}$ which to $f$ associates $[f^*]$ is called the *fundamental functor* and $B: \mathbf{HMC} \to \mathbf{B}$ which to $[f]$ associates $[\underline{f}^*]$ the *Borsuk shape functor*.

**Definition**

Two compact metric spaces $X$ and $Y$ are said to *have the same shape* if there is an isomorphism between the images of $X$ and $Y$ in **B**.

**Remarks**

(1) It is clear that $\mathbf{B}[\ ] = \mathbf{F}$.

(2) The notion of the shape of a space $X$ is independent of the choice of the embedding of $X$ in an AR, $M$, for suppose that we make a different choice (for each space $X$ in **MC**), which we will denote by $N$, etc., and let $\mathbf{B}$, $\mathbf{B}'$ be the corresponding shape categories defined using the $M$s and the $N$s respectively. Denoting $[\underline{1}_X^*]_{M,N}$ the homotopy class of the fundamental sequence generated by $1_X$ relative to $(M, N)$, we can put, for each $[f] \in \mathbf{B}(X, Y)$,

$$\theta([f]) = [\underline{1}_Y^*]_{M',N'}[f][\underline{1}_M^*]_{M,N} \in \mathbf{B}'(X, Y).$$

Thus defined, $\theta$ gives a functor from **B** to $\mathbf{B}'$ such that $\theta$ induces a bijection from $\mathbf{B}(X, Y)$ onto $\mathbf{B}'(X, Y)$. With the obvious notation, one has $\theta\mathbf{F} = \mathbf{F}'$ and $\theta\mathbf{B} = \mathbf{B}'$. Now it is easily checked that a homotopy class $[\underline{f}]$ in **B** is an isomorphism between $X$ and $Y$ if and only if $\theta([f])$ is one between $X$ and $Y$ within $\mathbf{B}'$. This proves what was claimed.

Initially, Borsuk defined the shape category by considering all compact metric spaces as being embedded in the Hilbert cube $I^\infty$; this category is, in fact, sufficient as a basis for a shape theory of compact metric spaces, but the more general approach can have advantages in some of the applications.

As a first illustration of the uses of the notion of shape, we will give a partial proof of a classification of connected compact subsets of $\mathbb{R}^2$.

**Theorem 1**

*Two connected compact subsets of $\mathbb{R}^2$ have the same shape if and only if they have the same first Betti number.*

*Proof*

One direction depends on a result (Theorem 2 of section 3.2) to be proved later to the effect that Čech homology is shape invariant. If we accept this, it is clear that if $X$, $Y$ are the two spaces in question and they have the same shape, then they have the same first Betti number.

Conversely, suppose $X$, $Y$ are connected compact subsets of $\mathbb{R}^2$ having the same Betti number. Alexander–Pontrjagin duality (see [96]) then gives that $\mathbb{R}^2 - X$ and

$\mathbb{R}^2 - Y$ have the same number of components. From now on we suppose that this number is infinite; for the finite case the proof is simpler.

We arrange the components of $\mathbb{R}^2 - X$ in a sequence $A_0, A_1, \ldots$, and the components of $\mathbb{R}^2 - Y$ in a sequence $B_0, B_1, \ldots$ such that $A_0$ and $B_0$ are the unbounded components. Then there exist, for every $k = 1, 2, \ldots$, two open subsets $U_k$ and $V_k$ of $\mathbb{R}^2$ satisfying the following conditions.

(1) The sets $U_k$, $V_k$ are both connected, $X \subset U_k$ and $Y \subset V_k$. The boundary of $U_k$ in $\mathbb{R}^2$ is the union of $k+1$ disjoint simple closed curves $C_0, C_1, \ldots, C_k$ such that $C_i \subset A_i$ for $i = 0, 1, \ldots, k$. The boundary of $V_k$ in $\mathbb{R}^2$ is the union of $k+1$ disjoint simple closed curves $D_0, D_1, \ldots, D_k$ such that $D_i \subset B_i$ for $i = 0, 1, \ldots, k$.
(2) There is a sequence $(\varepsilon_n)_{n \in \mathbb{N}}$ of positive real numbers convergent to zero such that if $x \in A_i \cap U_k$, $y \in B_i \cap V_k$ for $i = 0, 1, \ldots, k$ then $d(x, X) \leqslant \varepsilon_k$, $d(y, Y) \leqslant \varepsilon_k$.
(3) $\bar{U}_{k+1} \subset U_k$, $\bar{V}_{k+1} \subset V_k$ for $k = 1, 2$.

(1) implies that there is a sequence of homeomorphisms

$$h_k \colon \mathbb{R}^2 \to \mathbb{R}^2 \qquad k = 1, 2, \ldots$$

preserving the orientation of $\mathbb{R}^2$ and satisfying the conditions

(i) $h_k(C_i) = D_i$ for $i = 0, 1, \ldots, k$;
( (ii) $h_{k+1} | \mathbb{R}^2 - U_k = h_k | \mathbb{R}^2 - U_k$;
(iii) $h_k(U_k) = V_k$; and
(iv) $h_k | U_k \simeq h_{k+n} | U_k$ within $V_k$ for $k = 1, 2, \ldots$, and $n = 0, 1, 2, \ldots$.

Thus $\underline{h} = (h_n)_{n \in \mathbb{N}}$ is a fundamental sequence from $X$ to $Y$. Clearly $\underline{h}^{-1} = (h_n^{-1})_{n \in \mathbb{N}}$ is also a fundamental sequence and $[\underline{h}]$ and $[\underline{h}^{-1}]$ are mutually inverse to each other. (Note how convenient it is to take, as we have here, the chosen ARs to be $M = N = \mathbb{R}^2$.)

**Corollary**

*Let $X$ be a connected compactum in $\mathbb{R}^2$; then $X$ has the shape of a point, or a bouquet of $n$ circles (for some finite $n$) or a bouquet of an infinite number of circles.*

## 1.3 FUNDAMENTAL SEQUENCES AND INVERSE SYSTEMS OF ANRs

In Borsuk's approach outlined above, one approximates to a compact metric space $X$ by its system of neighbourhoods within some AR; a fundamental sequence is then a sequence of maps linking approximations to $X$ with approximations to $Y$. The second system of ideas which uses this idea of approximation is the ANR systems approach of Mardešić and Segal [76]. The idea behind this approach is already outlined in Christie's thesis [17]. It uses the notion of inverse systems of ANRs, and so we shall start by recalling the definition of an inverse system over a directed set.

**Convention**

*To simplify the exposition, we shall, in this section, suppose that all compact metric spaces are embedded in the Hilbert cube $I^\infty$.*

**Definition**

A *directed set* $(A, <)$ is a pair consisting of a set $A$ together with a relation $<$ on $A$ satisfying

(i) $\alpha \nless \alpha$ for all $\alpha \in A$;
(i) $\alpha < \beta$ and $\beta < \gamma$ implies $\alpha < \gamma$;
(iii) given $\alpha$, $\beta \in A$ there is some $\gamma \in A$ with $\alpha < \gamma$ and $\beta < \gamma$.

A subset $A'$ of $A$ is said to be *cofinal* in $A$ if, for each $\alpha \in A$, there is an $\alpha' \in A'$ with $\alpha' > \alpha$. (We will normally suppress the relation $<$ in the notation and refer to $A$ as a directed set.)

**Definition**
Let **C** be an arbitrary category. An *inverse system* in **C** is a collection $\underset{\sim}{X} = (X_\alpha, p_\alpha^\beta, A)$ where

(i) $A$ is a directed set;
(ii) for each $\alpha \in A$, $X_\alpha$ is a object of **C** and whenever $\beta > \alpha$,

$$p_\alpha^\beta : X_\beta \to X_\alpha \text{ is a morphism of } \mathbf{C};$$

(iii) whenever $\alpha < \beta$ and $\beta < \gamma$, $p_\alpha^\gamma = p_\alpha^\beta p_\beta^\gamma$ and, by convention, $p_\alpha^\alpha =$ identity on $X_\alpha$.

**Remarks**
(1) The $p_\alpha^\beta$ are called the *bonding* or *structure* maps of the system.
(2) It is, of course, only a short step to pass from the above description to one involving a functor from a category associated with $A$ to **C**. This transition, in fact, took quite some time to occur and so many important papers used the above definition. Because it is not our intention to handle all aspects of the theory and as it is often advantageous or even necessary for a reader to refer to the original papers, we shall for the moment use this more 'primitive' definition, dealing with the more technically elegant one later.

**Definition**
Let $X = (X_\alpha, p_\alpha, A)$ and $\underset{\sim}{Y} = (Y_\beta, q_\beta, B)$ be two inverse systems in our category **C**. A *morphism* from $\underset{\sim}{X}$ to $\underset{\sim}{Y}$ is a family $\underset{\sim}{\phi} = ((\phi_\beta)_{\beta \in B}, \theta)$ where

(i) $\theta$ is an increasing function from $B$ to $A$ (thus if $\beta < \beta'$, then $\theta(\beta) \leqslant \theta(\beta')$);

(ii) for each $\beta \in B$, $\phi_\beta : X_{\theta(\beta)} \to Y_\beta$ is a morphism in **C** such that

for all $\beta$, $\beta'$ with $\beta > \beta'$, the diagram below is commutative:

$$\begin{array}{ccc} X_{\theta(\beta)} & \xrightarrow{\ \phi_\beta\ } & Y_\beta \\ {\scriptstyle p_{\theta(\beta)}^{\theta(\beta')}} \uparrow & & \uparrow {\scriptstyle q_\beta^{\beta'}} \\ X_{\theta(\beta')} & \xrightarrow{\ \phi_{\beta'}\ } & Y_{\beta'} \end{array}$$

[that is $q_\beta^{\beta'} \phi_{\beta'} = \phi_\beta p_{\theta(\beta)}^{\theta(\beta')}$].

**Remark**
Our initial examples of morphisms of inverse systems will be in the case where **C** is a category of topological spaces and homotopy classes of maps. Then it is usual to

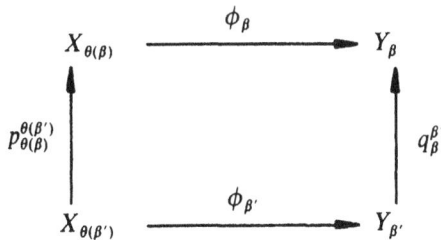

specify continuous bonding maps and continuous $\phi_\beta$ in the above and to require that

$$q_\beta^{\beta'}\phi_{\beta'} \simeq \phi_\beta p_{\theta(\beta)}^{\theta(\beta')}.$$

In this context two such morphisms $\underline{\phi}$ and $\underline{\phi}'$ from $\underline{X}$ and $\underline{Y}$ are said to be *homotopic* if

for all $\beta \in B$, there is an $\alpha \geq \theta(\beta)$, $\alpha \geq \theta'(\beta)$ such that

$$\phi_\beta p_{\theta(\beta)}^{\alpha} \simeq \phi_\beta p_{\theta'(\beta)}^{\alpha}.$$

We now return to our compact metric space, $X$, embedded in the Hilbert cube, $I^\infty$. It is well known [9] that given any open neighbourhood $V$ of $X$ in $I^\infty$ there is a $V'$ which is a compact ANR and a neighbourhood of $X$ such that $X \subset V' \subset V$. Consequently there is a decreasing sequence $(X_n)_{n\in\mathbb{N}}$ of compact ANR neighbourhoods of $X$ such that $X = \bigcap_{n\in\mathbb{N}} X_n$.

Let $\underline{X} = (X_n, i_m^n, \mathbb{N})$ be the resulting inverse system of ANRs where $i_m^n\colon X_n \to X_m$ is the inclusion. (Clearly one need only specify $i_n^{n+1}$ for each $n$ to specify all $i_m^n$ because of property (iii) of bonding maps.) Then $X = \operatorname{Lim} \underline{X}$ and the projection, $p_n$, from the limit, $X$, to $X_n$ is the canonical inclusion of $X$ in its neighbourhood $X_n$. (We refer the reader in need of illumination on the properties of inverse limits to MacLane [71].)

Now let $Y$ be another metric compactum with a decreasing sequence, $(Y_n)_{n\in\mathbb{N}}$, of compact ANRs with $j_n\colon Y_{n+1} \to Y_n$, the inclusions defining the bonding maps. Let $\underline{f} = (f_n)_{n\in\mathbb{N}}$ be a fundamental sequence from $X$ to $Y$. As $Y_0$ is a neighbourhood of $Y$, there is an open neighbourhood $U_0$ of $X$ and an index, $n_0$, such that for all $n \geq n_0$,

$$f_n|U_0 \simeq f_{n_0}|U_0 \text{ in } Y.$$

Let us consider a $\theta(0) \in \mathbb{N}$ such that $\theta(0) \geq n_0$ and $X_{\theta(0)} \subset U_0$ and set $\phi_0 = f_{\theta(0)}|X_{\theta(0)}$. Similarly for $Y$, there are a $U_1$ and an $n_1$ such that for all $n \geq n_1$,

$$f_n|U_1 \simeq f_{n_1}|U_1 \text{ within } Y_1;$$

let $\theta(1) \in \mathbb{N}$ be such that $\theta(1) \geq \max(n_1, \theta(0))$ and $X_{\theta(1)} \subset U_1 \cap X_{\theta(0)}$ and set $\phi_1 = f_{\theta(1)}|X_{\theta(1)}$.

As $f_{\theta(1)}|X_{\theta(0)} \simeq f_{\theta(0)}|X_{\theta(0)}$ within $Y_0$, we have

$$\phi_0 \phi_{\theta(0)}^{\theta(1)} = f_{\theta(0)}|X_{\theta(1)} \simeq f_{\theta(1)}|X_{\theta(1)} \text{ within } Y_0,$$

but $f_{\theta(1)}|X_{\theta(1)}$ is simply $j_0^1\phi_1$.

Continuing this process, we see that a fundamental sequence gives rise to a morphism of inverse systems; we will, in fact, show that the notion of a fundamental sequence, and thus of the shape of a space, is linked to that of morphisms of inverse systems, and thus to the approximation of the space by an inverse system of ANRs.

**Lemma 1**

*Let $(f_n)_{n\in\mathbb{N}}$ be a sequence of continuous maps from $I^\infty$ to itself, $\theta$ an increasing function from $\mathbb{N}$ to $\mathbb{N}$ and $\phi_n$ a continuous map from $X_{\theta(n)}$ to $Y_n$, where $(X_n)_{n\in\mathbb{N}}$ and $(Y_n)_{n\in\mathbb{N}}$ are two decreasing sequences of compact ANR neighbourhoods of $X$ and $Y$ respectively. Suppose that for all $n \in \mathbb{N}$, the following conditions hold:*

(1) *for all $m, m' \geq \theta(n)$, $f_m|X_{\theta(n)} \simeq f_{m'}|X_{\theta(n)}$ within $Y_n$,*
(2) *$\phi_n = f_{\theta(n)}|X_{\theta(n)}$.*

*Then $\underline{f} = (f_n)_{n\in\mathbb{N}}$ is a fundamental sequence from $X$ to $Y$ and $\underline{\phi} = ((\phi_n)_{n\in\mathbb{N}}, \theta)$ is a morphism of inverse systems.*

*Proof*
Let $V$ be a neighbourhood of $Y$ in $I^\infty$; then there is some $n \in \mathbb{N}$ such that $Y_n \subset V$. Let $n_0 = \theta(n)$ and $U = X_{\theta(n)}$, then for all $n, n' \geqslant n_0$,

$$f_n|U \simeq f_{n'}|U \text{ within } Y_n \subset V$$

so $\underline{f}$ is a fundamental sequence. $((\phi_n)_{n\in\mathbb{N}}, \theta)$ is clearly a morphism of inverse systems.

**Definition**
If $\underline{f} = (f_n)_{n\in\mathbb{N}}$, a sequence of mapping from $I^\infty$ to itself, and $\underline{\phi} = ((\phi_n)_{n\in\mathbb{N}}, \theta)$ satisfy the conditions of the preceding lemma, we will say that $\underline{f}$ and $\underline{\phi}$ are *linked*.

**Definition**
A morphism of inverse systems will be said to be *regular* if the function $\theta$ is strictly increasing.

**Lemma 2**
*Let $\underline{\phi} = ((\phi_n)_{n\in\mathbb{N}}, \theta)$ be a morphism of inverse systems from $\underline{X} = (X_n, p^n_m, \mathbb{N})$ to $\underline{Y} = (Y_n, q^n_m, \mathbb{N})$ then there is a regular morphism $\underline{\phi}'$ from $\underline{X}$ to $\underline{Y}$ homotopic to $\underline{\phi}$.*

*Proof*
Let $\theta': \mathbb{N} \to \mathbb{N}$ be defined by $\theta'(n) = \theta(n) + n$ and let $\phi'_n = \phi_n p^{\theta'(n)}_{\theta(n)}$; then $\underline{\phi}'$ is a regular morphism. We claim $\underline{\phi}'$ is homotopic to $\underline{\phi}$, since

$$\begin{aligned}\phi'_n p^{\theta'(n')}_{\theta(n)} &= \phi_n p^{\theta'(n)}_{\theta(n)} p^{\theta'(n')}_{\theta'(n)} = \phi_n p^{\theta'(n')}_{\theta(n)} \\ &= \phi_n p^{\theta(n')}_{\theta(n)} p^{\theta'(n')}_{\theta(n')} = q^{n'}_n \phi_{n'} p^{\theta'(n')}_{\theta(n')} \\ &= q^{n'}_n \phi'_{n'}.\end{aligned}$$

**Proposition 1**
*Let $\underline{\phi}$ be a regular morphism of inverse systems from $\underline{X}$ to $\underline{Y}$ where $\underline{X} = (X_n, i^n_m, \mathbb{N})$, $\underline{Y} = (Y_n, j^n_m, \mathbb{N})$, $X = \bigcap X_n$, $Y = \bigcap Y_n$ in $I^\infty$; then there is a fundamental sequence $\underline{f}$ from $X$ to $Y$ such that $\underline{f}$ and $\underline{\phi}$ are linked.*

*Proof*
Let $\underline{\phi} = ((\phi_n)_{n\in\mathbb{N}}, \theta)$ be the regular morphism from $\underline{X}$ to $\underline{Y}$. We will define by induction on $n$, maps $f_n: I^\infty \to I^\infty$:

—If $\theta(1) = 1$ take $f_1$ to be an extension of $\phi_1$ from $I^\infty$ to $I^\infty$
—If $\theta(1) > 1$, take $f_1$ to be any map from $I^\infty$ to itself.

Suppose now that we have defined $f_k: I^\infty \to I^\infty$ for all $k$, $1 \leqslant k \leqslant n$, so that the following conditions are satisfied:

$(1)_n$ For all $k$, $1 \leqslant k \leqslant n$, such that $\theta(k) \leqslant n$, and all $m, m'$ with $\theta(k) \leqslant m, m' \leqslant n$,

$$f_m|X_{\theta(k)} \simeq f_{m'}|X_{\theta(k)} \text{ in } Y_k.$$

$(2)_n$ For all $k$, $1 \leqslant k \leqslant n$, such that $\theta(k) \leqslant n$,

$$\phi_k = f_{\theta(k)}|X_{\theta(k)}.$$

Clearly $(1)_1$ and $(2)_1$ are satisfied since $\theta(k) \leqslant 1$ implies $k = 1$ and $\theta(1) = 1$. We shall define $f_{n+1}: I^\infty \to I^\infty$ so that $(1)_{n+1}$ and $(2)_{n+1}$ are satisfied:

If $\theta(1) > n+1$, take for $f_{n+1}$ any continuous map from $I^\infty$ to itself, then $(1)_{n+1}$ and $(2)_{n+1}$ are trivially satisfied since $\theta(k) \leqslant n+1$ implies $\theta(k) < \theta(1)$.

If $\theta(1) = n+1$, take for $f_{n+1}$ an extension (from $I^\infty$ to itself) of $\phi_1$; as $\theta(k) \leqslant n+1$ implies $k=1$, one has

$$f_{\theta(1)}|X_{\theta(1)} = f_{n+1}|X_{n+1} = \phi_1$$

and as $\theta(1) = n+1 \leqslant m, m' \leqslant n+1$, $(2)_{n+1}$ is trivially satisfied.

If, however, $\theta(1) < n+1$, we construct $f_{n+1}$ by induction. For this we shall make the convention that $X_0 = Y_0 = I^\infty$ and $\theta(0) = 0$; let $l$ be the largest of those $k \leqslant n$ with $\theta(k) \leqslant n$; we will construct a sequence of mappings $f^j_{n+1}: X_{\theta(l-j)} \to Y_{l-j}$ for $j = 0, \ldots, l$, each one being obtained from the preceding one by extension, $f^l_{n+1}: X_0 \to Y_0$ being the map we are after.

As $\theta(l+1) > n$, we have to consider two cases to define $f^0_{n+1}$:

—If $\theta(l+1) > n+1$, take $f^0_{n+1} = f_n|X_{\theta(l)}$
—If $\theta(l+1) = n+1$, consider $\phi_{l+1}: X_{\theta(l+1)} \to Y_{l+1}$.

As $j_l^{l+1}\phi_{l+1} \simeq \phi_l\, i_{\theta(l)}^{\theta(l+1)}$ within $Y_l$, there is some extension $f^0_{n+1}$ of $\phi_{l+1}$, $f^0_{n+1}: X_{\theta(l)} \to Y_l$, such that $f^0_{n+1} \simeq \phi_l$ (use the homotopy extension property recalling that $Y_l$ is an ANR).

In both cases we now have a map $f^0_{n+1}$ such that $f^0_{n+1} \simeq f_n|X_{\theta(l)}$ within $Y_l$ (since if $\theta(l+1) = n+1$, as $\theta(l) < n$ and $(1)_n$ and $(2)_n$ are satisfied, we have

$$f^0_{n+1} \simeq \phi_l = f_{\theta(l)}|X_{\theta(l)} \simeq f_n|X_{\theta(l)}.)$$

As $\phi_{l-1} i^{\theta(l)}_{\theta(l-1)} \simeq j^l_{l-1}\phi_l \simeq j^l_{l-1} f^0_{n+1}$ in $Y_{l-1}$, another use of the homotopy extension property gives the existence of an extension, $f^1_{n+1}: X_{\theta(l-1)} \to Y_{l-1}$, of $f^0_{n+1}$ such that

$$f^1_{n+1} \simeq \phi_{l-1} = f_{\theta(l-1)}|X_{\theta(l-1)} \simeq f_n|X_{\theta(l-1)}$$

within $Y_{l-1}$ and so on .... Thus one constructs a sequence $f^j_{n+1}$ for $0 \leqslant j \leqslant l-1$ having the properties

$(3)_j$ $f^j_{n+1} \simeq \phi_{l-j} \simeq f_n|X_{\theta(l-j)}$ within $Y_{l-j}$.
$(4)_j$ $f^{j+1}_{n+1}|X_{\theta(l-j)} = f_j|X_{n+1}$.

Taking for $f^l_{n+1}$ an extension over the whole $I^\infty$ of $f^{l-1}_{n+1}$, we set $f_{n+1} = f^l_{n+1}$ and we are left to verify that $(1)_{n+1}$ and $(2)_{n+1}$ are satisfied.

Let $k \leqslant n+1$ be such that $\theta(k) \leqslant n+1$.

—If $k \leqslant n$ and $\theta(k) \leqslant n$, it is clear that $(2)_n$ implies this case of $(2)_{n+1}$. Let $\theta(k) \leqslant m, m' \leqslant n+1$; if $m, m' \leqslant n$, again this case of $(1)_{n+1}$ is satisfied because of $(1)_n$. Let $m' = n+1$, $m \leqslant n$, then as $k = l-j$, for some $j$,

$$f_{n+1}|X_{\theta(k)} = f_{n+1} \simeq f_n|X_{\theta(k)} \simeq f_m|X_{\theta(k)}$$

(by $(3)_j$, $(4)_j$ and $(1)_n$).

—If $k \leqslant n$ and $\theta(k) = n+1$, $k > l$, and as $\theta(k) = n+1$ less than or equal to $\theta(l+1)$ implies $k \leqslant l+1$, one must have $k = l+1$ and $(2)_{n+1}$ is satisfied in this case, since

$$f_{\theta(k)}|X_{\theta(k)} = f_{n+1}|X_{\theta(l+1)} = f^l_{n+1}|X_{\theta(l+1)} = f^0_{n+1}|X_{\theta(l+1)} = \phi_{l+1}.$$

$(1)_{n+1}$ is satisfied again trivially here as $\theta(k) \leqslant m, m' \leqslant n+1$ implies $m = m' = n+1$.

—If $k = n + 1$, then $\theta(k) = n + 1 = \theta(l + 1)$, since $l + 1 \leqslant n + 1$ implies $n + 1 \leqslant \theta(l + 1) \leqslant \theta(n + 1) \leqslant n + 1$. Thus $(1)_{n+1}$ and $(2)_{n+1}$ are satisfied as before.

This completes the proof of the proposition.

**Remarks**
This proof only works if $\theta$ is strictly increasing.

Combining the various results of this section together, we find that the passage from fundamental sequences to the morphism of an inverse system is a two-way process. We will investigate this more fully in Chapter 3.

## HISTORICAL NOTE

In 1928, Alexandroff [1] proved that any compact metric space could be approached by an inverse sequence of finite simplicial complexes. Lefschetz [66] continued this study in depth using neighbourhoods of an embedded copy of the space and also the Alexandroff–Čech construction of a nerve of an open covering of the space. His student, Christie [17], started the study of the homotopy properties of these inverse systems, and showed that the homotopy properties of a compact metric space found using the neighbourhood systems of an embedded copy and those found by the nerve construction were equivalent. In 1968, Borsuk [10], wishing to study the global homotopy properties of a compact metric space by neglecting the local properties, introduced the notion of shape using the neighbourhoods of an embedded copy; the link with inverse systems was then rediscovered and further developed by Mardešić–Segal [76].

# 2

# Categorical Shape Theory

## 2.1 SHAPE THEORY

If **A** is a category, we will denote by $|\mathbf{A}|$ its class of objects, and if $X, Y \in |\mathbf{A}|$, $\mathbf{A}(X, Y)$ will be the set of morphisms from $X$ to $Y$ in **A**.

Let **K** be a functor from **A** to **B** and $X \in |\mathbf{B}|$. The comma category of **K**-objects under $X$, denoted $X \downarrow \mathbf{K}$, is the category whose objects are the pairs $(f, P)$ where $f \in \mathbf{B}(X, \mathbf{K}P)$ and where the morphisms $k: (f, P) \to (f', P')$ are the morphisms $k \in \mathbf{A}(P, P')$ which satisfy $\mathbf{K}(k)f = f'$. We denote by $\boldsymbol{\delta}_X$ the codomain functor from $X \downarrow \mathbf{K}$ to **A** defined by $\boldsymbol{\delta}_X(f, P) = P$, $\boldsymbol{\delta}_X(k) = k$.

We have just seen in Chapter 1 that the shape of a compact metric space, $X$, is linked to the approximation of $X$ by compact ANRs; how can one 'categorise' this idea? We have at our disposal the homotopy category of compact metric spaces, **HMC**, and **W** the full subcategory of **HMC** having as objects the compact ANRs. If $X \in |\mathbf{HMC}|$, the approximation of $X$ by ANRs is the 'set' of morphisms $\{X \xrightarrow{f} P \mid P \in |\mathbf{W}|\}$ and one can evidently consider this 'set' as the class of objects of the comma category $X \downarrow K$ where $\mathbf{K}: \mathbf{W} \to \mathbf{HMC}$ is the inclusion. Equally well it can be thought of as the functor $\mathbf{HMC}(X, \mathbf{K}-): \mathbf{W} \to \mathbf{Sets}$. If $h: X \to Y$ in **HMC**, then $h$ induces a functor

$$h^*: Y \downarrow \mathbf{K} \to X \downarrow \mathbf{K}$$

given by composition, such that $\boldsymbol{\delta}_X h^* = \boldsymbol{\delta}_Y$, so we have a natural notion of morphisms between approximations, namely those functors $\mathbf{V}: Y \downarrow \mathbf{K} \to X \downarrow \mathbf{K}$ which satisfy $\boldsymbol{\delta}_X \mathbf{V} = \boldsymbol{\delta}_Y$. Using the other notion of approximation, namely the functors $\mathbf{HMC}(X, -)$ from **W** to **Sets**, $h$ induces a natural transformation

$$h^*: \mathbf{HMC}(Y, \mathbf{K}-) \to \mathbf{HMC}(X, \mathbf{K}-)$$

again by composition and this time $\mathrm{Nat}(\mathbf{HMC}(Y, \mathbf{K}-), \mathbf{HMC}(X, \mathbf{K}-))$ will be the natural candidate for the set of morphisms between the approximations.

In this case $\mathbf{K}: \mathbf{W} \to \mathbf{HMC}$ is an inclusion, but clearly the constructions above do not depend on this fact, so instead of restricting ourselves to considering the shape theory of a category, $\mathbf{B}$, relative to subcategory $\mathbf{A}$ of $\mathbf{B}$, we will give a general definition of a shape theory for an arbitrary functor $\mathbf{K}: \mathbf{A} \to \mathbf{B}$.

Let $\mathbf{K}$ be a functor from $\mathbf{A}$ to $\mathbf{B}$, and $\mathbf{T}$ a functor from $\mathbf{B}$ into a category $\mathbf{C}$.

**Definition**

Let $X \in |\mathbf{B}|$ and $D \in |\mathbf{C}|$; we denote by $\mathrm{Func}(X \downarrow \mathbf{K}, D \downarrow \mathbf{TK})$ the set of functors $\mathbf{V}: X \downarrow \mathbf{K} \to D \downarrow \mathbf{TK}$ such that $\boldsymbol{\delta}_D \mathbf{V} = \boldsymbol{\delta}_X$.

If $f \in \mathbf{B}(X, \mathbf{K}P)$, we denote by $\mathbf{V}[f, P] \in \mathbf{C}(D, \mathbf{TK}P)$ the morphism such that $\mathbf{V}(f, P) = (\mathbf{V}[f, P], P)$.

One says that $\mathbf{T}$ is $\mathbf{K}$-*continuous* in $X$ if, for all $D \in |\mathbf{C}|$ and for all $\mathbf{V} \in \mathrm{Func}(X \downarrow \mathbf{K}, D \downarrow \mathbf{TK})$, there is a unique $g \in \mathbf{C}(D, \mathbf{T}X)$ such that the diagram

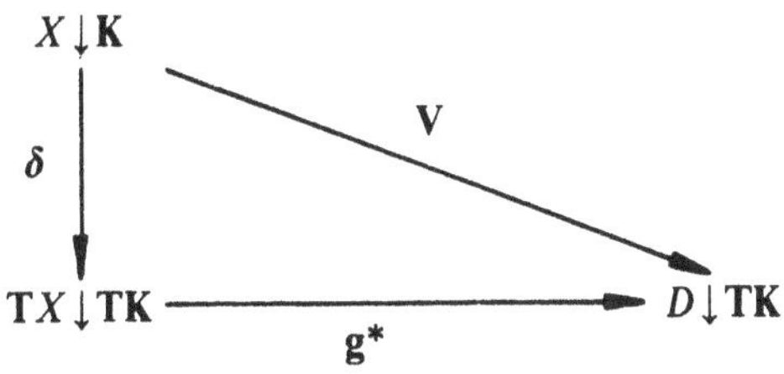


commutes where $\mathbf{g}^*: \mathbf{T}X \downarrow \mathbf{TK} \to D \downarrow \mathbf{TK}$ is the functor given by

$$\mathbf{g}^*(h, P) = (hg, P) \quad \text{and} \quad \mathbf{g}^*(k) = k$$

and where $\boldsymbol{\delta}$ is the functor from $X \downarrow \mathbf{K}$ to $\mathbf{T}X \downarrow \mathbf{TK}$ defined by

$$\boldsymbol{\delta}(f, P) = (\mathbf{T}f, P) \quad \text{and} \quad \boldsymbol{\delta}(k) = k.$$

The functor $\mathbf{T}$ will be said to be $\mathbf{K}$-*continuous* if $\mathbf{T}$ is $\mathbf{K}$-continuous in $X$ for all $X$ in $|\mathbf{B}|$.

**Remark**

Normally one would say that $\mathbf{T}: \mathbf{B} \to \mathbf{C}$ was continuous if it preserved all small limits. The $\mathbf{K}$-relative version given above is forced on us by the fact that, in the principal examples neither $\mathbf{A}$ nor $\mathbf{B}$ has small limits. The comma category then takes the place of the limiting cone in the more usual notion.

**Definition**

A pair $(\mathbf{S}, S)$ is said to be a *shape theory for* $\mathbf{K}$ where $\mathbf{K}: \mathbf{A} \to \mathbf{B}$ is a functor, and $S$ is a functor from $\mathbf{B}$ to $\mathbf{S}$ if

(S1) $|\mathbf{S}| = |\mathbf{B}|$ and for all $X$ in $|\mathbf{B}|$, $SX = X$.
(S2) If $s \in \mathbf{S}(X, \mathbf{K}Q)$, there is a unique $f \in \mathbf{B}(X, \mathbf{K}Q)$ such that $Sf = s$.
(S3) $S$ is $\mathbf{K}$-continuous.

**Remark**

There is also a notion of a coshape theory for $\mathbf{K}$ obtained by dualising all the definitions and results.

**Generic Example**

The Holsztyński shape theory ($\mathbf{S_K}$, $S_\mathbf{K}$) of a functor $\mathbf{K}: \mathbf{A} \to \mathbf{B}$.

Let $\mathbf{S_K}$ be the category with $|\mathbf{S_K}| = |\mathbf{B}|$ and $\mathbf{S_K}(X, Y) = \text{Func}(Y\downarrow\mathbf{K}, X\downarrow\mathbf{K})$ (we are taking for $\mathbf{T}$ the functor, $\mathbf{Id_B}$, the identity on $\mathbf{B}$). The assignment which to $s \in \mathbf{S_K}(X, Y)$ associates $\bar{s} \in \text{Nat}(\mathbf{B}(Y, \mathbf{K}-), \mathbf{B}(X, \mathbf{K}-))$ defined by

$$\bar{s}(f) = s[f, P],$$

where $f \in \mathbf{B}(X, \mathbf{K}P)$, is a bijection of $\mathbf{S_K}(X, Y)$ onto $\text{Nat}(\mathbf{B}(Y, \mathbf{K}-), \mathbf{B}(X, \mathbf{K}-)) = \mathbf{S'_K}(X, Y)$, say. It extends to give an isomorphism of $\mathbf{S_K}$ with the category $\mathbf{S'_K}$ defined by $|\mathbf{S'_K}| = |\mathbf{B}|$ and with the $\mathbf{S'_K}(X, Y)$ as hom-sets, the composition being the composition of transformations.

The functor $S_\mathbf{K}: \mathbf{B} \to \mathbf{S_K}$ is defined by $\mathbf{S_K}(h) = \mathbf{B}(h, \mathbf{K}-)$ (on identifying $\mathbf{S_K}$ and $\mathbf{S'_K}$). We say that $S_\mathbf{K}$ (resp. $\mathbf{S_K}$) is the *shape functor* (resp. *shape category*) *of Holsztyński* for $\mathbf{K}$. From now on we shall adopt both of the definitions of ($\mathbf{S_K}$, $S_\mathbf{K}$) and will change from one to another when convenient. There is one point to note, however:

—If $s \in \mathbf{S_K}(X, Y) = \text{Nat}(\mathbf{B}(Y, \mathbf{K}-), \mathbf{B}(X, \mathbf{K}-))$, we will denote by $s(f)$ the element of $\mathbf{B}(X, \mathbf{K}P)$ corresponding to $f \in \mathbf{B}(Y, \mathbf{K}P)$.

—If $s \in S_\mathbf{K}(X, Y) = \text{Func}(Y\downarrow\mathbf{K}, X\downarrow\mathbf{K})$, it will be expedient to denote the corresponding element by $s[f, P]$.

**Remark**

If $\mathbf{A}$ and $\mathbf{B}$ are in a given set theoretic universe $U$, in general $\mathbf{S}$ is in a bigger universe than $U$; however, in certain particular cases, one can show that $\mathbf{S}$ belongs to $U$.

We will next examine what conditions on $\mathbf{K}$ imply that ($\mathbf{S_K}$, $S_\mathbf{K}$) is indeed a shape theory for $\mathbf{K}$ in the sense of the definition.

**Proposition 1**

*If $S_\mathbf{K}$ satisfies (S2) then ($\mathbf{S_K}$, $S_\mathbf{K}$) is a shape theory for $\mathbf{K}: \mathbf{A} \to \mathbf{B}$.*

*Proof* (We will write $\mathbf{S} = \mathbf{S_K}$, $S = S_\mathbf{K}$ for short.)

Let $Y \in |\mathbf{B}|$ and $\mathbf{V} \in \text{Func}(X\downarrow\mathbf{K}, Y\downarrow\mathbf{SK})$; if $k$ is a morphism from $(f, P)$ to $(f', P')$ in $X\downarrow\mathbf{K}$ then

$$\mathbf{SK}(k)\mathbf{V}[f, P] = \mathbf{V}[K(k)f, P'].$$

Let $f \in \mathbf{B}(X, \mathbf{K}P)$. As $\mathbf{V}[f, P] \in \mathbf{S}(Y, \mathbf{K}P)$, there is, by (S2), a unique $h \in \mathbf{B}(Y, \mathbf{K}P)$ such that $S(h) = \mathbf{V}[f, P]$. Let $s$ be the mapping from $\mathbf{B}(X, \mathbf{K}-)$ to $\mathbf{B}(Y, \mathbf{K}-)$ defined as follows:

—if $f \in \mathbf{B}(X, \mathbf{K}P)$, $s(f) = h$, where $h$ is the unique morphism such that $\mathrm{S}(h) = \mathbf{V}[f, P]$.

Then $s \in \text{Nat}(\mathbf{B}(X, \mathbf{K}-), \mathbf{B}(Y, \mathbf{K}-))$ since if $k \in \mathbf{A}(P, P')$ and $f \in \mathbf{B}(X, \mathbf{K}P)$,

$$\begin{aligned} S(\mathbf{K}(k)s(f)) &= S\mathbf{K}(k)\mathbf{V}[f, P] \\ &= \mathbf{V}[\mathbf{K}(k)f, P'] \\ &= S(s(\mathbf{K}(k)f)) \end{aligned}$$

and by (S2), one has $\mathbf{K}(k)s(f) = s(\mathbf{K}(k)f)$.

We next check that $\mathbf{s}^*\boldsymbol{\delta} = \mathbf{V}$.

Let $(f, P) \in |X \downarrow \mathbf{K}|$, $\mathbf{s}^*\delta(f, P) = (S(f)s, P)$ and as $S(f)s \in \mathbf{S}(Y, \mathbf{K}P)$, there is a unique $g \in \mathbf{B}(X, \mathbf{K}P)$ such that $S(g) = S(f)s$. Thus one has

$$S(g)(\mathrm{Id}_{\mathbf{K}P}) = g = S(f)s(\mathrm{Id}_{\mathbf{K}P}) = s(f)$$

and hence

$$(S(f)s, P) = (S(s(f)), P) = (\mathbf{V}[f, P], P)$$

as required.

Let $s' \in \mathbf{S}(Y, X)$ be such that $\mathbf{s}'^*\delta = \mathbf{V}$. Since $\mathbf{s}'^*\delta(f, P) = (S(f)s', P) = (S(s'(f)), P) = (\mathbf{V}[f, P], P) = (S(s(f)), P)$ with $f \in \mathbf{B}(X, \mathbf{K}P)$, $s(f) = s'(f)$ again by (S2) and so $s = s'$, and (S3) is satisfied.

**Corollary**

*If* $\mathbf{K}$ *is a full functor, thus in particular if* $\mathbf{K}$ *is the insertion of a full subcategory* $\mathbf{A}$ *into* $\mathbf{B}$, $(\mathbf{S_K}, S_\mathbf{K})$ *is a shape theory for* $\mathbf{K}$.

*Proof*

Let $s \in \mathbf{S_K}(X, \mathbf{K}P)$ with $P$ in $\mathbf{A}$ and let $f = s(\mathrm{Id}_{\mathbf{K}P}) \in \mathbf{B}(X, \mathbf{K})$. If $h \in \mathbf{B}(\mathbf{K}P, \mathbf{K}Q)$ where $Q \in |\mathbf{A}|$, then $h = \mathbf{K}(k)$ for some $k \in \mathbf{A}(P, Q)$, since $\mathbf{K}$ is full and one has

$$S(f)(h) = hf = \mathbf{K}(k)f = \mathbf{K}(k)s(\mathrm{Id}_{\mathbf{K}P}) = s(\mathbf{K}(k)) = s(h)$$

since $s$ is a natural transformation from $\mathbf{B}(\mathbf{K}P, \mathbf{K}-)$ to $\mathbf{B}(X, \mathbf{K}-)$. Thus $S(f) = s$.

Now let $f' \in \mathbf{B}(X, \mathbf{K}P)$ be such that $S(f') = s$, one has

$$f' = S(f')(\mathrm{Id}_{\mathbf{K}P}) = s(\mathrm{Id}_{\mathbf{K}P}) = f$$

so $f' = f$ and $f$ is unique as required.

We next consider the uniqueness of shape theories for a given $\mathbf{K}$.

**Proposition 2**

*Suppose given* $\mathbf{K}: \mathbf{A} \to \mathbf{B}$ *as before and suppose* $\mathbf{C}$ *is a category and* $\mathbf{F}: \mathbf{B} \to \mathbf{C}$ *a functor satisfying* (S1) *and* (S2). *If* $\mathbf{T}: \mathbf{B} \to \mathbf{C}$ *is a* $\mathbf{K}$*-continuous functor, then there is a unique functor* $\mathbf{R}: \mathbf{C} \to \mathbf{D}$ *such that* $\mathbf{RF} = \mathbf{T}$.

*Proof*

For $X \in |\mathbf{B}|$, we set $\mathbf{R}X = \mathbf{T}X$. Let $c \in \mathbf{C}(X, Y)$ and let $\mathbf{V}$ be the functor $\mathbf{V}: Y \downarrow \mathbf{K} \to X \downarrow \mathbf{K}$ defined as follows:

> if $f \in \mathbf{B}(Y, \mathbf{K}P)$, $\mathbf{V}[f, P]$ is the unique element of $\mathbf{B}(X, \mathbf{K}P)$ such that $\mathbf{F}(\mathbf{V}[f, P]) = \mathbf{F}(f)c$.

Now $\mathbf{V} \in \mathrm{Func}(Y \downarrow \mathbf{K}, X \downarrow \mathbf{K})$ since if $k \in \mathbf{A}(P, Q)$, $\mathbf{F}(\mathbf{K}(k)\mathbf{V}[f, P]) = \mathbf{FK}(k)\mathbf{F}(f)c = \mathbf{F}(\mathbf{K}(k)f)c = \mathbf{F}(\mathbf{V}[\mathbf{K}(k)f, Q])$ and from (S2) we can conclude that

$$\mathbf{K}(k)\mathbf{V}[f, P] = \mathbf{V}[\mathbf{K}(k)f, Q].$$

Let $\delta: X \downarrow \mathbf{K} \to \mathbf{T}X \downarrow \mathbf{TK}$, $\delta': Y \downarrow \mathbf{K} \to \mathbf{T}Y \downarrow \mathbf{TK}$. As $\delta\mathbf{V}$ is in $\mathrm{Func}(Y \downarrow \mathbf{K}, \mathbf{T}X \downarrow \mathbf{TK})$ and $\mathbf{T}$ is $\mathbf{K}$-continuous, hence $\mathbf{K}$-continuous in $Y$, there is a unique $g \in \mathbf{D}(\mathbf{T}X, \mathbf{T}Y)$ such that

$$\mathbf{g}^*\delta' = \delta\mathbf{V}.$$

Thus we let $\mathbf{R}: \mathbf{C} \to \mathbf{D}$ be defined on morphisms by $\mathbf{R}(c) = g$. By the uniqueness of $g$, $\mathbf{R}$ is a functor. Let $h \in \mathbf{B}(X, Y)$ and $\mathbf{V}[f, P]$, the unique element such that $\mathbf{F}(\mathbf{V}[f, P]) = \mathbf{F}(f)\mathbf{F}(h)$ where $f \in \mathbf{B}(Y, \mathbf{K}P)$; $\mathbf{RF}(h)$ is then the unique morphism such that $\mathbf{RF}(h)^* \delta' = \delta \mathbf{V}$ and one has

$$\mathbf{RF}(h)^* \delta'(f, P) = (\mathbf{T}(f)\mathbf{RF}(h), P) = \mathbf{T}(\mathbf{V}[f, P], P)$$

for all $(f, P)$ and as $\mathbf{V}[f, P] = fh$,

$$\mathbf{T}(f)\mathbf{RF}(g) = \mathbf{T}(f)\mathbf{T}(h)$$

from which we conclude

$$\mathbf{RF}(h) = \mathbf{T}(h).$$

Now suppose $\mathbf{R}'$ is such that $\mathbf{R}'\mathbf{F} = \mathbf{T}$ and let $c \in \mathbf{C}(X, Y)$, $f \in \mathbf{B}(Y, \mathbf{K}P)$ as before;

$$\mathbf{T}(f)\mathbf{R}'(c) = \mathbf{R}'\mathbf{F}(f)\mathbf{R}'(c) = \mathbf{R}'(\mathbf{FV}[f, P]) = \mathbf{T}(\mathbf{V}[f, P]).$$

As $\mathbf{R}(c)$ is the unique morphism such that $\mathbf{T}(f)\mathbf{R}(c) = \mathbf{T}(\mathbf{V}[f, P])$, we have $\mathbf{R}(c) = \mathbf{R}'(c)$ as required.

**Corollary**

*If* $(\mathbf{S}, S)$ *and* $(\mathbf{S}', S')$ *are two shape theories for* $\mathbf{K}$, *where* $\mathbf{K}: \mathbf{A} \to \mathbf{B}$, *then there is a unique isomorphism* $\mathbf{R}$ *such that* $\mathbf{R}S' = \mathbf{S}$.

**Remark**

Let $\mathbf{K}$ be a functor from $\mathbf{A}$ to $\mathbf{B}$ such that the shape functor of Holsztyński, $S_{\mathbf{K}}: \mathbf{B} \to \mathbf{S_K}$ satisfies (S2); then a pair $(\mathbf{S}, S)$ is a shape theory for $\mathbf{K}$ if and only if (S1) and (S2) are satisfied and if

> (S3)′ for each functor $\mathbf{F}: \mathbf{B} \to \mathbf{C}$ satisfying (S1) and (S2), there is a unique functor $\mathbf{R}$ such that $\mathbf{RF} = S$.

## 2.2 EXAMPLES OF SHAPE THEORIES

As we shall be returning later to the topological examples occurring in 'classical' shape theory, we shall limit the examples here to algebraic and categorical ones which will illustrate some of the points that will be met later.

**(a) $\mathbf{K} = \mathrm{U}$: Groups → Sets, *the forgetful functor***

We first consider, for a set $X$, the comma category $X \downarrow \mathbf{U}$. Each $X \to \mathbf{U}G$, $G$ a group, corresponds uniquely, by adjointness, to a group morphism $\mathbf{F}(X) \to G$, where $\mathbf{F}(X)$ is the free group on $X$. Thus $X \downarrow \mathbf{U}$ is isomorphic to $\mathbf{F}(X) \downarrow \mathbf{Groups}$, the category of groups under $\mathbf{F}(X)$. This has an initial object, namely the identity on $\mathbf{F}(X)$. Thus if $\mathbf{V}: Y \downarrow \mathbf{U} \to X \downarrow \mathbf{U}$ is a functor satisfying $\delta_X \mathbf{V} = \delta_Y$, we have that it is completely determined by the element $\mathbf{V}(\eta_Y, \mathbf{F}(Y))$, where $\eta_Y: Y \to \mathbf{UF}(Y)$ is the unit of the adjunction. So we obtain $\mathbf{V}[\eta_Y, \mathbf{F}(Y)]: X \to \mathbf{UF}(Y)$ which, in turn, corresponds to a unique morphism $v: \mathbf{F}(X) \to \mathbf{F}(Y)$. All $v \in \mathbf{Groups}(\mathbf{F}(X), \mathbf{F}(Y))$ arise in this way, giving an isomorphism between $\mathbf{S_K}$ and the full subcategory of the category of groups determined by the free groups. Under this isomorphism, $S_{\mathbf{K}}(X)$ becomes $\mathbf{F}(X)$.

By Proposition 1 of section 2.1, to prove that $(\mathbf{S_K}, S_{\mathbf{K}})$ is a shape theory for $\mathbf{K}$, it suffices to verify (S2). Note that $\mathbf{K}$ is certainly not full so we cannot get this from the argument used earlier.

However, we know that the pair $(\mathbf{S_K}, S_K)$ is isomorphic to (**Free Groups, F**) so suppose $s \in \mathbf{S}(X, \mathrm{U}(G))$; this corresponds to $s\colon \mathbf{F}(X) \to \mathbf{FU}(G)$, and writing $\varepsilon(G)\colon \mathbf{FU}(G) \to G$ for the canonical projection (counit of the adjunction), one obtains

$$\varepsilon(G)s\colon \mathbf{F}(X) \to G$$

and the corresponding morphism in **Sets** (via the adjunction) denoted

$$(\overline{\varepsilon(G)s})\colon X \to \mathrm{U}(G),$$

say. It is then routine to check that $\mathbf{F}(\overline{\varepsilon(G)s}) = s$.

***(b) The general case of a functor* K: A → B *with a left adjoint***

The argument used above in example (a) at no point depended on any special characteristics of the category of groups; in fact a similar argument works whenever **K** has a left adjoint. To make this precise and concise, it will be convenient to recall here the construction of the Kleisli category of a monad, as this will take the place of the category of free groups in the previous example (see MacLane [71] for further details on monads and the Kleisli construction).

Any functor $\mathbf{G}\colon \mathbf{A} \to \mathbf{B}$ with a left adjoint, $\mathbf{F}\colon \mathbf{B} \to \mathbf{A}$ say, determines a monad on **B**, that is, an endofunctor $\mathbf{T}\colon \mathbf{B} \to \mathbf{B}$ together with natural transformations

$$\eta\colon 1 \to T \quad \text{and} \quad \mu\colon \mathbf{T}^2 \to \mathbf{T}$$

such that the diagrams

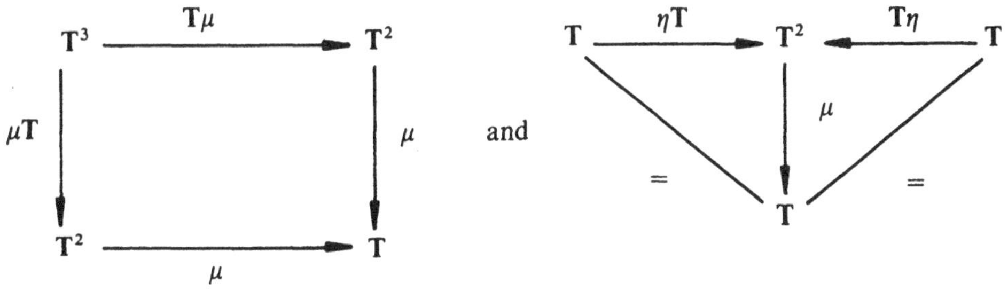


commute. In our case, $\mathbf{T} = \mathbf{GF}$, $\eta$ is the unit of the adjunction, $\eta\colon \mathbf{Id} \to \mathbf{GF}$ and $\mu = \mathbf{G}\varepsilon\mathbf{F}$ where $\varepsilon\colon \mathbf{FG} \to \mathbf{Id}$ is the corresponding counit.

Given any monad $\mathbf{T} = (\mathbf{T}, \eta, \mu)$ on **B**, one constructs its Kleisli category $\mathbf{B_T}$ as follows:

—$|\mathbf{B_T}| = |\mathbf{B}|$;
—if $X, Y \in |\mathbf{B_T}|$, $\mathbf{B_T}(X, Y) = \mathbf{B}(X, \mathbf{T}Y)$;
—if $f\colon X \to \mathbf{T}Y$, $g\colon Y \to \mathbf{T}Z$ are morphisms in **B**, we can form a composite

$$g \circ f = \mu\mathbf{T}(g)f\colon X \to \mathbf{T}Z;$$

this defines the composition in $\mathbf{B_T}$.

$\mathbf{B_T}$ comes equipped with functors $\mathbf{F_T}\colon \mathbf{B} \to \mathbf{B_T}$, $\mathbf{G_T}\colon \mathbf{B_T} \to \mathbf{B}$ defined by

—if $k\colon X \to Y$, $\mathbf{F_T}(k)$ is the morphism $\eta_Y k\colon X \to \mathbf{T}Y$;
—if $f\colon X \to \mathbf{T}Y$ in **B** considered as a map in $\mathbf{B_T}$, then

$$\mathbf{G_T}(f) = \mu_Y \mathbf{T}(f)\colon \mathbf{T}Y \to \mathbf{T}^2 Y \to \mathbf{T}Y.$$

In the case we considered in (a), $\mathbf{F}$ = free group, $\mathbf{G}$ = forget and $\mathbf{Sets_T}$ is isomorphic to the full subcategory of groups determined by the free groups. With this identification $\mathbf{F_T}$ is $\mathbf{F}$ with its codomain restricted, and $\mathbf{G_T}$ is the 'underlying set of the free group on __' functor. $\mathbf{B_T}$ is in general interpreted as being '**Free T-algebras**'.

Returning to our functor $\mathbf{K}: \mathbf{A} \to \mathbf{B}$, suppose it has a left adjoint $\mathbf{L}: \mathbf{B} \to \mathbf{A}$. To illustrate the equivalence of the comma category and natural transformation approaches we shall use the latter here, as in (a) we used comma categories. (We leave it to the reader to rephrase the results in terms of comma categories.)

We first note that there is a natural isomorphism of functors from $\mathbf{A}$ to $\mathbf{Sets}$, given by the adjunction

$$\mathbf{B}(X, \mathbf{K}-) \cong \mathbf{A}(\mathbf{L}(X), -)$$

and by the Yoneda Lemma

$$\begin{aligned}\mathrm{Nat}(\mathbf{B}(Y, \mathbf{K}-), \mathbf{B}(X, \mathbf{K}-)) &\cong \mathrm{Nat}(\mathbf{A}(\mathbf{L}(Y), -), \mathbf{A}(\mathbf{L}(X), -)) \\ &\cong \mathbf{A}(\mathbf{L}(X), \mathbf{L}(Y))\end{aligned}$$

which, in turn, is naturally isomorphic to $\mathbf{B}(X, \mathbf{KL}(Y))$.

Hence $\mathbf{S_K}$ is naturally isomorphic to a category with $|\mathbf{B}|$ as objects and with $\mathbf{B}(X, \mathbf{KL}(Y))$ as the set of morphisms between $X$ and $Y$.

To show that $\mathbf{S_K}$ is isomorphic to $\mathbf{B_T}$ for $\mathbf{T}$ generated by the adjunction between $\mathbf{K}$ and $\mathbf{L}$, it remains to show that the composition is the same.

We will use explicitly the natural isomorphism

$$\phi_{X,P}: \mathbf{B}(X, \mathbf{K}P) \cong \mathbf{A}(\mathbf{L}X, P)$$

so that

$$\begin{cases}\eta_X: X \to \mathbf{KL}(X) & \text{is given by } \phi_{X,\mathbf{L}(X)}{}^{-1}(\mathrm{Id}_{\mathbf{L}X}) \\ \varepsilon_P: \mathbf{LK}(P) \to P & \text{is given by } \phi_{\mathbf{K}(P),P}(\mathrm{Id}_{\mathbf{K}(P)}).\end{cases}$$

To a natural transformation $f: \mathbf{B}(Y, \mathbf{K}-) \to \mathbf{B}(X, \mathbf{K}-)$, we have associated a morphism $\tilde{f}: X \to \mathbf{KL}(Y)$. Following $f$ through the isomorphisms we find

$$\tilde{f} = f(\mathbf{L}(Y))(\eta_Y).$$

Now suppose we are given $f: \mathbf{B}(Y, \mathbf{K}-) \to \mathbf{B}(X, \mathbf{K}-)$ and $g: \mathbf{B}(Z, \mathbf{K}-) \to \mathbf{B}(Y, \mathbf{K}-)$; the composite natural transformation, $fg$, represents a map in $\mathbf{S_K}$ from $X$ to $Z$ and corresponds to

$$(\widetilde{fg}) = (fg)(\mathbf{L}(Z))(\eta_Z).$$

As $g(\mathbf{L}(Z))(\eta_Z) = \mathbf{K}(\phi_{Y,\mathbf{Y}(Z)}(g(\mathbf{L}(Z))(\eta_Z))) \circ \eta_Y$ and $f$ is a natural transformation, we obtain

$$(\widetilde{fg}) = \mathbf{K}(\phi_{Y,\mathbf{L}(Z)}(g(\mathbf{L}(Z))(\eta_Z))) \circ (f(\mathbf{L}(Y))(\eta_Y)). \quad (*)$$

On the other hand, we have, within the Kleisli category, the composite

$$X \xrightarrow{\tilde{f}} \mathbf{KL}(Y) \xrightarrow{\mathbf{KL}(\tilde{g})} \mathbf{KLKL}(Z) \xrightarrow{\mu(Z)} \mathbf{KL}(Z)$$

or more precisely

$$\mathbf{K}\varepsilon\mathbf{L}(Z) \circ \mathbf{KL}(g(\mathbf{L}(Z))(\eta_Z)) \circ (f(\mathbf{L}(Y))(\eta_Y)). \quad (**)$$

Comparing (*) and (**) we see that it is sufficient to prove that

$$\varepsilon\mathbf{L}(Z)\circ\mathbf{L}(g(\mathbf{L}(Z))(\eta_Z))=\phi_{Y,\mathbf{L}(Z)}(g(\mathbf{L}(Z))(\eta_Z)). \qquad (***)$$

Now $\phi$ is a natural transformation so with respect to $\tilde{g}\colon Y\to\mathbf{KL}(Z)$ applied in the first variable, we obtain a commutative square

$$\begin{array}{ccc}
\mathbf{B}(Y,\mathbf{KL}(Z)) & \xrightarrow[\phantom{xx}]{\phi_{Y,\mathbf{L}(Z)}\ \cong} & \mathbf{A}(\mathbf{L}(Y),\mathbf{L}(Z)) \\
\uparrow{\scriptstyle \tilde{g}^*} & & \uparrow{\scriptstyle \mathbf{L}(\tilde{g})^*} \\
\mathbf{B}(\mathbf{KL}(Z),\mathbf{KL}(Z)) & \xrightarrow[\cong]{\phi_{\mathbf{KL}(Z),\mathbf{L}(Z)}} & \mathbf{A}(\mathbf{LKL}(Z),\mathbf{L}(Z))
\end{array}$$

Starting with $\mathrm{Id}_{\mathbf{KL}(Z)}$ in the bottom left-hand corner and following it around anti-clockwise gives

$$\varepsilon\mathbf{L}(Z)\circ\mathbf{L}(g(\mathbf{L}(Z))(\eta_Z));$$

following it clockwise gives

$$\phi_{Y,\mathbf{L}(Z)}(g(\mathbf{L}(Z))(\eta_Z))$$

so (***) is satisfied and we have proved that $\mathbf{S_K}\cong\mathbf{B_T}$. The functor $S_\mathbf{K}\colon\mathbf{B}\to\mathbf{S_K}$ is, of course, just the functor $\mathbf{F_T}\colon\mathbf{B}\to\mathbf{B_T}$.

We have not yet proved that $(\mathbf{S_K}, S_\mathbf{K})$ is a shape theory for $\mathbf{K}$, but as this merely requires that (S2) is satisfied this is fairly quickly done; in fact, given $s\in\mathbf{S_K}(X,\mathbf{K}Q)$, we have that it is completely determined by $\tilde{s}\colon X\to\mathbf{KLK}(Q)$ in $\mathbf{B}$. We set $f=\mathbf{K}\varepsilon(Q)\circ s$, and note that

$$\mathbf{F_T}(f)=\eta_{\mathbf{K}Q}\circ\mathbf{K}(\varepsilon(Q))\circ s=s$$

since the composite $\eta_{\mathbf{K}(Q)}\circ\mathbf{K}(\varepsilon(Q))$ is the identity.

**Remark**

This 'meta-example' includes not only example (a) but, of course, many other algebraic and quasialgebraic situations. However, even in some simple algebraic cases, the condition it requires is not satisfied. We next look at one of the simplest of such cases: $\mathbf{K}$ = inclusion: **Fin. Groups** → **Groups** where **Fin. Groups** is the category of finite groups.

**(c) K = '*inclusion*': Fin. Groups → Groups**

We first see how near $\mathbf{K}$ is to having a left adjoint. Suppose $G$ is a group and $H$ a finite group; we want to study the set $\mathbf{Groups}(G,\mathbf{K}(H))$, i.e. the set of group homomorphisms from $G$ to $H$. If $f\colon G\to H$, then clearly $\operatorname{Ker} f$ is of finite index in $G$ and $f$ can be factored as

$$G\xrightarrow{\text{Proj}}\mathbf{K}(G/\operatorname{Ker} f)\to H.$$

As $f$ varies in **Groups**$(G, \mathbf{K}(H))$, Ker $f$ will vary through the normal subgroups of finite index of $G$. Thus we are led to consider

$$\hat{G} = \{G/N \mid N \lhd G,\ G/N \text{ finite}\}.$$

Although we have given this as a set, it clearly has more structure because if $N_i \lhd G$, $i = 1, 2$, and $N_1 \subset N_2$, then there is a homomorphism

$$\frac{G}{N_1} \to \frac{G}{N_2}.$$

There is, thus, a diagram, $\hat{G}$, of finite groups with objects the finite quotients of $G$ and indexing category the normal subgroups of finite index of $G$. This indexing category is associated with a directed set (since given $N_1, N_2$ normal of finite index in $G$ then $N_1 \cap N_2$ is normal of finite index in $G$), hence we can consider $\hat{G}$ as an inverse system. We will shortly see how to define a category of inverse (or projective) systems, **pro(C)**, for an arbitrary category **C** in such a way that $\hat{G}$ is an object in **pro(Fin. Groups)** and one has a natural isomorphism

$$\mathbf{Groups}(G, \mathbf{K}(H)) \cong \mathbf{pro(Fin.\,Groups)}(\hat{G}, H)$$

where one considers the finite group $H$ as an inverse system in the obvious trivial way. (We shall see that $G \to \hat{G}$ defines a functor $\widehat{\ }$: **Groups** → **pro(Fin. Groups)** and will say that $\widehat{\ }$ is proadjoint to **K**.)

Now looking at the natural transformation description of shape categories, one obtains in this situation

$$\begin{aligned} \mathbf{S_K}(G_1, G_2) &= \mathrm{Nat}(\mathbf{Groups}(G_2, \mathbf{K}-), \mathbf{Groups}(G_1, \mathbf{K}-)) \\ &\cong \mathrm{Nat}(\mathbf{pro(Fin.\,Groups)}(\hat{G}_2, -), \mathbf{pro(Fin.\,Groups)}(\hat{G}_1, -)) \\ &\cong \mathbf{pro(Fin.\,Groups)}(\hat{G}_1, \hat{G}_2). \end{aligned}$$

And one can identify $\mathbf{S_K}$ with a full subcategory of **pro(Fin. Groups)**.

Clearly such a description will work for any $\mathbf{K}\colon \mathbf{A} \to \mathbf{B}$ which has a 'proadjoint'. Since inverse systems came into the Mardešić–Segal description of Borsuk's classical shape, one thus expects the original topological **K** to have a proadjoint, but before examining this, we need a technical section on procategories, proadjoints and related concepts. As this material is not well represented in the shape theoretic literature, we will go into it in slightly more detail and in slightly more generality than is necessary for what follows.

## 2.3 PRO-OBJECTS, PROCATEGORIES AND PROADJOINTS

We start by recalling some standard ideas of elementary category theory.

**Definition**

(1) A category **J** is called *connected* is given any two objects $j, k \in \mathbf{J}$, there is a finite sequence of arrows (i.e. morphisms)

$$j = j_0 \to j_1 \leftarrow j_2 \to \cdots \to j_{2n-1} \leftarrow j_{2n} = k$$

(both directions possible) joining $j$ to $k$.

(2) A functor $\mathbf{L}: \mathbf{J}' \to \mathbf{J}$ is *final* if for each $k \in \mathbf{J}$, the comma category $(k \downarrow \mathbf{L})$ is non-empty and connected.
(3) A functor $\mathbf{L}: \mathbf{J}' \to \mathbf{J}$ is *initial* if $\mathbf{L}^{\mathrm{op}}: \mathbf{J}'^{\mathrm{op}} \to \mathbf{J}^{\mathrm{op}}$, its dual functor, is final.

**Proposition 1** (see MacLane [71], p. 213)
*If* $\mathbf{L}: \mathbf{J}' \to \mathbf{J}$ *is final and* $\mathbf{F}: \mathbf{J} \to \mathbf{C}$ *is a functor such that* $X = \mathrm{Colim}\,\mathbf{FL}$ *exists then* $\mathrm{Colim}\,\mathbf{F}$ *exists and the canonical map*

$$h: \mathrm{Colim}\,\mathbf{FL} \to \mathrm{Colim}\,\mathbf{F}$$

*is an isomorphism.*

*Proof*
Given any $k \in |J|$, there is some arrow $u: k \to \mathbf{L}(j')$; pick one and define

$$\tau_k: \mathbf{F}(k) \to X$$

to be the composite

$$\mathbf{F}(k) \xrightarrow{\mathbf{F}(u)} \mathbf{F}(\mathbf{L}(j')) \xrightarrow{\mu(j')} X$$

where $\mu(j')$ is the $j$th projection onto the colim (i.e. is part of the colimiting cone

$$\mu: \mathbf{FL} \to \mathbf{k}_X = \text{Constant functor with value } X).$$

As $k \downarrow \mathbf{L}$ is connected and $\mu$ is a natural transformation, $\tau_k$ is independent of the choice of $u$. In fact if $u = u_0: k \to \mathbf{L}(j'_0)$ and $u' = u_1: k \to \mathbf{L}(j'_1)$ with a morphism $f: (u_0, j'_0) \to (u_1, j'_1)$ in $(k \downarrow \mathbf{L})$

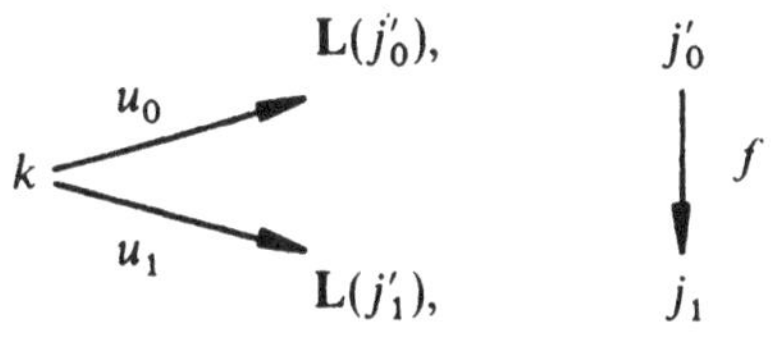


with $u_1 = \mathbf{L}(f)u_0$

then one has a possible $\tau'_k$ defined to be $\mu(j'_1)\mathbf{F}(u')$, but

$$\begin{aligned} \tau'_k &= \mu(j'_1)\mathbf{F}(u_1) \\ &= \mu(j'_1)\mathbf{FL}(f)\mathbf{F}(u_0) \\ &= \mu(j'_0)\mathbf{F}(u_0) \\ &= \tau_k. \end{aligned}$$

This also shows that the $\{\tau_k: \mathbf{F}(k) \to X\}$ combine to give a natural transformation

$$\tau: \mathbf{F} \to \mathbf{k}_X.$$

If $\lambda: \mathbf{F} \to \mathbf{k}_Y$ is another cone, then $\lambda\mathbf{L}: \mathbf{FL} \to \mathbf{k}_Y$ is a cone so one obtains a unique $g: X \to Y$ with $\mathbf{k}_g\mu = \lambda\mathbf{L}$ and hence with $\mathbf{k}_g\tau = \lambda$. Thus $\tau$ has a colimiting cone and so $X = \mathrm{Colim}\,\mathbf{F}$, making $h$ an ismorphism.

**Corollary: (Dual of Proposition 1)**
*If* $\mathbf{L}: \mathbf{J}' \to \mathbf{J}$ *is initial and* $\mathbf{F}: \mathbf{J} \to \mathbf{C}$ *is a functor such that* $X = \text{Lim}\,\mathbf{FL}$ *exists then* Lim $\mathbf{F}$ *exists and the canonical map*

$$h: \text{Lim}\,\mathbf{F} \to \text{Lim}\,\mathbf{FL}$$

*is an isomorphism.*

*Proof*
Apply the proposition to $\mathbf{L}^{\mathbf{op}}: \mathbf{J}'^{\mathbf{op}} \to \mathbf{J}^{\mathbf{op}}$ and $\mathbf{F}^{\mathbf{op}}: \mathbf{J}^{\mathbf{op}} \to \mathbf{C}^{\mathbf{op}}$.
In fact this property is characteristic of final functors and one has the following.

**Proposition 2**
*Let* $\mathbf{L}: \mathbf{J}' \to \mathbf{J}$ *be a functor (between small categories) such that for every* $\mathbf{F}: \mathbf{J} \to \mathbf{C}$ *with* $\mathbf{C}$ *cocomplete (i.e. having all colimits) the canonical map*

$$\text{Colim}\,\mathbf{FL} \to \text{Colim}\,\mathbf{F}$$

*is an isomorphism. Then* $\mathbf{L}$ *is final.*

*Proof*
We suppose $k \in |\mathbf{J}|$ and examine the case when $\mathbf{F} = \mathbf{J}(k, -): \mathbf{J} \to \mathbf{Sets}$ is the hom-functor. First we prove a lemma.

**Lemma**
Colim $\mathbf{J}(k, -)$ *is a one-point set.*

*Proof of Lemma*
We use the adjointness of Colim and $\mathbf{k}$, the 'constant-functor' functor and also the Yoneda Lemma. Let $X$ be a set,

$$\begin{aligned} \mathbf{Sets}(\text{Colim}\,\mathbf{J}(k, j), X) &\cong \mathbf{Sets}^{\mathbf{J}}(\mathbf{J}(k, -), \mathbf{k}_X) \\ &= \text{Nat}(\mathbf{J}(k, -), \mathbf{k}_X) \\ &\cong \mathbf{k}_X(k) \\ &= X. \end{aligned}$$

Thus Colim $\mathbf{J}(k, -)$ must have exactly one element.
Returning to the proof of the proposition, we have to show that $k \downarrow \mathbf{L}$ is non-empty and connected. Since Colim $\mathbf{J}(k, \mathbf{L}-)$ is non-empty, $\mathbf{J}(k, \mathbf{L}(i))$ must be non-empty for some $i \in |\mathbf{J}'|$. Next suppose

$$\alpha: k \to \mathbf{L}(i)$$

$$\beta: k \to \mathbf{L}(i')$$

are two objects in $k \downarrow \mathbf{L}$. One can construct Colim $\mathbf{J}(k, \mathbf{L}-)$ by means of the coequaliser diagram

$$\coprod_{\gamma} \mathbf{J}(k, \mathbf{L}(\text{dom}(\gamma))) \overset{f}{\underset{g}{\rightrightarrows}} \coprod_{i \in |J'|} \mathbf{J}(k, \mathbf{L}(i)) \longrightarrow \text{Colim}\,\mathbf{J}(k, \mathbf{L}-)$$

where the $\gamma$s indexing the left-hand coproduct are the morphisms of $\mathbf{J}'$, and $f, g$ are respectively given by

$$fi_\gamma = i_{\mathrm{dom}(\gamma)}$$

$$gi_\gamma = i_{\mathrm{codom}(\gamma)}\mathbf{J}(k, \gamma)$$

(with $i_\gamma$ the coproduct inclusions).

Thus Colim $\mathbf{J}(k, L(-))$ is a quotient of $\coprod_{i \in |\mathbf{J}'|} \mathbf{J}(k, \mathbf{L}(i))$ by the equivalence relation generated by

if $\alpha_1 : k \to \mathbf{L}(i_1)$, $\alpha_2 : k \to \mathbf{L}(i_2)$, there is a $\gamma : i_1 \to i_2$ in $\mathbf{J}'$ such that $\alpha_2 = \mathbf{L}(\gamma)\alpha_1$.

As Colim $\mathbf{J}(k, L-)$ consists of one point only, the equivalence classes of $\alpha$ and $\beta$ must be the same, hence they are linked by a chain of related elements. However, this is exactly the condition that $k \downarrow \mathbf{L}$ be connected, so $\mathbf{L}$ is final.

**Definition**

A (non-empty) category $\mathbf{I}$ is said to be *filtering* if it satisfies the following two axioms.

(a) Given any two objects $i, i'$ in $\mathbf{I}$, there is a $j \in \mathbf{I}$ and morphisms $i \to j$, $i' \to j$.

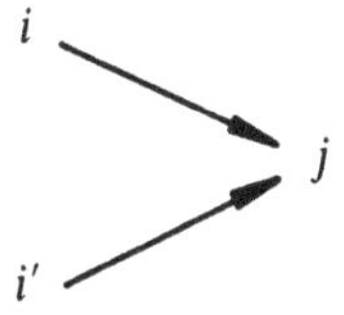


(b) To any parallel morphisms $u, v : i \to j$ in $\mathbf{I}$ there is a $k \in |\mathbf{I}|$ and a morphism $w : j \to k$ such that $wu = wv$, i.e.

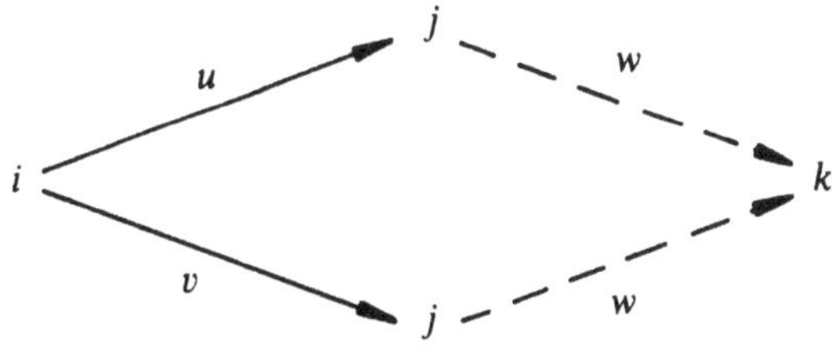


is commutative.

A category $\mathbf{I}$ is *cofiltering* if $\mathbf{I}^{\mathrm{op}}$ is filtering.

**Remark and Example**

Let $(A, <)$ be a directed set (see section 1.3); then one may form an associated category (also denoted $A$) with objects the elements of $A$ and with:

—$A(a_1, a_2)$ a one-element set if $a_1 = a_2$ or $a_1 < a_2$
—$A(a_1, a_2)$ empty otherwise.

$(A, <)$ is directed implies that $A$ is filtering. An inverse system in a category $\mathbf{C}$ can now be given as a functor $\mathbf{X} : A^{\mathrm{op}} \to \mathbf{C}$.

A cofinal subset $A'$ of $A$ is one for which the inclusion functor $A' \to A$ is final. ('Cofinal' is the older term but the 'co-' seems out of place here.)

**Proposition 3**

*Let* $\mathbf{L}: \mathbf{J}' \to \mathbf{J}$ *be a functor, and suppose* $\mathbf{J}'$ *is filtering. Then* $\mathbf{L}$ *is final if and only if it satisfies the conditions:*

(F1) *Given any object* $k$ *of* $\mathbf{J}$, *there is a* $j'$ *in* $\mathbf{J}'$ *and a morphism* $k \to \mathbf{L}(j')$;

(F2) *For any parallel morphisms* $u, v: k \to \mathbf{L}(j')$ *in* $\mathbf{J}$, *there is a morphism* $h: j' \to j''$ *in* $\mathbf{J}$ *such that*

$$\mathbf{L}(h)u = \mathbf{L}(h)v;$$

*Moreover if* $\mathbf{L}$ *is final,* $\mathbf{J}$ *is filtering.*

*Proof*

Suppose $\mathbf{L}$ satisfies (F1) and (F2), and consider $k\downarrow\mathbf{L}$ for some $k \in |\mathbf{J}'|$.

Suppose we have two objects $(u_1, j'_1)$, $(u_2, j'_2)$ in $k\downarrow\mathbf{L}$, so $u_i: k \to \mathbf{L}(\mathbf{j}'_i)$, $i = 1, 2$. Since $\mathbf{J}'$ is filtering, there is a $j'$ and maps $\alpha_i: \mathrm{j}'_1 \to j'$ in $\mathbf{J}'$. Set $u = \mathbf{L}(\alpha_1)u_1$, $v = \mathbf{L}(\alpha_2)u_2$ to obtain two parallel morphisms $u, v: k \to \mathbf{L}(j')$. (F2) tells us there is a $j''$ and $h: j' \to j''$ such that $\mathbf{L}(h)u = \mathbf{L}(h)v$. We thus have a commutative diagram

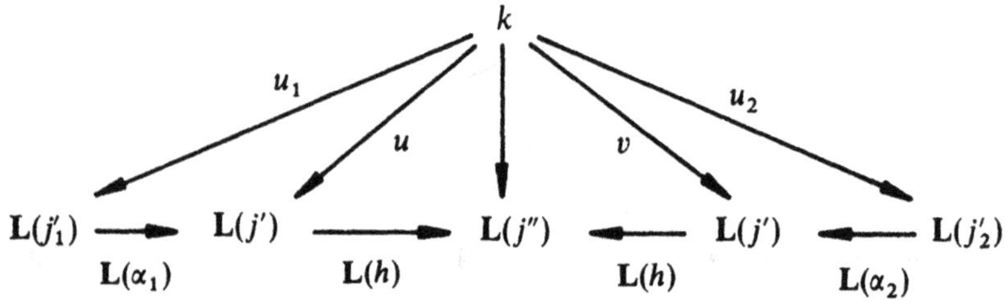


so $k\downarrow\mathbf{L}$ is connected, and $\mathbf{L}$ is final.

Now suppose $\mathbf{L}$ is final. $k\downarrow\mathbf{L}$ is non-empty for each $k$ so (F1) is satisfied. If we consider $u, v: k \to \mathbf{L}(j')$ as objects $(u, j')$, $(v, j')$ of $k\downarrow\mathbf{L}$, we have that there is a chain of maps

$$(u, j') = (u_0, j'_0) \xrightarrow[\alpha_0]{} (u_1, j'_1) \xleftarrow[\beta_1]{} \cdots \xrightarrow[\alpha_{n-1}]{} (u_{2n-1}, j'_{2n-1}) \xleftarrow[\beta_n]{} (u_{2n}, j'_{2n}) = (v, j').$$

For any $i$, $1 < i < n$, consider the segments

$$(u_{2i-1}, j'_{2i-1} \xleftarrow[\beta_i]{} (u_{2i}, j'_{2i}) \xrightarrow[\alpha_1]{} (u_{2i+1}, j'_{2i+1})$$

and its image in $\mathbf{J}'$ (via the codomain functor $k\downarrow\mathbf{L} \to \mathbf{J}'$)

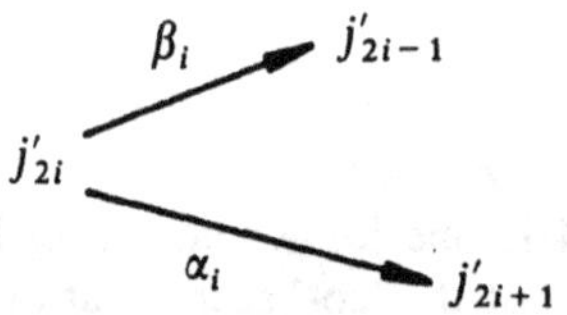

Since $\mathbf{J}'$ is filtering, we can use properties (a) to obtain a diagram

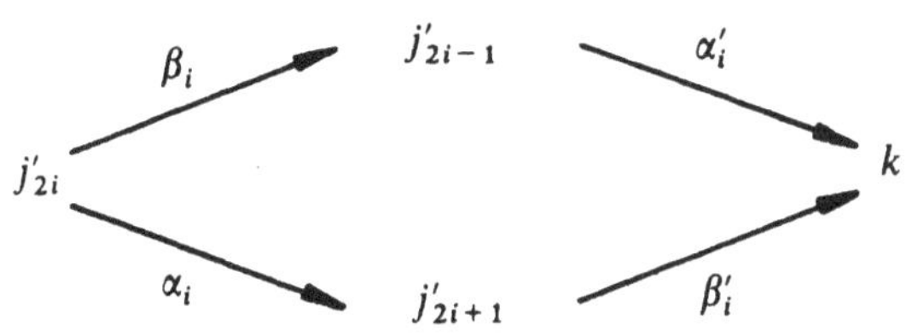


but this need not commute, so use property (b) of filtering categories to find a $k \xrightarrow{\gamma} k'$ such that

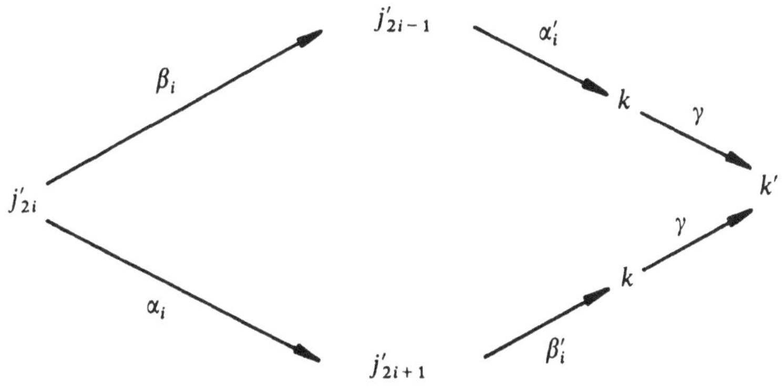


commutes. Looking at adjacent parts of the chain, we can replace

$$(u_{2i-2}, j'_{2i-2}) \xrightarrow[\alpha_{i-1}]{} (u_{2i-1}, j'_{2i-1}) \xleftarrow[\beta_i]{} (u_{2i}, j'_{2i}) \xrightarrow[\alpha_i]{} (u_{2i+1}, j'_{2i+1}) \xleftarrow[\beta_{i+1}]{} (u_{2i+2}, j'_{2i+2})$$

by

$$(u_{2i-2}, j'_{2i-2}) \xrightarrow[\alpha_{i-1}]{} (u_{2i-1}, j'_{2i-1}) \xrightarrow[\gamma\alpha'_i]{} (\mathbf{L}(\gamma\alpha'_i)u_{2i-1}, k') \xleftarrow[\gamma\beta'_i]{} (u_{2i+1}, j'_{2i+1}) \xleftarrow[\beta_{i+1}]{} (u_{2i+2}, j_{2i+2})$$

and hence on composing,

$$(u_{2i-2}, j_{2i-2}) \xrightarrow[\gamma_{\alpha'_i\alpha_{i-1}}]{} (\mathbf{L}(\gamma\alpha'_i)u_{2i-1}, k') \xleftarrow[\gamma\beta'_{i+1}]{} (u_{2i+2}, j_{2i+2})$$

thus shortening the length of the chain by two. Continuing in this way we can reduce the chain joining $(u, j')$ and $(v, j')$ to one of length two, i.e.

$$(u, j') \xrightarrow{\alpha} (w, j'') \xleftarrow{\beta} (v, j').$$

Again we project into $\mathbf{J}'$ to obtain the two parallel maps

$$\alpha, \beta: j' \to j''.$$

Using the filtering property (b), we can find a $j'''$ and a map $\delta$, say, $\delta: j'' \to j'''$ such that $\delta\alpha = \delta\beta$. Thus we finally arrive at a diagram

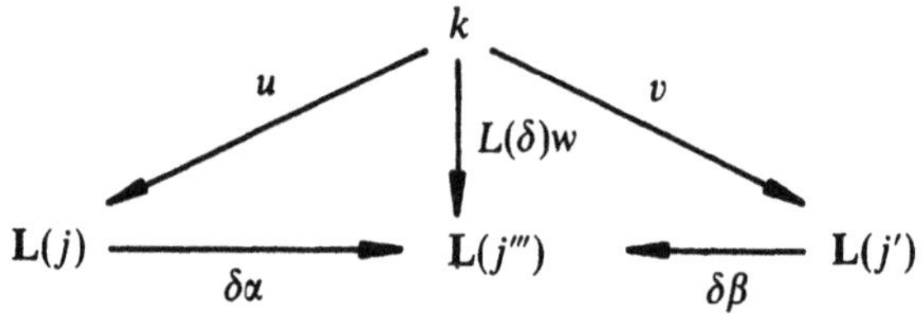


We set $h = \delta\alpha = \delta\beta$ to obtain the verification of property (F2).

Finally if $\mathbf{L}$ is final, then given $i, i'$ in $\mathbf{J}$, there are $j, j'$ in $\mathbf{J}'$ with morphisms $i \to \mathbf{L}(j)$, $i' \to \mathbf{L}(j')$. Now $\mathbf{J}'$ is filtering, so there is some $k$ with $j \to k$, $j' \to k$. One has a diagram

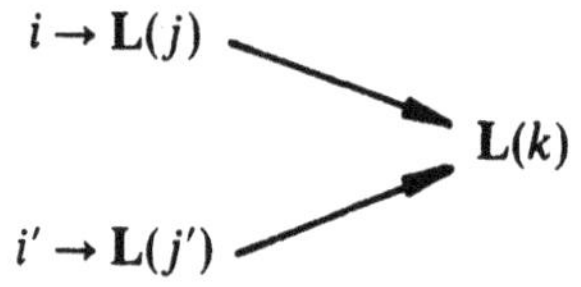


as required.

If $u, v: i \to i'$ in $\mathbf{J}$ then we can find a $j'$ in $\mathbf{J}'$ and $i' \xrightarrow{\alpha} \mathbf{L}(j')$. Now $\alpha u, \alpha v$ can be fed into property (F2) to find a $j' \xrightarrow{h} k$ in $\mathbf{J}'$ with $\mathbf{L}(h)\alpha u = \mathbf{L}(h)\alpha v$. Taking $w = \mathbf{L}(h)\alpha$ completes the proof.

As mentioned earlier, the classical theory of projective or inverse systems used directed sets. If one uses filtering categories instead of directed sets in the definition, it would seem that the notion was more general. There has been some discussion of this point in several shape theory texts. Here we note and prove a theorem of P. Deligne which shows that the two notions of projective system are equivalent.

**Proposition 4**

*Let $\mathbf{I}$ be a small filtering category. Then there is a (small) directed set $(E, <)$ and a final functor $\phi: E \to \mathbf{I}$ where $E$ also denotes the category associated to $(E, <)$.*

*Proof*

First we note that given $\mathbf{I}$, we can preorder $|\mathbf{I}|$ by

$$x \leqslant y \Leftrightarrow I(x, u) \neq \varnothing.$$

Now suppose that $(|\mathbf{I}|, \leqslant)$ has no greatest element. (We will reduce the general case to this one later.)

By a subdiagram of $\mathbf{I}$, we shall mean a pair $D = (Q, F)$ where $F \subseteq$ Morphisms of $I$, $Q \subseteq |I|$ such that for each $f \in F$, both domain and codomain of $f$ are in $Q$. An element $e$ of $Q$ will be said to be a final object of $D$ if for all $x \in D$, $\mathbf{I}(x, e) \cap F$ consists of exactly one element $f_x$ such that for each $g: x \to y$ in $D$, $f_x = f_y\, g$ and also $f_e = id$.

Let $E$ be the set of all finite subdiagrams $D$ of $\mathbf{I}$ having a unique final element $\phi(D)$. We order $E$ by inclusion. If $D, D' \in E$ and $D \subset D'$, then there is a unique morphism $\phi(D', D)\colon \phi(D) \to \phi(D')$ in $D'$, and if one has $D \subset D' \subset D''$ in $E$, then one clearly has

$$\phi(D'', D) = \phi(D'', D')\phi(D', D).$$

Similarly $\phi(D, D) = \mathrm{Id}_{\phi(D)}$. Thus $\phi\colon E \to \mathbf{I}$ is a functor and it remains to prove that $E$ is filtering and that $\phi$ is final. We start by checking conditions (F1) and (F2) of Proposition 3.

**Condition (F1)**
If $i \in |\mathbf{I}|$, take $D = (\{i\}, \{\mathrm{Id}_i\})$, then $i \leqslant \phi(D) = i$. Note also that this shows that $E \neq \varnothing$.

**Condition (F2)**
For each $i \in |\mathbf{I}|$, $D = (\underset{\sim}{O}, F) \in E$ and pairs of arrows

$$i \rightrightarrows \phi(D)$$

we need to find a diagram $D' \in E$, with $D \subset D'$, such that

$$\phi(D', D)\colon \phi(D) \to \phi(D')$$

equalises the given pair of arrows. Now $\mathbf{I}$ is filtering so there is a $j \in |\mathbf{I}|$ and an arrow

$$f\colon \phi(D) \to j$$

equalising the given pair. As $(|\mathbf{I}|, \leqslant)$ has no greatest element, we can suppose that $j$ is not among the objects of the subdiagram $D$. Thus we can take $D' = (\underset{\sim}{O}', F')$ to be the subdiagram of $\mathbf{I}$ with $\underset{\sim}{O}' = \underset{\sim}{O} \cup \{j\}$ and $F' = F \cup \{ff_x\colon x \in \underset{\sim}{O}\} \cup \{\mathrm{Id}_j\}$ so that $D' \in E$, $\phi(D') = j$ and $D'$ is the required element of $E$.

Finally we have to verify that $E$ is filtering.

Suppose given two elements, $D, D' \in E$, $D = (\underset{\sim}{O}, F)$, $D' = (\underset{\sim}{O}', F')$, then, as $\mathbf{I}$ is filtering, there is an object $j$ of $\mathbf{I}$ and arrows,

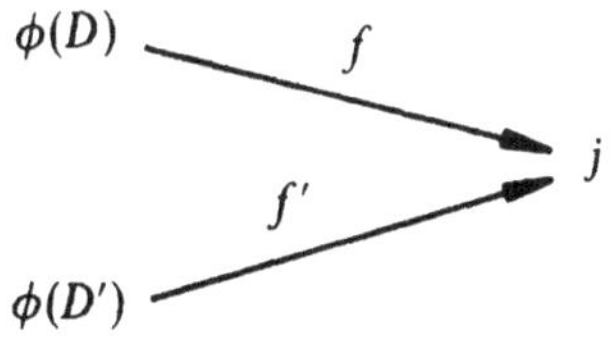


and we may assume that neither arrow is an identity. If $x \in \underset{\sim}{O} \cap \underset{\sim}{O}'$, then $ff_x$ and $f'f'_x$ have the same domain and codomain and so we cannot yet be sure that $j$ is a final object in some diagram $D''$ containing $D, D'$ and $j$. However, there are only finitely many $x \in \underset{\sim}{O} \cap \underset{\sim}{O}'$ so we can number them, $x_1, \ldots, x_n$, say, find $g_1\colon j \to j_1$ equalising $ff_{x_1}$ and $f'f'_{x_1}$, then $g_2\colon j_1 \to j_2$ equalising $g_1 ff_{x_2}$ and $g_1 f'f'_{x_2}$ and so on to obtain finally some $j'$ and $g\colon j \to j'$ equalising $ff_x$ and $f'f'_x$ for all $x \in \underset{\sim}{O} \cap \underset{\sim}{O}$. We now take $D'' = (\underset{\sim}{O}'', F'')$ with

$$\underset{\sim}{O}'' = \underset{\sim}{O} \cup \underset{\sim}{O}' \cup \{j'\}$$
$$F'' = F \cup F' \cup \{gff_x\colon x \in \underset{\sim}{O}\} \cup \{gf'f'_x\colon x \in \underset{\sim}{O}'\} \cup \{\mathrm{Id}_{j'}\}.$$

$j'$ is a final object of $D''$ and $D'' \in E$. (As $E$ is associated with an ordering, the second axiom for a filtering category is automatically satisfied.)

We now turn to the case when $(|\mathbf{I}|, \leqslant)$ has a largest element. Clearly, writing $\mathbf{N}$ for the filtering category of natural numbers, the product category $\mathbf{N} \times \mathbf{I}$ has no largest element so by the above there is a final functor $E \to \mathbf{N} \times \mathbf{I}$; however, the projection $\mathbf{N} \times \mathbf{I} \to \mathbf{I}$ is also final so we obtain a final functor $E \to \mathbf{I}$ as required.

**Remarks**

(1) It is worth noting that for each $D \in E$, there are only finitely many $D'$ in $E$ with $D' \leqslant D$.

(2) The above result is clearly dualisable to one on cofiltering categories. In the above form it appears in [51], whereas in its cofiltering form it is in [36].

If $\mathbf{C}$ is an arbitrary (small) category, it is customary to write $\hat{\mathbf{C}}$ for the category $\mathbf{Sets}^{\mathbf{C}^{\mathrm{op}}}$, often called the category of presheaves on $\mathbf{C}$. We will need below a dual version of this and will write

$$\check{\mathbf{C}} = (\widehat{\mathbf{C}^{\mathrm{op}}}) = \mathbf{Sets}^{\mathbf{C}}.$$

(For more than sufficient information on presheaf categories, see [51] Exposé I.)

**Definition**

A *pro-object* in $\mathbf{C}$ is a functor

$$\mathbf{F}: \mathbf{I} \to \mathbf{C}$$

where $\mathbf{I}$ is a cofiltering category.

**Notation**

(It is sometimes useful to write $(\mathbf{I}, \mathbf{F})$ to emphasise the *indexing category* $\mathbf{I}$ in the notation; another usual notation is $(X_i)_{i \in |\mathbf{I}|}$ where $X_i = \mathbf{F}(i)$; this latter, however, suppresses the *transition* or *bonding* morphisms of the pro-object, hence it must be used with care. To any pro-object $(\mathbf{I}, \mathbf{F})$, we can associate a functor, $\mathbf{L}(\mathbf{F}): \mathbf{C} \to \mathbf{Sets}$, by defining

$$\mathbf{L}(\mathbf{F})(X) = \operatorname*{Lim}_{i \in \mathbf{I}} \mathbf{C}(\mathbf{F}(i), X).$$

We define the *procategory*, $\mathbf{pro}(\mathbf{C})$, to be the category with objects the pro-objects in $\mathbf{C}$ and with

$$\mathbf{pro}(\mathbf{C})(\mathbf{F}, \mathbf{G}) = \check{\mathbf{C}}(\mathbf{L}(\mathbf{G}), \mathbf{L}(\mathbf{F})).$$

Using the definition of morphisms in $\check{\mathbf{C}}^{\mathrm{op}}$, we obtain

$$\mathbf{pro}(\mathbf{C})((\mathbf{I}, \mathbf{F}), (\mathbf{J}, \mathbf{G})) = \operatorname*{Lim}_{j \in |\mathbf{J}|} \operatorname*{Colim}_{i \in |\mathbf{I}|} \mathbf{C}(\mathbf{F}(i), \mathbf{G}(j)).$$

Since the category $\mathbf{1}$ with one morphism is a cofiltering category, within $\mathbf{pro}(\mathbf{C})$ we have all pro-objects of the form

$$\mathbf{F}: \mathbf{1} \to \mathbf{C}.$$

Clearly any such functor is determined completely by the one object of **C** to which it corresponds. We therefore obtain a functor

$$\mathbf{c}: \mathbf{C} \rightarrow \mathbf{pro}(\mathbf{C})$$

which embeds the category **C** in **pro(C)**. (As **pro(C)** is equivalent to a subcategory of $\check{\mathbf{C}}^{\mathbf{op}}$, we thus obtain an embedding of **C** into $\check{\mathbf{C}}^{\mathbf{op}}$ which is precisely the Yoneda embedding.) Pro-objects isomorphic to objects of the form $\mathbf{c}(X)$ for $X$ in **C** are called *essentially constant* or *stable*. It is usual to consider **C** as a subcategory of **pro(C)** and **c** as an inclusion. We shall sometimes omit **c** when this does not lead to confusion.

***Alternative description of morphisms in* pro(C)**

We have a formula

$$\mathbf{pro}(\mathbf{C})(\mathbf{F}, \mathbf{G}) = \operatorname*{Lim}_{j \in |\mathbf{J}|} \operatorname*{Colim}_{i \in |\mathbf{I}|} \mathbf{C}(\mathbf{F}(i)), \mathbf{G}(j)).$$

Interpreting this via the usual constructions of Lim and Colim over filtering and cofiltering categories, we obtain a very useful alternative description of $\mathbf{pro}(\mathbf{C})(\mathbf{F}, \mathbf{G})$:

$\mathbf{pro}(\mathbf{C})(\mathbf{F}, \mathbf{G})$ will be the set of equivalence classes of maps of systems were we make the following definitions:

A *map of systems* from $(\mathbf{I}, \mathbf{F})$ to $(\mathbf{J}, \mathbf{G})$ is a pair $\underline{h} = ((h_j)_{j \in |\mathbf{J}|}, \theta)$ where $\theta$ is a map from $|\mathbf{J}|$ to $|\mathbf{I}|$ and $h_j \in \mathbf{C}(\mathbf{F}(\theta(j)), \mathbf{G}(j))$, such that if $v: j \rightarrow j'$ in $\mathbf{J}$, there are $i \in \mathbf{I}$, $v: i \rightarrow \theta(j)$, and $v': i \rightarrow \theta(j')$ such that the following diagram is commutative:

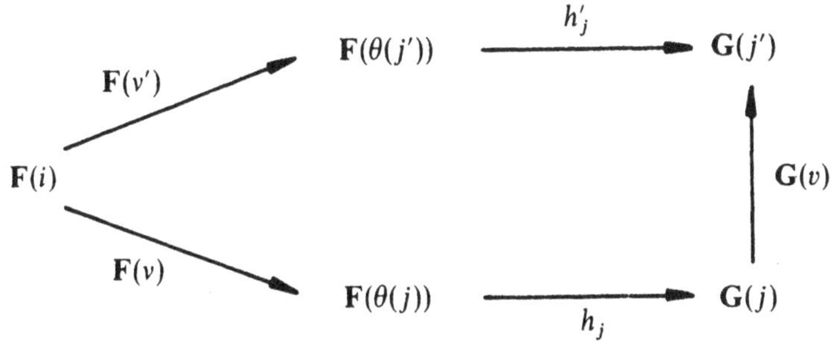


*Equivalence of systems*: $\underline{h}$ and $\underline{h}' = ((h'_j)_{j \in |\mathbf{J}|}, \theta')$ are said to be *equivalent* if for each $j \in |\mathbf{J}|$, there are an $i \in |\mathbf{I}|$, $v: i \rightarrow \theta(j)$ and $v': i \rightarrow \theta'(j)$ such that the following diagram commutes:

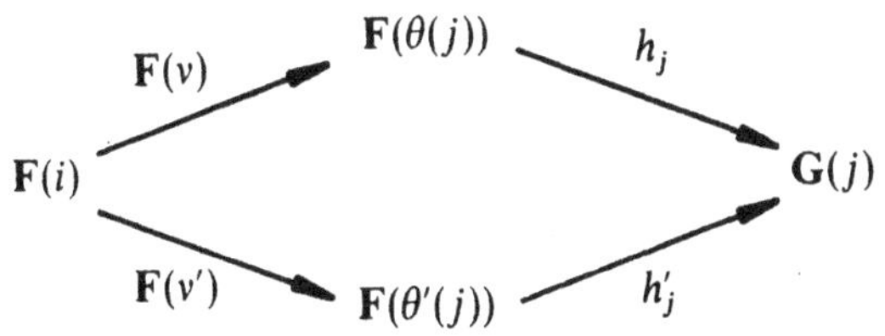


It will be convenient to denote by $[\underline{h}]$ the promorphism corresponding to $\underline{h}$ from $(\mathbf{I}, \mathbf{F})$ to $(\mathbf{J}, \mathbf{G})$. With this notation one can easily write down the composite of $[\underline{h}]$ with $[\underline{g}] = [(g_k)_{k \in |\mathbf{K}|}, \gamma]$ a promorphism, or promap, from $(\mathbf{J}, \mathbf{G})$ to $(\mathbf{K}, \mathbf{H})$. The composite is given by

$$[\underline{g}][\underline{h}] = [(g_k h_{\gamma(k)})_{k\in|\mathbf{K}|}, \theta\gamma].$$

The identity $\mathrm{Id}_{(\mathbf{I},\mathbf{F})}$ is then given by $((\mathrm{Id}_{\mathbf{F}(i)})_{i\in|\mathbf{I}|}, \mathrm{Id}_{|\mathbf{I}|})$.

**Proposition 5**

*Let* $(\mathbf{I}, \mathbf{F})$ *be a pro-object in* $\mathbf{C}$, $\mathbf{I}'$ *a cofiltering category and* $\phi\colon \mathbf{I}' \to \mathbf{I}$ *an initial functor then in* **pro**($\mathbf{C}$), $(\mathbf{I}, \mathbf{F})$ *is isomorphic to* $(\mathbf{I}', \mathbf{F}\phi)$.

*Proof*

We compare $\mathbf{L}(\mathbf{F})$ and $\mathbf{L}(\mathbf{F}\phi)$. For an arbitrary $X$ in $\mathbf{C}$

$$\mathbf{L}(\mathbf{F})(X) = \mathrm{Lim}\left(\mathbf{I} \xrightarrow{\mathbf{C}(\mathbf{F}(-),X)} \mathbf{Sets}\right)$$

$$\mathbf{L}(\mathbf{F}\phi)(X) = \mathrm{Lim}\left(\mathbf{I}' \xrightarrow{\phi} \mathbf{I} \xrightarrow{\mathbf{C}(\mathbf{F}(-),X)} \mathbf{Sets}\right).$$

As $\phi$ is initial, the Corollary (Dual to Proposition 1) implies that the natural transformation

$$\mathbf{L}(\mathbf{F}) \to \mathbf{L}(\mathbf{F}\phi)$$

is a natural isomorphism, hence an isomorphism in $\check{\mathbf{C}}$, i.e. $(\mathbf{I}, \mathbf{F})$ and $(\mathbf{I}', \mathbf{F}\phi)$ are isomorphic.

***Prorepresentable functors***

A functor $\mathbf{L}\colon \mathbf{C} \to \mathbf{Sets}$ is said to be *prorepresentable* if it is isomorphic in $\check{\mathbf{C}}^{\mathrm{op}}$ to a functor of the form $\mathbf{L}(\mathbf{F})$ for $\mathbf{F}$ in **pro**($\mathbf{C}$).

**Remarks**

If $\mathbf{F}$ is in **pro**($\mathbf{C}$), $\mathbf{L}(\mathbf{F}) = $ **pro**($\mathbf{C}$)($\mathbf{F}$, $\mathbf{c}(\quad)$) with $\mathbf{c}\colon \mathbf{C} \to$ **pro**($\mathbf{C}$) the canonical embedding.

Consider the comma category $(\mathbf{Yon}\downarrow\mathbf{L})$ where $\mathbf{Yon}\colon \mathbf{C} \to (\mathbf{C})^{\mathrm{op}}$ is the Yoneda Functor, $\mathbf{Yon}(X) = \mathbf{C}(X, -)$. (Using the Yoneda Lemma one has that a natural transformation, $\mathbf{u}\colon \mathbf{Yon}(X) \to \mathbf{L}$, corresponds to a $u \in \mathbf{L}(X)$; hence the objects of $(\mathbf{Yon}\downarrow\mathbf{L})$ can be thought of as pairs $(X, u)$ with $X \in |\mathbf{C}|$, $u \in \mathbf{L}(X)$ and one has a map $f$ from $(X, u)$ to $(Y, v)$ if $f\colon Y \to X$ in $\mathbf{C}$ is such that $\mathbf{L}(f)(v) = u$.)

There is a domain functor

$$(\mathbf{Yon}\downarrow\mathbf{L}) \xrightarrow{\partial_{\mathbf{L}}} \mathbf{C}^{\mathrm{op}}$$

and one has

$$\mathbf{L} \cong \mathrm{Lim}\left((\mathbf{Yon}\downarrow\mathbf{L})^{\mathrm{op}} \xrightarrow{\partial_{\mathbf{L}}{}^{\mathrm{op}}} \mathbf{C}^{\mathrm{op}}\right)$$

by general results on functor categories.

If $(\mathbf{Yon}\downarrow\mathbf{L})$ is filtering, then $(\mathbf{Yon}\downarrow\mathbf{L})^{\mathrm{op}}$ is cofiltering and $\mathbf{L}$ is prorepresented by $\partial_{\mathbf{L}^{\mathrm{op}}}\colon (\mathbf{Yon}\downarrow\mathbf{L})^{\mathrm{op}} \to \mathbf{C}$ (provided $(\mathbf{Yon}\downarrow\mathbf{L})^{\mathrm{op}}$ has an initial small subcategory, in which case, we say it is *essentially small*).

Now assume $\mathbf{L} \cong \mathbf{L}(\mathbf{F})$, $\mathbf{F}\colon \mathbf{I} \to \mathbf{C}$ a pro-object, then

$$\mathbf{L} \cong \mathrm{Lim}\ \mathbf{C}(\mathbf{F}(i), -)$$

so one has an initial functor

$$\mathbf{I} \xrightarrow{\cong} \{\mathbf{L} \to \mathbf{C}(\mathbf{F}(i), -)\}_{i \in |\mathbf{I}|} \subseteq (\mathbf{Yon} \downarrow \mathbf{L})^{\mathbf{op}}$$

and hence $(\mathbf{Yon} \downarrow \mathbf{L})^{\mathbf{op}}$ is cofiltering (by Proposition 3). We have thus proved

**Proposition 6**
*If* $\mathbf{L}: \mathbf{C} \to \mathbf{Sets}$ *is a functor then* $\mathbf{L}$ *is prorepresentable if and only if the comma category* $(\mathbf{Yon} \downarrow \mathbf{L})$ *is filtering and essentially small where* $\mathbf{Yon}: \mathbf{C} \to \check{\mathbf{C}}^{\mathbf{op}} = (\mathbf{Sets}^{\mathbf{C}})^{\mathbf{op}}$ *is the Yoneda functor.*

**Remarks**
If $\mathbf{C}$ is equivalent to a small category, $(\mathbf{Yon} \downarrow \mathbf{L})$ is automatically essentially small.

**Proposition 7**
*If* $\mathbf{C}$ *has all finite limits and is equivalent to a small category,* $\mathbf{L}$ *is prorepresentable if and only if it is left exact.*

*Proof*
(i) If $\mathbf{L}$ is prorepresentable

$$\begin{aligned} \mathbf{L} &\cong \operatorname*{Lim}_{\mathbf{I}} \mathbf{C}(\mathbf{F}(i), -) \text{ within } (\check{\mathbf{C}})^{\mathbf{op}} \\ &\cong \operatorname*{Colim}_{\mathbf{I}^{\mathbf{op}}} \mathbf{C}(\mathbf{F}(i), -) \text{ within } \check{\mathbf{C}}. \end{aligned}$$

Each $\mathbf{C}(\mathbf{F}(i), -)$ preserves finite limits, and filtered colimits commute with finite limits (see [71], Ch. IX §2). Thus $\mathbf{L}$ is left exact.

(ii) Now suppose $\mathbf{L}$ is left exact and let $(X, a)$, $(X', a')$ be two objects of $(\mathbf{Yon} \downarrow \mathbf{L})^{\mathbf{op}}$. One has a product diagram

$$\begin{array}{ccc} & & X \\ & \overset{\pi}{\nearrow} & \\ X \times X' & & \\ & \underset{\pi'}{\searrow} & \\ & & X' \end{array}$$

in $\mathbf{C}$, hence as $\mathbf{L}$ preserves finite limits, a product diagram

$$\begin{array}{ccc} & & \mathbf{L}(X) \\ & \overset{\mathbf{L}(\pi)}{\nearrow} & \\ \mathbf{L}(X) \times \mathbf{L}(X') \cong \mathbf{L}(X \times X') & & \\ & \underset{\mathbf{L}(\pi')}{\searrow} & \\ & & \mathbf{L}(X) . \end{array}$$

and $\mathbf{L}(\pi)(a, a') = a$, $\mathbf{L}(\pi')(a, a') = a'$.

Thus $\pi, \pi'$ are morphisms in $(\mathbf{Yon} \downarrow \mathbf{L})^{\mathbf{op}}$, hence verifying the first property of a cofiltering category.

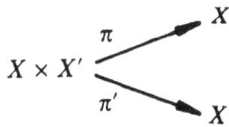

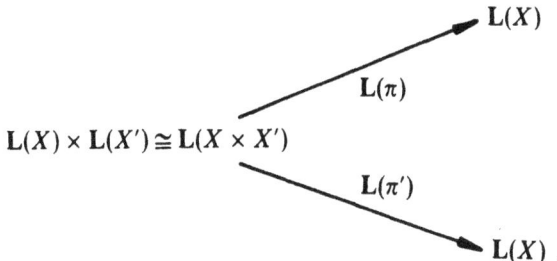

Next suppose $f, g\colon (X, a) \to (X', a')$ is a pair of morphisms in $(\mathbf{Yon}\downarrow\mathbf{L})^{\mathrm{op}}$. Let $E \xrightarrow{m} X$ be the equaliser of $f$ and $g$ in $\mathbf{C}$; then, as $\mathbf{L}$ preserves finite limits and

$$f(a) = g(a) = a',$$

one has $a \in \mathbf{L}(E)$. Thus

$$(E, a) \xrightarrow{m} (X, a) \overset{f}{\underset{g}{\rightrightarrows}} (X', a')$$

is a diagram in $(\mathbf{Yon}\downarrow\mathbf{L})^{\mathrm{op}}$ with $fm = gm$. Hence $(\mathbf{Yon}\downarrow\mathbf{L})^{\mathrm{op}}$ is cofiltering and $(\mathbf{Yon}\downarrow\mathbf{L})$ is filtering. As $\mathbf{C}$ is assumed to be equivalent to a small category, $(\mathbf{Yon}\downarrow\mathbf{L})$ is essentially small and so $h$ is prorepresentable by Proposition 5.

***Proadjoints***

Let $\mathbf{F}\colon \mathbf{C} \to \mathbf{C}'$ be a functor; then $\mathbf{F}$ induces a functor

$$\mathbf{F}^*\colon \check{\mathbf{C}}' \to \check{\mathbf{C}}$$

defined by composition:

$$\text{if } \mathbf{X}\colon \mathbf{C}' \to \mathbf{Sets},\ \mathbf{F}^*(\mathbf{X}) = \mathbf{X} \circ \mathbf{F}\colon \mathbf{C} \to \mathbf{Sets}.$$

**Definition**

We shall say that $\mathbf{F}$ *admits a proadjoint* if for any prorepresentable functor $\mathbf{X}$ in $\mathbf{C}'$, $\mathbf{F}^*(\mathbf{X})$ is prorepresentable.

**Proposition 8**

$\mathbf{F}$ *admits a proadjoint if and only if there is a functor*

$$\mathbf{g}\colon \mathbf{pro}(\mathbf{C}') \to \mathbf{pro}(\mathbf{C})$$

*and a natural isomorphism*

$$\mathbf{pro}(\mathbf{C})(\mathbf{g}(\underline{Y}'), X) \cong \mathbf{pro}(\mathbf{C}')(\underline{Y}', \mathbf{cF}(X)),$$

*where* $\mathbf{c}\colon \mathbf{C} \to \mathbf{pro}(\mathbf{C})$ *is the canonical embedding.*

*Proof*

Suppose $\mathbf{F}$ admits a proadjoint and let $\mathbf{L}(\underline{Y}')$ be a prorepresentable functor in $\mathbf{C}'$. If $X \in |\mathbf{C}|$, we have

$$\mathbf{F}^*(\mathbf{L}(\underline{Y}'))(X) = \mathbf{pro}(\mathbf{C})(\underline{Y}', \mathbf{cF}(X)).$$

As $\mathbf{F}$ has a proadjoint, we can restrict $\mathbf{F}^{*\mathrm{op}}$ to $\mathbf{pro}(\mathbf{C}')$:

$$\mathbf{g} = \mathbf{F}^{*\mathrm{op}}\colon \mathbf{pro}(\mathbf{C}') \to \mathbf{pro}(\mathbf{C})$$

(identifying $\mathbf{pro}(\mathbf{C})$ (etc.) with its image in $\check{\mathbf{C}}$).

We have thus, for each $\underline{Y}'$ in $\mathbf{pro}(\mathbf{C}')$, a natural isomorphism

$$\mathbf{F}^*(\mathbf{L}(\underline{Y}') \xrightarrow{\cong} \mathbf{L}(\mathbf{g}(\underline{Y}')),$$

hence a natural isomorphism,

$$\mathbf{pro}(\mathbf{C})(\mathbf{g}(\underline{Y}'), \mathbf{c}(X)) \cong \mathbf{pro}(\mathbf{C}')(\underline{Y}', \mathbf{cF}(X))$$

as required. As $\mathbf{g}$ is exactly $\mathbf{F}^{*\mathrm{op}}$ restricted to $\mathbf{pro}(\mathbf{C}') \subset \check{\mathbf{C}}'$, the opposite implication is clear.

**Corollary**

*Let* $\mathbf{F}: \mathbf{C} \to \mathbf{C}'$ *be a functor, then* $\mathbf{F}$ *admits a proadjoint if and only if the extension,* $\mathbf{pro}(\mathbf{F}): \mathbf{pro}(\mathbf{C}) \to \mathbf{pro}(\mathbf{C}')$, *of* $\mathbf{F}$ *to* $\mathbf{pro}(\mathbf{C})$ *admits a left adjoint.*

*Proof*

We shall first make explicit the description of $\mathbf{pro}(\mathbf{F})$. Suppose $\underline{Y}$ is a pro-object in $\mathbf{C}$; we can represent $\underline{Y}$ as a functor

$$\underline{Y}: \mathbf{I} \to \mathbf{C}$$

with $\mathbf{I}$ cofiltering, then $\mathbf{pro}(\mathbf{F})(\underline{Y}) = \mathbf{F} \circ \underline{Y}$. Equivalently if $\underline{Y} = (Y_i)_{i \in |\mathbf{I}|}$,

$$\mathbf{pro}(\mathbf{F})(\underline{Y}) = (\mathbf{F}(Y_i))_{i \in |\mathbf{I}|}.$$

Now for each $i \in \mathbf{I}$, we have an isomorphism of functors from $\mathbf{I}$ to $\mathbf{Sets}$.

$$\{\mathbf{pro}(\mathbf{C})(\mathbf{g}(\underline{Y}'), \mathbf{c}(Y_i)) \cong \mathbf{pro}(\mathbf{C}')(\underline{Y}', \mathbf{cF}(Y_i))\}_{i \in |\mathbf{I}|}.$$

Taking the limit of these functors (with respect to $i$) gives a natural isomorphism

$$\mathbf{pro}(\mathbf{C})(\mathbf{g}(\underline{Y}'), \underline{Y}) \cong \mathbf{pro}(\mathbf{C}')(\underline{Y}', \mathbf{pro}(\mathbf{F})(\underline{Y})),$$

i.e. $\mathbf{g}$ is left adjoint to $\mathbf{pro}(\mathbf{F})$.

The converse follows from this natural isomorphism on restricting $\underline{Y}$ to be of the form $\mathbf{c}(X)$, $X \in |\mathbf{C}|$ and then citing Proposition 8.

We also have the following explicit description of a proadjointness situation.

**Proposition 9**

*Let* $\mathbf{F}: \mathbf{C} \to \mathbf{C}'$ *be a functor. Then* $\mathbf{F}$ *admits a proadjoint if and only if the following three conditions are satisfied*:

(i) *for any* $Y$ *in* $\mathbf{C}'$, $X_1, X_2$ *in* $\mathbf{C}$ *and morphisms*

$$f_1: Y \to \mathbf{F}X_1$$

$$f_2: Y \to \mathbf{F}X_2$$

*there is an* $X$ *in* $\mathbf{C}$ *with morphisms* $f: Y \to \mathbf{F}X, f'_1: X \to X_1, f'_2: X \to X_2$ *such that the diagram*

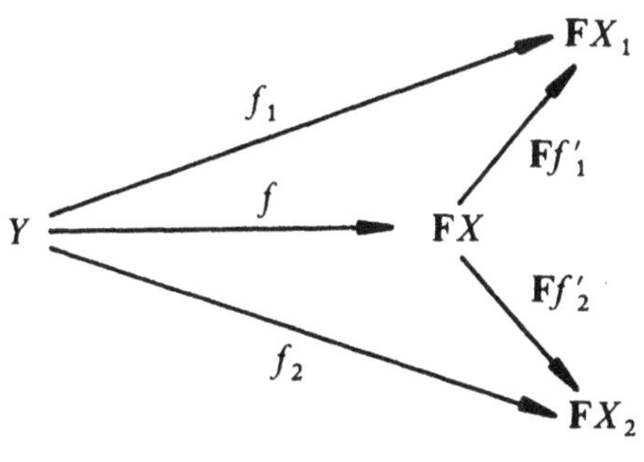


*commutes*;

(ii) *for any commutative diagram*

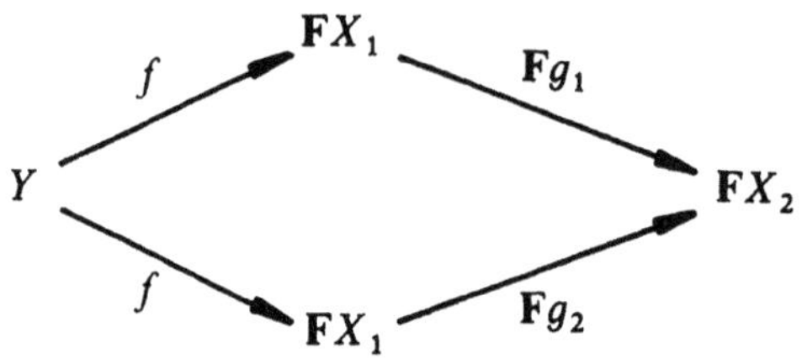

*with $Y$ in $\mathbf{C}'$, $X_1, X_2$ in $\mathbf{C}$, there is an $X$ in $\mathbf{C}$ and morphisms $f_1 \colon Y \to \mathbf{F}X$ in $\mathbf{C}'$, $f_2 \colon X \to X_1$ in $\mathbf{C}$ such that the diagrams*

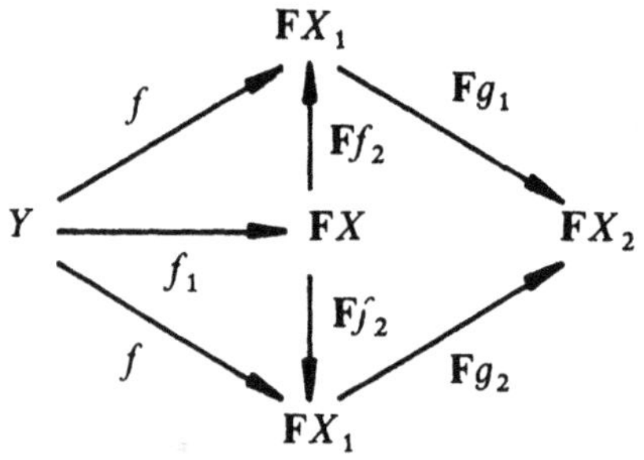

and

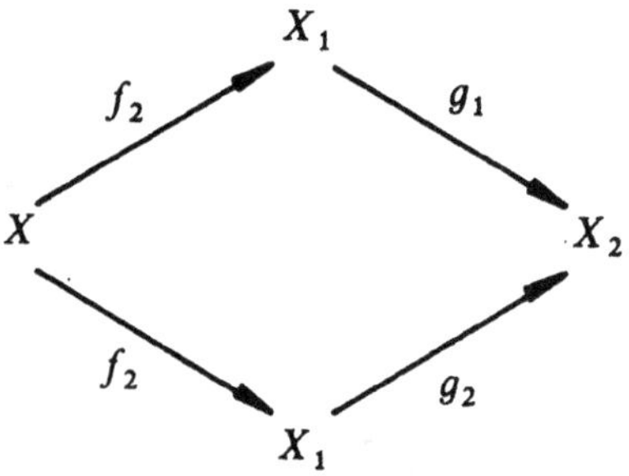

*commute*;

(iii) *for any $Y$ in $\mathbf{C}'$, there is a small subcategory $\mathbf{C}(Y)$ of $\mathbf{C}$ such that every morphism $f \colon Y \to \mathbf{F}X$, $X$ in $\mathbf{C}$, has a factorisation*

$$Y \xrightarrow{f_1} \mathbf{F}X' \xrightarrow{\mathbf{F}f_2} \mathbf{F}X$$

*with $X' \in |\mathbf{C}(Y)|$.*

*Proof*

We first remark that in Proposition 8, it suffices to prove that there is a functor

$$\mathbf{g} \colon \mathbf{C}' \to \mathbf{pro}(\mathbf{C})$$

such that

$$\mathbf{pro}(\mathbf{C})(\mathbf{g}(Y), X) \cong \mathbf{C}'(Y, \mathbf{F}(X))$$

since if $\underline{Y} \colon \mathbf{I} \to \mathbf{C}'$ is a pro-object then $\mathbf{g}(Y') \colon \mathbf{I} \to \mathbf{pro}(\mathbf{C})$ can also be considered as a pro-object in $\mathbf{C}$. Accepting this for the moment we have an isomorphism of functors

$$i \to \mathbf{pro}(\mathbf{C})(\mathbf{g}(\underline{Y}(i)), X) \cong \mathbf{C}'(\underline{Y}(i), \mathbf{F}(X))$$

from $\mathbf{I}$ to **Sets** and hence an isomorphism of their colimits, i.e.

$$\mathbf{pro}(\mathbf{C})(\mathbf{g}(\underline{Y}), X) \cong \mathbf{pro}(\mathbf{C}')(\underline{Y}, \mathbf{F}(X))$$

as required.

Returning to the result we assumed, we shall state it as a separate result, as it will often be useful later. It will be proved after the end of the proof of Proposition 9.

**Proposition 10**
*The category,* **pro(C)**, *has all projective limits; in fact if* **F**: **J** → **pro(C)** *is a pro-object in* **pro(C)**, *then* **F** *defines a new pro-object (essentially still* **F** *itself) which is the limit of* **F** *in* **pro(C)**.

*Continuation of Proof of Proposition* 9
The rest is now quite simple. Conditions (i) and (ii) state that the comma category $(Y \downarrow \mathbf{F})$ is cofiltering, whereas (iii) states it is essentially small (in fact has a small cofinal subcategory). Hence the codomain functor defines a pro-object

$$(Y \downarrow \mathbf{F}) \xrightarrow{\partial_Y} \mathbf{C}$$

which is to be $\mathbf{g}(Y)$.

Now assume given $f: Y \to \mathbf{F}(X)$. We can, by (iii), factor this via

$$Y \xrightarrow{f_1} \mathbf{F}(X_1) \xrightarrow{\mathbf{F}f_2} \mathbf{F}(X)$$

with $X'$ in $|\mathbf{C}(Y)|$. Taking colimits, $f_2$ represents an element of

$$\operatorname*{Colim}_{\mathbf{C}(Y)} \mathbf{C}(\partial_Y, X)$$

which is, by the cofilteringness of $Y \downarrow \mathbf{F}$, itself independent of the choice of factorisation. Thus

$$\begin{aligned}\mathbf{pro(C)}(\mathbf{g}(Y), X) &= \operatorname*{\mathbf{Colim}}_{\mathbf{C}(Y)} \mathbf{C}(\partial_Y, X) \\ &\cong \mathbf{C}'(Y, \mathbf{F}(X))\end{aligned}$$

as required.

*Proof of Proposition* 10
Suppose $\mathbf{F}: \mathbf{J} \to \mathbf{pro(C)}$ is a projective system in **pro(C)** and that for each $j \in |\mathbf{J}|$, $\mathbf{F}(j)$ is the pro-object represented by $\mathbf{F}(j): \mathbf{I}_j \to \mathbf{C}$. We may assume by Proposition 4 that **J**, and all the $\mathbf{I}_j$, are categories associated with directed sets. We define a new directed set as follows:

For each $\alpha: j_1 \to j_2$, there is a corresponding bonding morphism $\mathbf{F}(\alpha): \mathbf{F}(j_1) \to \mathbf{F}(j_2)$ in **F**. We pick a representing map of systems $((f_k), \phi)$ for $\mathbf{F}(\alpha)$. Now let **L** be the disjoint union of the sets $\mathbf{I}_j$ and denote elements of **L** by pairs $(i, j)$ where $\in \mathbf{I}_j$, $j \in \mathbf{J}$. We order the elements of **L** by $(i_1, j_1) \leqslant (i_2, j_2)$ if and only if $j_1 \leqslant j_2$ and, if $((f_k), \phi)$ is the chosen representative of the map $F(j_1) \to F(j_2)$, then $i_1 \leqslant \phi(i_2)$. We leave, as an exercise, the verification that this makes **L** into a cofiltered ordered set.

Define a functor $\bar{F}: \mathbf{L} \to \mathbf{C}$ by

$$\bar{\mathbf{F}}(i, j) = \mathbf{F}(j)(i)$$

and if $(i_1, j_1) \to (i_2, j_2)$ is a map in **L**, then

$$\bar{\mathbf{F}}(i_1, j_1) \longrightarrow \bar{\mathbf{F}}(i_2, j_2) = \mathbf{F}(j_1)(i_2) \longrightarrow \mathbf{F}(j_1)(\phi(i_2)) \xrightarrow{f_{i_2}} \mathbf{F}(j_2)(i_2)$$

(here the first map in the composite is a transition in $\mathbf{F}(\mathbf{j}_1)$ and the second is $f_{i_2}$ where $((f_i), \phi)$ is the transition promorphism $\mathbf{F}(j_1) \to \mathbf{F}(j_2)$). We claim that $\bar{\mathbf{F}}$ is the (projective) limit of $\mathbf{F}$. Clearly, $\bar{\mathbf{F}} \in |\mathbf{pro}(\mathbf{C})|$ and the limiting cone is given by the promorphisms

$$\mu_j \colon \bar{\mathbf{F}} \to \mathbf{F}(j)$$

represented by the maps of systems $(\{\mathrm{id}\}, \mathrm{inc})$ where $\mathrm{inc}\colon |\mathbf{I}_j| \to |\mathbf{L}|$ is the inclusion, and the maps $\bar{\mathbf{F}}(\mathrm{inc}(j)) \to \mathbf{F}(j)$ are the relevant identities. Note that $\bar{\mathbf{F}}(\mathrm{inc}(j))$ is the same as $\mathbf{F}(j)$ under the obvious identification.

If $\mathbf{G}\colon \mathbf{K} \to \mathbf{C}$ and $\{\lambda_j \colon \mathbf{G} \to \mathbf{F}(j)\}$ gives another cone on $\mathbf{F}\colon \mathbf{J} \to \mathbf{pro}(\mathbf{C})$ then each $\lambda_j$ is represented by some $((l_j), \psi_j)$, where $\psi_j \colon |\mathbf{I}_j| \to |\mathbf{K}|$ and

$$l_j(i) \colon \mathbf{G}(\psi_j(i)) \to \mathbf{F}(j)(i) = \bar{\mathbf{F}}(i, j)$$

is a morphism in $\mathbf{C}$. Since $\mathbf{L}$ is a disjoint union, we can define

$$\lambda \colon |\mathbf{L}| \to |\mathbf{K}|$$

by $\lambda(i, j) = \psi_j(i)$ and then the morphisms $l_j$ give morphisms

$$l_j(i) \colon \mathbf{G}(\psi(i, j)) \to \bar{\mathbf{F}}(i, j)$$

which represent the unique map $\lambda \colon \mathbf{G} \to \bar{\mathbf{F}}$ in $\mathbf{pro}(\mathbf{C})$. Thus $\bar{\mathbf{F}}$ is a limit of $\mathbf{F}$.

**Remarks**

(a) If $\mathbf{F}\colon \mathbf{C} \to \mathbf{Sets}$ has a proadjoint then it is prorepresentable. The converse is true if and only if $\mathbf{pro}(\mathbf{C})$ is closed under (small) products taken in $\mathbf{C}$. Thus proadjointness is the natural generalisation of prorepresentability. Clearly, if an arbitrary $\mathbf{F}\colon \mathbf{C} \to \mathbf{C}'$ has a proadjoint, then $\mathbf{F}$ must be left exact (i.e. preserve all finite limits). If $\mathbf{C}$ is small, this condition is also sufficient.

(b) More results on prorepresentability and proadjointness can be found in SGA4 [51] Exposé I. Of particular note are those relating to strict pro-objects, i.e. those with epimorphic bondings.

Our motivation for introducing procategories was to explain the example on profinite groups and to prepare the way for the next section where procategories come in in an essential way. We finish this section with the (now obvious) calculation of $\mathbf{S}_\mathbf{K}$ when $\mathbf{K}$ has a proadjoint.

We suppose $\mathbf{K}\colon \mathbf{A} \to \mathbf{B}$ has a proadjoint, hence that $\mathbf{pro}(\mathbf{K})$ has a left adjoint $\mathbf{L}$. Thus for any $Y$ in $\mathbf{B}$, we can make the following natural identifications:

$$\begin{aligned} \mathbf{B}(Y, \mathbf{K}-) &\cong \mathbf{pro}(\mathbf{B})(\mathbf{c}Y, \mathbf{cK}-) \\ &\cong \mathbf{pro}(\mathbf{A})(\mathbf{L}(Y), -). \end{aligned}$$

Hence

$$\mathrm{Nat}(\mathbf{B}(Y, \mathbf{K}-), \mathbf{B}(X, \mathbf{K}-)) \cong \mathbf{pro}(\mathbf{A})(\mathbf{L}(X), \mathbf{L}(Y))$$

by the Yoneda Lemma. (Here we have explicitly considered objects in $\mathbf{A}$ and $\mathbf{B}$ as being constant pro-objects; then, having made that point, we have discarded the '$\mathbf{c}$'.)

Thus $\mathbf{S}_\mathbf{K}$ is isomorphic to the full subcategory of $\mathbf{pro}(\mathbf{A})$ determined by the image of $\mathbf{L}$ restricted to $\mathbf{B}$. Being a left adjoint, $\mathbf{L}$ may be thought of as being a 'profree' functor so we can further identify $\mathbf{S}_\mathbf{K}$ as being the full subcategory of the Kleisli category of the promonad generated by **pro(K) L** on **pro(B)** determined by the objects of **B** (as always, considered as constant pro-objects).

Later we will see that this sort of 'pro-Kleisli' category description can be made completely general.

## 2.4 SHAPE THEORY AND PROCATEGORIES

As always let $\mathbf{K}: \mathbf{A} \to \mathbf{B}$ be a functor and let $X \in |\mathbf{B}|$.

**Definition**

A pro-object $(\mathbf{I}, \mathbf{F})$ in $\mathbf{B}$ is said to be $\mathbf{K}$-*associated to* $X$ *with pseudoprojections* $p_i$ for $i \in |\mathbf{I}|$ if:

(B1) for all $i$ in $\mathbf{I}$, $p_i \in \mathbf{B}(X, \mathbf{F}(i))$ and if $v \in \mathbf{I}(i, i')$,

$$\mathbf{F}(v)p_i = p_{i'};$$

(B2) for all $P$ in $\mathbf{A}$ and $f \in \mathbf{B}(X, \mathbf{K}P)$, there is some $i \in |\mathbf{I}|$ and $f_i \in \mathbf{B}(\mathbf{F}(i), \mathbf{K}P)$ such that $f_i p_i = f$;

(B3) for all $P$ in $\mathbf{A}$ and $f, f' \in \mathbf{B}(\mathbf{F}(i), \mathbf{K}P)$ such that $fp_i = f'p_i$, there is a $v \in \mathbf{I}(i', i)$ with $f\mathbf{F}(v) = f'\mathbf{F}(v)$.

**Remarks**

(a) $(1, X)$ is always $\mathbf{K}$-associated to $X$ where 1 is the category with one morphism.

(b) It is instructive to compare (B1)–(B3) with the criteria for existence of a proadjoint given in Proposition 9. It should be clear how closely they are related. (In fact if $\mathbf{F}: \mathbf{I} \to \mathbf{B}$ factors through $\mathbf{K}$, say $\mathbf{F} = \mathbf{K}\mathbf{F}'$, where $\mathbf{F}': \mathbf{I} \to \mathbf{A}$, then $\mathbf{F}'$ satisfies the isomorphism

$$\mathbf{B}(X, \mathbf{K}P) \cong \mathbf{pro}(\mathbf{A})(\mathbf{F}', P)$$

if $\mathbf{K}$ is full.)

(c) Related to (b) is the following: suppose given a pro-object, $(\mathbf{I}, \mathbf{F})$, and maps $p_i \in \mathbf{B}(X, \mathbf{F}(i))$, $i \in |\mathbf{I}|$ satisfying (B1); then $(\mathbf{I}, \mathbf{F})$ is $\mathbf{K}$-associated to $X$ if and only if for each $P$ in $\mathbf{A}$, the function $R$ from $\operatorname{Colim}_{i \in \mathbf{I}} \mathbf{B}(\mathbf{F}(i), \mathbf{K}P)$ to $\mathbf{B}(X, \mathbf{K}P)$ defined by $R[f_i] = f_i p_i$ is a bijection, $[f_i]$ being the equivalence class of $f_i \in \mathbf{B}(\mathbf{F}(i), \mathbf{K}P)$ in $\operatorname{Colim} \mathbf{B}(\mathbf{F}-, \mathbf{K}P)$, the relation being defined by $f_i \sim f'_i$ where $f'_i \in \mathbf{B}(F(i'), \mathbf{K}P)$ if there is some $i_0 \in |\mathbf{I}|$ and $v: i_0 \to i$, $v': i_0 \to i'$ such that $f_i\mathbf{F}(v) = f'_i\mathbf{F}(v')$.

Thus we can rephrase the above definition as follows:

**Proposition 1**

*Let* $X$ *be an object of* $\mathbf{B}$, $(\mathbf{I}, F)$ *a pro-object in* $\mathbf{B}$, *and suppose we are given*

$$\{p_i \in \mathbf{B}(X, \mathbf{F}(i)) : i \in |\mathbf{I}|\}.$$

*Then* $(\mathbf{I}, \mathbf{F})$ *is* $\mathbf{K}$-*associated to* $X$ *with pseudoprojections* $p_i$ *if and only if the* $p_i$ *form a promorphism* $X \xrightarrow{p} \mathbf{F}$ *inducing a natural isomorphism*

$$\mathbf{B}(X, \mathbf{K}-) \xleftarrow{\cong} \mathbf{pro}(\mathbf{B})(\mathbf{F}, \mathbf{K}-).$$

*Proof*

We need only observe that since $X$ as pro-object is indexed by a one-object category, any map $X \xrightarrow{f} G$ in **pro(B)** is given by a family of maps satisfying (B1). Thus $\{p_i : i \in |\mathbf{I}|\}$ satisfy (B1) if and only if they give a map of systems and hence a promap. The remainder of the proposition results from the remark.

**Remark**

Let $S_{\mathbf{K}}$ be the shape functor (of Holsztyński) of **K** and suppose $(\mathbf{I}, \mathbf{F})$ **K**-associated to $X$; then if for all $v \in \mathbf{I}(i, i')$, $S_{\mathbf{K}}(\mathbf{F}(v))$ is invertible in $S_{\mathbf{K}}$, $S_{\mathbf{K}}(p_i)$ is invertible for all $i \in |\mathbf{I}|$. This can be seen in various ways; perhaps one of the simplest is the following:

Using $S_{\mathbf{K}}$, we put **B** inside $\mathbf{S_K}$ and look at the diagram

$$\begin{array}{ccc}
\mathbf{B}(X, \mathbf{K}-) & \xleftarrow[\;p\;]{\cong} & \mathbf{pro(B)(F, K}-) \\
\downarrow & & \downarrow \\
\mathbf{S_K}(X, \mathbf{K}-) & \xleftarrow[\;p\;]{} & \mathbf{pro(S_K)(F, K}-)
\end{array}$$

(Considering $\mathbf{pro(S_K)}$ might raise eyebrows with set theorists, but we can handle this, if challenged, by recourse to bigger universes.)

As we are looking at morphisms into things in the image of **K**, the vertical morphisms are isomorphisms (i.e. bijections). Considered as a pro-object in $\mathbf{S_K}$, **F** is essentially constant, as all the transition maps are invertible; in other words, for any $P$,

$$\{\mathbf{S_K}(\mathbf{F}(i), \mathbf{K}P) : i \in \mathbf{I}\}$$

is a diagram of isomorphisms, hence $\mathbf{pro(S_K)}(\mathbf{F}, \mathbf{K}P) \cong \mathbf{S_K}(\mathbf{F}(i), \mathbf{K}P)$ for any $i$, but then $\mathbf{S_K}(\mathbf{F}(i), \mathbf{K}P) \cong \mathbf{B}(\mathbf{F}(i), \mathbf{K}P)$ so we obtain that

$$\mathbf{B}(p_i, \mathbf{K}) : \mathbf{B}(\mathbf{F}(i), \mathbf{K}-) \rightarrow \mathbf{B}(\mathbf{X}, \mathbf{K}-)$$

which is an isomorphism of functors, i.e. $S_{\mathbf{K}}(p_i)$ is invertible.

We next link up the above with the notion of a **K**-continuous functor introduced earlier (section 2.1).

**Proposition 2**

*Let* $(\mathbf{I}, \mathbf{F})$ *be* **K**-*associated to* $X$ *with pseudoprojections* $p_i$; *then if* $\mathbf{T} : \mathbf{B} \rightarrow \mathbf{C}$ *is* **K**-*continuous,* $\mathbf{T}X = \operatorname{Lim} \mathbf{TF}$ **with projection** $\mathbf{T}(p_i)$.

*Proof*

Let $D \in |\mathbf{C}|$ and let $\{g_i : D \rightarrow \mathbf{TF}(i) | i \in |\mathbf{I}|\}$ be a cone on **TF**, i.e. the $g_i$ satisfy: if $v \in \mathbf{I}(i, i')$, $\mathbf{TF}(v) g_i = g_{i'}$.

Let $P \in |\mathbf{A}|$ and $f \in \mathbf{B}(X, \mathbf{K}P)$. By (B2), there is an $f_i : \mathbf{F}(i) \rightarrow \mathbf{K}P$ such that $f_i p_i = f$; let $V[f, P] = \mathbf{T}(f_i) g_i \in \mathbf{C}(D, \mathbf{TK}P)$; then $\mathbf{V} : X \downarrow \mathbf{K} \rightarrow D \downarrow \mathbf{TK}$ defined by

$$\mathbf{V} : (f, P) \mapsto (\mathbf{V}[f, P], P)$$

is in $\operatorname{Func}(X \downarrow \mathbf{K}, D \downarrow \mathbf{TK})$; in fact let $k \in \mathbf{A}(P, P')$; one has $\mathbf{V}[f, P] = \mathbf{T}(f_i) g_i$, whilst $\mathbf{V}[\mathbf{K}(k) f, P'] = \mathbf{T}(f'_i) g'_i$ where $f'_i p'_i = \mathbf{K}(k) f$.

As **I** is cofiltering, there is an $i_0 \in |\mathbf{I}|$, $v: i_0 \to i$, $v': i_0 \to i'$ with

$$\mathbf{K}(k)f_i\mathbf{F}(v)p_{i_0} = f'_i\mathbf{F}(v')p_{i_0}.$$

By (B3) there is an $i'_0$ and $v_0: i_0 \to i'_0$ such that $\mathbf{K}(k)f_i\mathbf{F}(v)\mathbf{F}(v_0) = f'_i\mathbf{F}(v')\mathbf{F}(v_0)$, so that one now has

$$\begin{aligned}\mathbf{TK}(k)\mathbf{V}[f, P] &= \mathbf{TK}(k)\mathbf{T}(f_i)g_i \\ &= \mathbf{TK}(k)\mathbf{T}(f_i)\mathbf{TF}(v)\mathbf{TF}(v_0)g_{i_0} \\ &= \mathbf{T}(f'_i)g'_i \\ &= \mathbf{V}[\mathbf{K}(k)f, P']\end{aligned}$$

as required.

As **T** is **K**-continuous in $X$, there is a unique $g \in \mathbf{C}(D, \mathbf{T}X)$ such that $\mathbf{g}^*\delta = \mathbf{V}$, i.e. such that for all $(f, P) \in |X \downarrow \mathbf{K}|$

$$\mathbf{T}(f)g = \mathbf{V}[f, P].$$

Let $i \in |\mathbf{I}|$ and we have to show that $\mathbf{T}(p_i)g = g_i$. Let $\mathbf{V}_i: \mathbf{F}(i)\downarrow\mathbf{K} \to D\downarrow\mathbf{TK}$ be defined by $\mathbf{V}_i: (f_i, P) \to (\mathbf{T}(f_i)g_i, P)$. As $\mathbf{V}_i \in \mathrm{Func}(\mathbf{F}(i)\downarrow\mathbf{K}, D\downarrow\mathbf{TK})$, there is a unique $\bar{g}_i \in \mathbf{C}(D, \mathbf{TF}(i))$ such that

$$\bar{\mathbf{g}}_i^*\delta_i = \mathbf{V}_i$$

where $\delta_i: \mathbf{F}(i)\downarrow\mathbf{K} \to \mathbf{TF}(i)\downarrow\mathbf{TK}$, that is, such that $\mathbf{T}(f_i)\bar{g}_i = \mathbf{T}(f_i)g_i$. Thus $\bar{g}_i = g_i$ and as $\mathbf{T}(f_i)\mathbf{T}(p_i)g = \mathbf{T}(f_i)g_i$, we have $\mathbf{T}(p_i)g = g_i$; hence $g$ is the required morphism. It remains only to check that it is unique with this property.

Let $g' \in \mathbf{C}(D, \mathbf{T}X)$ be such that for all $i \in |\mathbf{I}|$, $\mathbf{T}(p_i)g' = g_i$; then one has for all $P \in |\mathbf{A}|$ and $f \in \mathbf{B}(X, \mathbf{K}P)$,

$$\mathbf{T}(f)g' = \mathbf{T}(f_i)\mathbf{T}(p_i)g' = \mathbf{T}(f_i)g_i = \mathbf{V}[f, P]$$

where $f = f_i p_i$ (using (B2)). By the uniqueness of $g$ as satisfying $\mathbf{T}(f)g = \mathbf{V}[f, P]$, we see that $g = g'$.

**Definition**

Let $\mathbf{T}: \mathbf{B} \to \mathbf{C}$ be a functor. **T** is said to be *continuous* if for each $(\mathbf{I}, \mathbf{F})$ in $\mathbf{pro}(\mathbf{B})$ and $X = \mathrm{Lim}\,\mathbf{F}$ with projection $p_i$, then $\mathbf{T}X = \mathrm{Lim}\,\mathbf{TF}$ with projection $\mathbf{T}(p_i)$. As an easy consequence of this one has:

**Proposition 3**

*Let* **K** *be a functor from* **A** *to* **B**, **D** *a category and* $\mathbf{L}: \mathbf{D} \to \mathbf{B}$ *a functor such that* $|\mathbf{D}| = |\mathbf{B}|$, $\mathbf{L}X = X$ *for all* $X \in |\mathbf{B}|$ *and furthermore such that the following condition* (P) *is satisfied*:

*If* **I** *is a cofiltering category*, $\mathbf{F}: \mathbf{I} \to \mathbf{D}$ *a functor and* $X = \mathrm{Lim}\,\mathbf{F}$ *with projection* $p_i$, *then* $(\mathbf{I}, \mathbf{LF})$ *is* **K**-*associated to* **X** *with pseudoprojection* $\mathbf{L}(p_i)$.

*Then if* $\mathbf{T}: \mathbf{B} \to \mathbf{C}$ *is a functor*, **T** *is* **K**-*continuous implies* **TL** *is continuous.*

**Remark**

In Proposition 3, think of **B** as the homotopy category, **D** as the category of spaces and **L** as the quotient functor.

**Theorem 1**

*Let* **K** *be a full functor from* **A** *to* **B** *such that for all* $X \in |\mathbf{B}|$, *there is a pro-object* (**I**, **F**) *in* **A** *such that* (**I**, **KF**) *is* **K**-*associated to* **X** *with pseudoprojection* $p_i$. *Then a functor* $\mathbf{T}: \mathbf{B} \to \mathbf{C}$ *is* **K**-*continuous if and only if* $\mathbf{TX} = \operatorname{Lim} \mathbf{TKF}$ *with projection* $\mathbf{T}(p_i)$ *for each* $X \in |\mathbf{B}|$.

*Proof*

Proposition 3 gives us the result in one direction. So now assume $\mathbf{T}X = \operatorname{Lim} \mathbf{TKF}$ and let $X \in |\mathbf{B}|$, (**I**, **F**) such that (**I**, **KF**) is **K**-associated to $X$ and $\mathbf{V} \in \operatorname{Func}(X \downarrow \mathbf{K}, D \downarrow \mathbf{TK})$ for some $D \in |\mathbf{C}|$. As $p_i \in \mathbf{B}(X, \mathbf{KF}(i))$ for $i \in |\mathbf{I}|$, $\mathbf{V}[p_i, \mathbf{F}(i)] \in \mathbf{C}(D, \mathbf{TKF}(i))$ and as for $v \in \mathbf{I}(i, i')$ one has

$$\mathbf{TKF}(v)\mathbf{V}[p_i, \mathbf{F}(i)] = \mathbf{V}[\mathbf{KF}(v)p_i, \mathbf{F}(i)];$$

thus there is exactly one $g \in \mathbf{C}(D, \mathbf{T}X)$ such that $\mathbf{T}(p_i)g = \mathbf{V}[p_i, \mathbf{F}(i)]$ for all $i \in |\mathbf{I}|$.

One has

$$\mathbf{g}^*\boldsymbol{\delta} = \mathbf{V};$$

in fact let $P \in |\mathbf{A}|$ and $f \in \mathbf{B}(X, \mathbf{K}P)$; by (B2) there is an $f_i$ such that $f_i p_i = f$ where $f_i \in \mathbf{B}(\mathbf{KF}(i), \mathbf{K}P)$ and as **K** is full, there is some $k_i \in \mathbf{A}(\mathbf{F}(i), P)$ such that $\mathbf{K}(k_i) = f_i$. This implies

$$\begin{aligned} \mathbf{V}[f, P] &= \mathbf{V}[\mathbf{K}(k_i)p_i, P] \\ &= \mathbf{TK}(k_i)\mathbf{V}[p_i, F(i)] \\ &= \mathbf{TK}(k_i)\mathbf{T}(p_i)g \\ &= \mathbf{T}(f)g \end{aligned}$$

as required.

We are left to verify uniqueness of $g$. Suppose $g' \in \mathbf{C}(D, \mathbf{T}X)$ is such that $\mathbf{g}'^*\boldsymbol{\delta} = \mathbf{V}$; then one has for all $(f, P) \in |X \downarrow \mathbf{K}|$,

$$\mathbf{V}[f, P] = \mathbf{T}(f)g';$$

thus in particular

$$\mathbf{V}[p_i, \mathbf{F}(i)] = \mathbf{T}(p_i)g' \quad \text{for all } i \in \mathbf{I};$$

as a result one has $g = g'$.

**Remark**

We can rephrase the result in the light of our experience with proadjoints. Suppose $\mathbf{K}: \mathbf{A} \to \mathbf{B}$ has a proadjoint $\mathbf{L}: \mathbf{B} \to \mathbf{pro}(\mathbf{A})$ and $\mathbf{T}: \mathbf{B} \to \mathbf{C}$; then **T** is **K**-continuous if and only if $\mathbf{T}X \cong \operatorname{Lim} \mathbf{TKL}(X)$ where the natural isomorphism is induced by the unit of the adjunction, $\mathrm{Id} \to \mathbf{KL}$.

**Corollary**

*Let* **A** *be a full subcategory of* **B** *and* **K** *the inclusion of* **A** *in* **B**, **L** *a functor from* **D** *to* **B**, *where* $|\mathbf{D}| = |\mathbf{B}|$ *and* **L** *is the identity on objects*, i.e. $\mathbf{L}X = X$ *for all* $X \in |\mathbf{B}|$. *Suppose furthermore that* **L** *satisfies the condition* (**P**) *of Proposition* 3, *namely*:

*If* $(\mathbf{I}, \mathbf{F}) \in \mathbf{pro}(\mathbf{D})$ *is such that* $X = \operatorname{Lim} \mathbf{F}$ *with projection* $p_i$ *then* $(\mathbf{I}, \mathbf{LF}) \in \mathbf{pro}(\mathbf{D})$ *is* **K**-*associated to* **X** *with pseudoprojection* $\mathbf{L}(p_i)$.

*If, for each* $X \in |\mathbf{B}|$, *there is a pro-object* $(\mathbf{I}, \mathbf{F})$ *in* $\mathbf{pro}(\mathbf{D})$ *such that* $X = \operatorname{Lim} \mathbf{F}$ *and* $\mathbf{LF}(i) \in |\mathbf{A}|$ *for all* $i \in |\mathbf{I}|$, *then if* $\mathbf{T}$ *is a functor* $\mathbf{T}: \mathbf{B} \to \mathbf{C}$, $\mathbf{T}$ *is* $\mathbf{K}$*-continuous if and only if* $\mathbf{TL}$ *is continuous.*

**Remark**

(1) This corollary explains the sense in which **K**-continuity is a weak form of continuity. The homotopy category does not have limits, but any compact metric (or compact Hausdorff) space has been seen in Chapter 1 to be the intersection (and therefore the inverse limit) of its ANR neighbourhoods. To express the way the shape of a space $X$ depends 'continuously' on the homotopy types of its ANR neighbourhoods, one thus needs exactly such a concept as **K**-continuity.

(2) Let $\mathbf{A}$ be a full subcategory of $\mathbf{B}$ satisfying the conditions of the last corollary; then $(\mathbf{S}, S)$ is a shape theory for $K$ if (S1), (S2) and

$$(S3)'': \mathbf{SL} \text{ is a continuous functor}$$

are satisfied.

We have remarked that if for each $X \in |\mathbf{B}|$ there is a $(\mathbf{I}_X, \mathbf{F}_X)$, say, in $\mathbf{pro}(\mathbf{A})$ such that $(\mathbf{I}_X, \mathbf{KF}_X)$ is $\mathbf{K}$-associated to $X$, there is an isomorphism

$$\mathbf{B}(X, \mathbf{K}-) \cong \mathbf{pro}(\mathbf{A})(\mathbf{F}_X, -).$$

At present we know only that such an $\mathbf{F}_X$ exists for each $X$, not that $\mathbf{F}_X$ depends in any way functorially on $X$. However, it is clear that a morphism

$$f: X \to Y$$

induces a natural transformation

$$f^*: \mathbf{B}(Y, \mathbf{K}-) \to \mathbf{B}(X, \mathbf{K}-),$$

thus a natural transformation

$$f^*: \mathbf{pro}(\mathbf{A})(\mathbf{F}_Y, -) \to \mathbf{pro}(\mathbf{A})(\mathbf{F}_X, -).$$

By the Yoneda Lemma such a transformation corresponds uniquely to a morphism of pro-objects

$$\bar{f}: \mathbf{F}_X \to \mathbf{F}_Y$$

and this uniqueness in its turn implies that the correspondence

$$f \to \bar{f}$$

is functorial. Thus $X \to \mathbf{F}_X$ is a functor from $\mathbf{B}$ to $\mathbf{pro}(\mathbf{A})$, giving $\mathbf{K}$ a proadjoint. We shall state this formally in the following theorem.

**Theorem 2**

*Given* $\mathbf{K}: \mathbf{A} \to \mathbf{B}$ *then* $\mathbf{K}$ *has a proadjoint if and only if for each* $X \in |\mathbf{B}|$, *there is a* $(\mathbf{I}, \mathbf{F}_X)$ *in* $\mathbf{pro}(\mathbf{A})$ *such that* $(\mathbf{I}, \mathbf{KF}_X)$ *is* $\mathbf{K}$*-associated to* $X$.

We thus have from the final remarks of section 2.3 that in the above situation the Holsztyński shape category $(\mathbf{S_K}, S_\mathbf{K})$ is isomorphic to the category $\mathbf{C}$ defined as follows:

—$|\mathbf{C}| = |\mathbf{B}|$
—if $X, Y \in |\mathbf{B}|$,

$$\mathbf{C}(X, Y) = \mathbf{pro}(\mathbf{A})((\mathbf{I}_X, \mathbf{F}_X), (\mathbf{I}_Y, \mathbf{F}_Y)).$$

(Taking a different choice of $(\mathbf{I}_X, \mathbf{F}_X)$, say $(\mathbf{J}_X, \mathbf{F}'_X)$, gives an equivalent proadjoint to $\mathbf{K}$, hence an isomorphic category.)

Later we shall see that exactly this situation pertains in the canonical examples $\mathbf{B} = \mathbf{HMC}$, $\mathbf{A} =$ homotopy category of ANRs, and that this 'explains' why the Mardešić–Segal approach worked so well to capture Borsuk's geometric notion.

## 2.5 KAN EXTENSIONS AND HOLSZTÝNSKI'S SHAPE FUNCTOR

In the topological context of shape theory, we have traced the beginnings of the subject to the work of Alexandroff, Čech, Vietoris and others working with cohomological invariants in the 1930s. Then the idea was to extend invariants which were theoretically calculable for finite polyhedra to more general compact spaces. This idea of extending 'invariants' from $\mathbf{A}$ to $\mathbf{B}$ by using the approximating systems has been central to the applications of shape theoretic ideas. The simplest abstract version of such constructions comes in the theory of Kan extensions.

Let, as usual, $\mathbf{K}: \mathbf{A} \to \mathbf{B}$ and let $\mathbf{C}$ be a category. We denote by $\mathbf{C}^{\mathbf{B}}$ and $\mathbf{C}^{\mathbf{A}}$ the categories of natural transformations between functors from $\mathbf{B}$ to $\mathbf{C}$ and from $\mathbf{A}$ to $\mathbf{C}$ respectively. The functor $\mathbf{K}$ induces a functor

$$\mathbf{C}^{\mathbf{K}}: \mathbf{C}^{\mathbf{B}} \to \mathbf{C}^{\mathbf{A}}$$

by composition: if $t: \mathbf{R} \to \mathbf{R}'$, $\mathbf{C}^{\mathbf{K}}(t) = t\mathbf{K}: \mathbf{RK} \to \mathbf{R}'\mathbf{K}$.

**Definition**

Let $\mathbf{T}$ be a functor from $\mathbf{A}$ to $\mathbf{C}$. A *right Kan extension of* $\mathbf{T}$ along $\mathbf{K}$ is a pair $(\tilde{\mathbf{T}}, \varepsilon)$ where $\tilde{\mathbf{T}}$ is a functor from $\mathbf{B}$ to $\mathbf{C}$ and $\varepsilon$ is a natural transformation from $\tilde{\mathbf{T}}\mathbf{K}$ to $\mathbf{T}$ such that $\varepsilon$ is universal as a morphism from $\mathbf{C}^{\mathbf{K}}: \mathbf{C}^{\mathbf{B}} \to \mathbf{C}^{\mathbf{A}}$ to $\mathbf{T} \in |\mathbf{C}^{\mathbf{A}}|$.

Thus we have a diagram

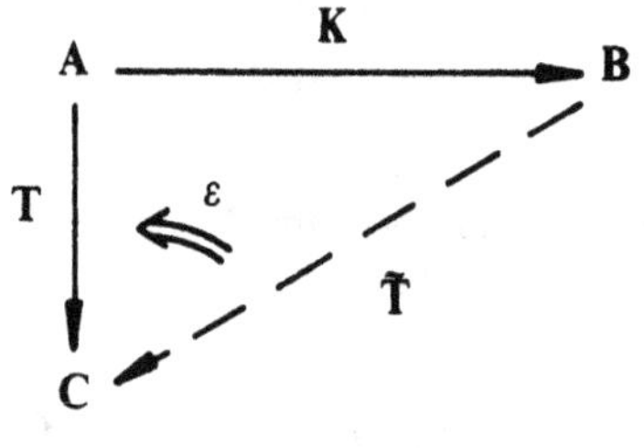


and if $T_1: B \to C$, $\varepsilon_1: \mathbf{T}_1\mathbf{K} \to \mathbf{T}$ is another pair then there is a unique natural transformation $\mathbf{T}_1 \xrightarrow{\phi} \tilde{\mathbf{T}}$ such that

$$\mathbf{T}_1\mathbf{K} \xrightarrow{\phi\mathbf{K}} \tilde{\mathbf{T}}\mathbf{K} \xrightarrow{\varepsilon} \mathbf{T} = \mathbf{T}_1\mathbf{K} \xrightarrow{\varepsilon_1} \mathbf{T}.$$

Clearly $\tilde{\mathbf{T}}$ is unique up to natural equivalence. We will often denote it by $\mathbf{Ran}_{\mathbf{K}}\,\mathbf{T}$.

**Remark**

The basic reference for the general theory of Kan extensions is MacLane [71], Chapter X.

We start by examining the links between **K**-continuity and Kan extensions.

**Proposition 1**

*Let* $\mathbf{T}\colon \mathbf{A} \to \mathbf{Sets}$, *a right Kan extension of* $\mathbf{T}$ *along* $\mathbf{K}$ *is given by the functor* $\tilde{\mathbf{T}}$ *defined by*

—$\tilde{\mathbf{T}}X = \mathrm{Nat}(\mathbf{B}(X, \mathbf{K}-), \mathbf{T})$ *for* $X \in |\mathbf{B}|$
—$\tilde{\mathbf{T}}h(t) = t\mathbf{B}(h, \mathbf{K}-)$ *for* $h \in \mathbf{B}(X, Y)$ *and* $t \in \mathrm{Nat}(\mathbf{B}(X, \mathbf{K}-), \mathbf{T})$
—*and* $\varepsilon\colon \tilde{\mathbf{T}}\mathbf{K} \to \mathbf{T}$ *defined by*

$$\varepsilon_P(t) = t_P(\mathrm{Id}_{\mathbf{K}P}) \quad \text{for } P \in |\mathbf{A}| \quad \text{and} \quad t\colon \mathbf{B}(\mathbf{K}P, \mathbf{K}-) \to \mathbf{T}.$$

*Proof*

Firstly $\varepsilon$ is a transformation from $\tilde{\mathbf{T}}\mathbf{K}$ to $\mathbf{T}$; in fact let $k \in \mathbf{A}(P, Q)$ and $t\colon \mathbf{B}(\mathbf{K}P, \mathbf{K}-) \to \mathbf{T}$ then

$$\mathbf{T}(k)\varepsilon_P(t) = \mathbf{T}(t)t_P(\mathrm{Id}_{\mathbf{K}P}) = t_Q\mathbf{K}(k),$$

$t$ being a transformation, and

$$\varepsilon_Q\tilde{\mathbf{T}}\mathbf{K}(k)(t) = \varepsilon_Q(t\mathbf{B}(\mathbf{K}(k), \mathbf{K}-)) = t_Q(\mathbf{B}(\mathbf{K}(k), \mathbf{K}Q)(\mathrm{Id}_{\mathbf{K}Q})$$
$$= t_Q\mathbf{K}(k) \quad \text{as required.}$$

We now check universality.

Let $\mathbf{U}\colon \mathbf{B} \to \mathbf{Sets}$ be a functor and $\eta\colon \mathbf{UK} \to \mathbf{T}$; let $\eta_X\colon \mathbf{U}X \to \tilde{\mathbf{T}}X$ be defined by

$$\bar{\eta}_X(x)_P(f) = \eta_P\mathbf{U}(f)(x) \quad \text{for } x \in \mathbf{U}X,\ P \in |\mathbf{A}| \quad \text{and} \quad f \in \mathbf{B}(Y, \mathbf{K}P).$$

$\bar{\eta}_X(x)$ is a natural transformation from $\mathbf{B}(X, \mathbf{K}-)$ to $\mathbf{T}$.

Let $\bar{\eta}\colon \mathbf{U} \to \tilde{\mathbf{T}}$ be defined by $\bar{\eta}_X$ for $X \in |\mathbf{B}|$; then $\bar{\eta}$ is a natural transformation; in fact if $h \in \mathbf{B}(X, Y)$ and $x \in \mathbf{U}X$, one has

$$\tilde{\mathbf{T}}(h)\bar{\eta}_X(x) = \bar{\eta}_X(x)\mathbf{B}(h, \mathbf{K}-)$$

and if $P \in |\mathbf{A}|$ and $f \in \mathbf{B}(Y, \mathbf{K}P)$

$$\bar{\eta}_X(x)_P\mathbf{B}(h, \mathbf{K}P)(f) = \bar{\eta}_X(x)_P(fh) = \eta_P\mathbf{U}(fh)(x)$$

whilst

$$\bar{\eta}_Y\mathbf{U}(h)(x)_P(f) = \eta_Y(\mathbf{U}(h)(x)_P)(f)$$
$$= \eta_P\mathbf{U}(f)(\mathbf{U}(h)(x)) = \eta_P\mathbf{U}(fh)(x)$$

as required.

It remains to check the uniqueness of $\bar{\eta}$.

Consider $\varepsilon\bar{\eta}\mathbf{K}$. Let $x \in \mathbf{UK}P$ for $P \in |\mathbf{A}|$; then one has

$$(\varepsilon\bar{\eta}\mathbf{K})_P(x) = \varepsilon_P\bar{\eta}\mathbf{K}_P(x) = \bar{\eta}_{\mathbf{K}P}(x)_P(\mathrm{Id}_{\mathbf{K}P})$$
$$= \eta_P\mathbf{U}(\mathrm{Id}_{\mathbf{K}P})(x) = \eta_P(x).$$

If $\bar{\eta}'\colon \mathbf{U} \to \tilde{\mathbf{T}}$ is a natural transformation such that $\varepsilon\bar{\eta}'\mathbf{K} = \eta$, then, one has for $x \in \mathbf{U}X$ and $f \in \mathbf{B}(Y, \mathbf{K}P)$,

$$\begin{aligned}\bar{\eta}'_X(x)_P(f) &= \bar{\eta}'_X(x)_P\mathbf{B}(f, \mathbf{K}P)(\mathrm{Id}_{\mathbf{K}P})\\ &= (\bar{\mathbf{T}}(f)\bar{\eta}'_X(x)_P)(\mathrm{Id}_{\mathbf{K}P})\\ &= (\bar{\eta}'_{\mathbf{K}P}\mathbf{U}(f)(x)_P)(\mathrm{Id}_{\mathbf{K}P})\\ &= \varepsilon_P(\bar{\eta}'_{\mathbf{K}P}\mathbf{U}(f)(x))\\ &= \eta_P\mathbf{U}(f)(x)\\ &= \bar{\eta}_X(x)_P(f).\end{aligned}$$

Consequently $\bar{\eta}$ is indeed the unique transformation such that $\varepsilon\bar{\eta}\mathbf{K} = \eta$.

The importance of this result to categorical shape theory may perhaps be more evident after the following result.

**Corollary**

*Let* $\mathbf{T}$ *be a functor from* $\mathbf{B}$ *to* $\mathbf{C}$. $\mathbf{T}$ *is* $\mathbf{K}$*-continuous if and only if for each* $D \in |\mathbf{C}|$, $\mathbf{C}(D, \mathbf{T}-)$ *is a Kan extension of* $\mathbf{C}(D, \mathbf{TK}-)$ *along* $\mathbf{K}$, $\varepsilon$ *being the identity transformation.*

*Proof*

Let $\mathbf{T}_1 = \mathbf{C}(D, \mathbf{TK}-)$ from $\mathbf{A}$ to **Sets**. As $\mathbf{T}$ is $\mathbf{K}$-continuous if and only if for all $D \in |\mathbf{C}|$ and $X \in |\mathbf{B}|$, there is a bijection from $\mathbf{C}(D, \mathbf{T}X)$ to

$$\mathrm{Func}(X \downarrow \mathbf{K}, D \downarrow \mathbf{TK}) \cong \mathrm{Nat}(\mathbf{B}(X, \mathbf{K}-), \mathbf{T}_1) = \bar{\mathbf{T}}_1(X)$$

which is natural in $X$, $\mathbf{T}$ is $\mathbf{K}$-continuous if and only if $\mathbf{C}(D, \mathbf{T}-)$ is a Kan extension of $\mathbf{C}(D, \mathbf{TK}-)$.

Apart from this link with $\mathbf{K}$-continuity, there is another even more obvious connection between Kan extensions and shape. Proposition 1 tells us that

$$\bar{\mathbf{T}}Y = \mathrm{Nat}(\mathbf{B}(Y, \mathbf{K}-), \mathbf{T}) \quad \text{for } Y \in |\mathbf{B}|$$

but if $T = \mathbf{B}(X, \mathbf{K}-)$, this tells us that

$$\begin{aligned}\bar{\mathbf{T}}Y &= \mathrm{Nat}(\mathbf{B}(Y, \mathbf{K}-), \mathbf{B}(X, \mathbf{K}-))\\ &= \mathbf{S_K}(X, Y)\\ &= \mathbf{S_K}(X, S_\mathbf{K}-)(Y).\end{aligned}$$

Similarly on morphisms—if $h \in \mathbf{B}(X, Y)$,

$$\begin{aligned}\mathbf{S_K}(X, S_\mathbf{K}(h))(t) &= S_\mathbf{K}(h)t\\ &= t\mathbf{B}(h, \mathbf{K}-)\\ &= \bar{\mathbf{T}}(h)(t).\end{aligned}$$

Thus we have

**Proposition 2**

*If* $(\mathbf{S_K}, S_\mathbf{K})$ *is the shape theory of Holsztyński of* $\mathbf{K}$, *then* $\mathbf{S_K}(X, S_\mathbf{K}-)$ *is a (right) Kan extension of* $\mathbf{B}(X, K-)\colon \mathbf{A} \to \mathbf{Sets}$ *along* $\mathbf{K}$, *with the transformation* $\varepsilon$ *as in Proposition* 1.

The final link with $\mathbf{K}$-continuity is the following:

**Proposition 3**
*Let* $\mathbf{T}: \mathbf{B} \to \mathbf{C}$ *be a* $\mathbf{K}$*-continuous functor; then* $\mathbf{T}$ *is a (right) Kan extension of* $\mathbf{TK}$ *along* $\mathbf{K}$ *with* $\varepsilon$ *the identity transformation on* $\mathbf{TK}$.

*Proof*
Let $\mathbf{T}': \mathbf{B} \to \mathbf{C}$, $t: \mathbf{T}'\mathbf{K} \to \mathbf{TK}$ be a functor and natural transformation to test for universality of $\mathbf{T}$.

Let $\mathbf{V}: X \downarrow \mathbf{K} \to \mathbf{T}'X \downarrow \mathbf{TK}$ be defined by

$$\mathbf{V}[f, P] = t_P\mathbf{T}'(f) \quad \text{for } P \in |\mathbf{W}| \quad \text{and} \quad f \in \mathbf{B}(X, \mathbf{K}P).$$

(As it is a fair time since we defined the notation of $[f, P]$, it is probably worth recalling that if $(f, P) \in |X \downarrow \mathbf{K}|$, $\mathbf{V}[f, P]$ is defined by

$$\mathbf{V}(f, P) = (\mathbf{V}[f, P], P).)$$

This presupposes that such a definition works, i.e. that $\mathbf{V}$ is in Func$(X \downarrow \mathbf{K}, \mathbf{T}'X \downarrow \mathbf{TK})$, but if $k \in \mathbf{A}(P, Q)$,

$$\begin{aligned}\mathbf{V}[\mathbf{K}(k)f, Q] &= t_Q\mathbf{T}'(\mathbf{K}(k)f)\\ &= \mathbf{TK}(k)t_P\mathbf{T}'(f)\\ &= \mathbf{TK}(k)\mathbf{V}[f, P]\end{aligned}$$

as required. As $\mathbf{T}$ is $\mathbf{K}$-continuous in $\mathbf{X}$, there is a unique $\bar{t}_X \in \mathbf{C}(\mathbf{T}'X, \mathbf{T}X)$ such that for all $P \in |\mathbf{A}|$ and $f \in \mathbf{B}(X, \mathbf{K}P)$,

$$\mathbf{T}(f)\bar{t}_X = \mathbf{V}[f, P] = t_P\mathbf{T}'(f).$$

Let $\bar{t}: \mathbf{T}' \to \mathbf{T}$ be defined by $\bar{t}_X: \mathbf{T}'X \to \mathbf{T}X$ for each $X \in |\mathbf{B}|$; $\bar{t}$ is a transformation from $\mathbf{T}'$ to $\mathbf{T}$; indeed if $h \in \mathbf{B}(X, Y)$ and $\mathbf{V}': Y \downarrow \mathbf{K} \to \mathbf{T}'X \downarrow \mathbf{TK}$ is defined by

$$\mathbf{V}'[g, P] = \mathbf{T}(g)\mathbf{T}(h)\bar{t}_X \quad \text{for } P \in |\mathbf{A}| \quad \text{and} \quad g \in \mathbf{B}(Y, \mathbf{K}P)$$

then as $\mathbf{V}' \in$ Func$(Y \downarrow \mathbf{K}, \mathbf{T}'X \downarrow \mathbf{TK})$, there is a unique element, $\tau$, of $\mathbf{C}(\mathbf{T}'X, \mathbf{T}Y)$ satisfying the equality

$$\mathbf{T}(g)c = \mathbf{V}[g, P];$$

however,

$$\mathbf{T}(g)\mathbf{T}(h)\bar{t}_X = t_P\mathbf{T}'(gh) = \mathbf{T}(g)\bar{t}_Y\mathbf{T}'(h)$$

so $\mathbf{T}(h)\bar{t}_X = \bar{t}_Y\mathbf{T}'(h)$ and $\bar{t}$ is a transformation.

Now let $\mathbf{K}P = X$ and $f = \mathrm{Id}_{\mathbf{K}P}$, to obtain

$$\bar{t}\mathbf{K}P = \bar{t}_{\mathbf{K}P} = \mathbf{T}(\mathrm{Id}_{\mathbf{K}P})\bar{t}_{\mathbf{K}P} = t_P\mathbf{T}'(\mathrm{Id}_{\mathbf{K}P}) = t_P,$$

i.e. $\bar{t}\mathbf{K} = t$.

If $\bar{t}: \mathbf{T}' \to \mathbf{T}$ is a transformation such that $\bar{t}'\mathbf{K} = t$, then for all $P \in |\mathbf{A}|$ and $f \in \mathbf{B}(X, \mathbf{K}P)$,

$$\mathbf{T}(f)\bar{t}'_X = \bar{t}'_{\mathbf{K}P}\mathbf{T}'(f) = t_P\mathbf{T}'(f) = \mathbf{T}(f)\bar{t}_X$$

and as $\bar{t}_X$ is the unique element satisfying this equality $\bar{t}'_X = \bar{t}_X$ for all $X$ in $|\mathbf{B}|$, i.e. $\bar{t}' = \bar{t}$.

**Corollary**
*If* $S$ *satisfies (S2),* $S$ *is a Kan extension of* $S\mathbf{K}$ *along* $\mathbf{K}$.

We now turn our attention to the link between the notion of Kan extension and the process of constructing shape invariants.

**Definition**
Let $\mathbf{K}: \mathbf{A} \to \mathbf{B}$ be a functor and $S$ the shape functor of Holsztyński of $\mathbf{K}$. Let $\mathbf{T}: \mathbf{B} \to \mathbf{C}$ be any functor; we say $\mathbf{T}$ is a *shape invariant functor* if $\mathbf{T}$ factorises via $S$, i.e. there is a $\bar{\mathbf{T}}: \mathbf{S} \to \mathbf{C}$ such that $\bar{\mathbf{T}}S = \mathbf{T}: \mathbf{B} \to \mathbf{C}$.

**Proposition 4**
*Let $\mathbf{C}$ be a complete category, and $\mathbf{T}: \mathbf{A} \to \mathbf{C}$ any functor. If $\tilde{\mathbf{T}}$ is the right Kan extension of $\mathbf{T}$ along $\mathbf{K}$, then $\tilde{\mathbf{T}}$ is a shape invariant.*

*Proof*
As $\mathbf{C}$ is complete, we can calculate $\tilde{\mathbf{T}}$ as follows:

$$\tilde{\mathbf{T}}X = \mathrm{Lim}(X \downarrow \mathbf{K} \xrightarrow{\delta_X} \mathbf{A} \to \mathbf{C})$$

with projection $\mathbf{V}[f, P]: \tilde{\mathbf{T}}X \to \mathbf{T}P$, $\delta_X$ being the codomain functor (see [71], p. 233). If $k \in \mathbf{A}(P, Q)$,

$$\mathbf{T}(k)\mathbf{V}[f, P] = \mathbf{V}[\mathbf{K}(k)f, Q].$$

If $h \in \mathbf{B}(X, X')$, $\tilde{\mathbf{T}}(h)$ is the unique morphism from $\tilde{\mathbf{T}}X$ to $\tilde{\mathbf{T}}X'$ such that $\mathbf{V}'[f, P]\tilde{\mathbf{T}}(h) = \mathbf{V}[fh, P]$.

We will construct directly a functor $\bar{\mathbf{T}}: \mathbf{S} \to \mathbf{C}$ such that $\bar{\mathbf{T}}S = \tilde{\mathbf{T}}$.

Of course $\bar{\mathbf{T}}X = \tilde{\mathbf{T}}X$. So consider $s \in \mathbf{S}(X, X')$; then for all $P \in |\mathbf{A}|$ and $f \in \mathbf{B}(X', \mathbf{K}P)$, we have

$$\mathbf{V}[s(f), P]: \tilde{\mathbf{T}}X \to \mathbf{T}\delta_{X'}(f, P) = \mathbf{T}P.$$

If $k \in \mathbf{A}(P, P')$ is such that $k: (f, P) \to (f', P)$ in $X' \downarrow \mathbf{K}$ then

$$\begin{aligned} \mathbf{T}(k)\mathbf{V}[s(f), P] &= \mathbf{V}[\mathbf{K}(k)s(f), P'] \\ &= \mathbf{V}[s(\mathbf{K}(k)f), P'] \\ &= \mathbf{V}[s(f'), P']; \end{aligned}$$

consequently there is a unique $\bar{\mathbf{T}}(s): \tilde{\mathbf{T}}X \to \tilde{\mathbf{T}}X'$ such that

$$\mathbf{V}'[f, P]\bar{\mathbf{T}}(s) = \mathbf{V}[s(f), P].$$

$\bar{\mathbf{T}}$ is a functor such that $\bar{\mathbf{T}}S = \tilde{\mathbf{T}}$ since if $h \in \mathbf{B}(X, Y)$, $\bar{\mathbf{T}}S(h) = \bar{\mathbf{T}}(\mathbf{B}(h, \mathbf{K}-))$ is the unique element of $\mathbf{C}(\bar{\mathbf{T}}X, \bar{\mathbf{T}}Y)$ such that

$$\mathbf{V}'[f, P]\bar{\mathbf{T}}(\mathbf{B}(h, \mathbf{K}-)) = \mathbf{V}[fh, P] = \mathbf{V}'[f, P]\tilde{\mathbf{T}}(h)$$

for all $(f, P)$ in $X' \downarrow \mathbf{K}$.

**Corollary**
*Let $\mathbf{C}$ be a cocomplete category and $\mathbf{T}: \mathbf{A}^{\mathrm{op}} \to \mathbf{C}$ a functor. If $\tilde{\mathbf{T}}$ is a Kan extension of $\mathbf{T}$ along $\mathbf{K}$, $\tilde{\mathbf{T}}$ is a shape invariant.*

*Proof*
$\check{\mathbf{T}}$ is a 'cofunctor' from **B** to **C** defined by

$$\check{\mathbf{T}}X = \operatorname{Colim}((X \downarrow \mathbf{K})^{\mathrm{op}} \xrightarrow{\delta_X} \mathbf{A}^{\mathrm{op}} \xrightarrow{\mathbf{T}} \mathbf{C})$$

and by duality $\check{\mathbf{T}}$ is a cofunctor from **S** to **C**.

We are next going to suppose that **K** is the inclusion of a full subcategory **A** into a category **B** such that for each $X$ in **B**, there is a pro-object $(\mathbf{I}(X), \mathbf{F}_X)$ in **A** such that $(\mathbf{I}(X), \mathbf{F}_X)$ considered as a pro-object in **B** is **K**-associated to $X$, i.e. the inclusion $\mathbf{K}: \mathbf{A} \to \mathbf{B}$ has a proadjoint, **F** (so we might say **A** is a *proreflective subcategory*).

Let **C** be a complete category and $\mathbf{T}: \mathbf{A} \to \mathbf{C}$ a functor. Let $\check{\mathbf{T}}$ be the functor from **A** to **C** defined by

—$\check{\mathbf{T}}C = \operatorname{Lim}_{\mathbf{I}(X)} \mathbf{T}F_X$ where $\bar{p}_i: \check{\mathbf{T}}X \to \mathbf{TF}_X$ corresponds to the $i$th pseudoprojection $p_i$.

—if $h \in \mathbf{B}(X, Y)$, and $\check{\mathbf{T}}Y = \operatorname{Lim}_{\mathbf{I}(Y)} \mathbf{TF}_Y$ with projection $\bar{q}_j$ where $(\mathbf{I}(Y), \mathbf{F}_Y)$ is **K**-associated to $Y$ with pseudoprojection $q_j$, then $\check{\mathbf{T}}h: \check{\mathbf{T}}X \to \check{\mathbf{T}}Y$ is the unique element of $\mathbf{C}(\check{\mathbf{T}}X, \check{\mathbf{T}}Y)$ such that

$$\bar{q}_j \check{\mathbf{T}}h = \mathbf{T}\mathrm{f}_j \bar{p}_{\theta(j)}$$

where $f_j: \mathbf{F}_X(\theta(j)) \to \mathbf{F}_Y(j)$ is the morphism such that $f_j p_{\theta(j)} = q_j h$.

**Definition**
The functor $\check{\mathbf{T}}: \mathbf{B} \to \mathbf{C}$ is called the *Čech extension* of **T** along **K**. Dually, if **T** is a cofunctor from **A** to **C** where **C** is cocomplete, then $\check{\mathbf{T}}$ is a cofunctor from **B** to **C** defined by

$$\check{\mathbf{T}}X = \operatorname{Colim} \mathbf{TF}_X$$

for $X \in |\mathbf{B}|$.

**Proposition 5**
*If* **T** *is a functor (resp. a cofunctor) from* **A** *to* **C** *where* **C** *is a complete (resp. cocomplete) category, then* $\check{\mathbf{T}}: \mathbf{B} \to \mathbf{C}$ *is a shape invariant.*

*Proof*
We will show that $\check{\mathbf{T}}$ is a Kan extension of **T** along **K**. To show this we make the following remark:

Let $(\mathbf{I}(X), \mathbf{F}_X)$ be **K**-associated to $X$ with pseudoprojection $p_i$ and using (B1) (see section 2.4), let $\bar{\mathbf{F}}_X: \mathbf{I}(X) \to X \downarrow \mathbf{K}$ be defined by

—$\bar{\mathbf{F}}_X(i) = (p_i, \mathbf{F}_X(i))$
—if $v \in \mathbf{I}(X)(i, i')$, $\bar{\mathbf{F}}_X(v) = \mathbf{F}_X(v)$;

then (B2) and (B3) are equivalent to the statement that $\bar{\mathbf{F}}_X$ is an initial functor from $\mathbf{I}(X)$ to $X \downarrow \mathbf{K}$.

(We have almost used this result several times, but until now have not needed to state it explicitly.)

Now let $\tilde{\mathbf{T}}$ be the Kan extension of $\mathbf{T}$ along $\mathbf{K}$. As $\bar{\mathbf{F}}_X$ is initial and $\boldsymbol{\delta}_X\bar{\mathbf{F}}_X = \mathbf{F}_X$, there is a unique invertible

$$t_X\colon \tilde{\mathbf{T}}X = \operatorname{Lim} \mathbf{T}\boldsymbol{\delta}_X \to \check{\mathbf{T}}X = \operatorname{Lim} \mathbf{T}\mathbf{F}_X$$

such that

$$\bar{p}_i t_X = \mathbf{V}[p_i, F_X(i))] \quad \text{for all } i \in |\mathbf{I}|.$$

Let $t\colon \tilde{\mathbf{T}} \to \check{\mathbf{T}}$ be defined by $t_X\colon \tilde{\mathbf{T}}X \to \check{\mathbf{T}}X$; then $t$ is a natural equivalence since, if $h \in \mathbf{B}(X, X')$ where $(\mathbf{I}(X'), \mathbf{F}_{X'})$ is $\mathbf{K}$-associated to $X'$ with pseudoprojection $q_j$, we have

$$\begin{aligned} \bar{q}_j \check{\mathbf{T}}(h) t_X &= \mathbf{T}f_j \bar{p}_{\theta(j)} t_X = \mathbf{T}f_j \mathbf{V}[p_{\theta(j)}, \mathbf{F}_X(\theta(j))] \\ &= \mathbf{V}[f_j p_{\theta(j)}, \mathbf{F}_{X'}(j)] = \mathbf{V}'[q_j, \mathbf{F}_{X'}(j)]\tilde{\mathbf{T}}h = \bar{q}_j t_{X'} \tilde{\mathbf{T}}(h) \end{aligned}$$

for every $j$ in $|\mathbf{I}(X')|$. Consequently, $\check{\mathbf{T}}(h) t_X = t_{X'} \tilde{\mathbf{T}}(h)$ so $\check{\mathbf{T}}$ being naturally equivalent to $\tilde{\mathbf{T}}$ is a Kan extension of $\mathbf{T}$ and thus a shape invariant.

## 2.6 CODENSITY MONADS

The use of Kan extensions allows us to extend the Kleisli category description of $\mathbf{S}_{\mathbf{K}}$ beyond the cases handled in section 2.2. To start with, however, we will look at some 'standard' results on Kan extensions.

We suppose as usual that $\mathbf{K}\colon \mathbf{A} \to \mathbf{B}$ and $\mathbf{T}\colon \mathbf{A} \to \mathbf{C}$, and that $\tilde{\mathbf{T}} = \mathbf{Ran}_{\mathbf{K}}\, \mathbf{T}$ exists with counit $\varepsilon\colon \tilde{\mathbf{T}}\mathbf{K} \to \mathbf{T}$. Now suppose that there is a functor $\mathbf{G}\colon \mathbf{C} \to \mathbf{D}$; we say $\mathbf{G}$ preserves this right Kan extension if $\mathbf{G} \circ \mathbf{Ran}_{\mathbf{K}}\, \mathbf{T}$ is a right Kan extension of $\mathbf{GT}$ along $\mathbf{K}$ with counit $\mathbf{G}\varepsilon\colon \mathbf{G}(\mathbf{Ran}_{\mathbf{K}}\, \mathbf{T})\mathbf{K} \to \mathbf{GT}$. This implies, of course, that

$$\mathbf{G}(\mathbf{Ran}_{\mathbf{K}}\, \mathbf{T}) \cong \mathbf{Ran}_{\mathbf{K}}(\mathbf{GT}).$$

**Proposition 1**
*If* $\mathbf{G}\colon \mathbf{C} \to \mathbf{D}$ *has a left adjoint* $\mathbf{F}$, *then it preserves all right Kan extensions which exist in* $\mathbf{C}$.

*Proof*
We have a natural isomorphism

$$\mathbf{C}(\mathbf{F}X, Y) \cong \mathbf{D}(X, \mathbf{G}Y) \quad \text{for } X \in |\mathbf{C}|,\ \mathrm{Y} \in |\mathbf{D}|.$$

Given any functors $\mathbf{H}\colon \mathbf{B} \to \mathbf{C}$, $\mathbf{L}\colon \mathbf{B} \to \mathbf{D}$, the naturality of this isomorphism gives a bijection

$$\operatorname{Nat}(\mathbf{FH}, \mathbf{L}) \cong \operatorname{Nat}(\mathbf{H}, \mathbf{GL}).$$

We can express the Kan extension property in general as follows: Given $\mathbf{K}\colon \mathbf{A} \to \mathbf{B}$, $\mathbf{T}\colon \mathbf{A} \to \mathbf{C}$, and, say, $\mathbf{S}\colon \mathbf{B} \to \mathbf{C}$, there is a natural bijection

$$\phi_{\mathbf{S}}\colon \operatorname{Nat}(\mathbf{S}, \mathbf{Ran}_{\mathbf{K}}\, \mathbf{T}) \cong \operatorname{Nat}(\mathbf{SK}, \mathbf{T})$$

and the counit $\varepsilon\colon (\mathbf{Ran}_{\mathbf{K}}\, \mathbf{T})\mathbf{K} \to \mathbf{T}$ is $\phi_{\mathbf{S}}(\mathrm{Id}_{\mathbf{Ran}_{\mathbf{K}}\mathbf{T}})$. (In turn if $\sigma\colon \mathbf{S} \to \mathbf{Ran}_{\mathbf{K}}\, \mathbf{T} \in \operatorname{Nat}(\mathbf{S}, \mathbf{Ran}_{\mathbf{K}}\, \mathbf{T})$ then $\phi_{\mathbf{S}}(\sigma) = \varepsilon \cdot \sigma \cdot \mathbf{K} \cdot )$.

Now suppose given $\mathbf{H}: \mathbf{B} \to \mathbf{C}$ as before, we have

$$\begin{aligned}\mathrm{Nat}(\mathbf{H}, \mathbf{G} \circ \mathbf{Ran_K}\, \mathbf{T}) &\cong \mathrm{Nat}(\mathbf{FH}, \mathbf{Ran_K}\, \mathbf{T}) \\ &\cong \mathrm{Nat}(\mathbf{FHK}, \mathbf{T}) \\ &\cong \mathrm{Nat}(\mathbf{HK}, \mathbf{GT}).\end{aligned}$$

Thus $\mathbf{G} \circ \mathbf{Ran_K} \cong \mathbf{Ran_K}\, \mathbf{GT}$ with counit $\mathbf{G}\varepsilon$ as required.

**Proposition 2**

*If* $\mathbf{G}: \mathbf{A} \to \mathbf{B}$ *has a left adjoint* $\mathbf{F}$ *with counit* $\varepsilon: \mathbf{GF} \to \mathbf{Id}$, *then* $\mathbf{Ran_G}(\mathbf{Id_A})$ *exists, is equal to* $\mathbf{F}$ *with counit* $\varepsilon$ *and is preserved by any functor (i.e. it is an absolute Kan extension).*

*Proof*

Suppose we have $\mathbf{G}$ with left adjoint $\mathbf{F}$, with unit $\eta: \mathbf{Id} \to \mathbf{GF}$ and counit $\varepsilon: \mathbf{FG} \to \mathbf{Id}$. We can construct, for any functor $\mathbf{H}: \mathbf{A} \to \mathbf{C}$, a bijection

$$\mathrm{Nat}(\mathbf{S}, \mathbf{HF}) \cong \mathrm{Nat}(\mathbf{SG}, \mathbf{H})$$

natural in $\mathbf{S}: \mathbf{B} \to \mathbf{C}$ by

$$\Phi: \{\sigma: \mathbf{S} \to \mathbf{HF}\} \longrightarrow \{\mathbf{SG} \xrightarrow{\sigma\mathbf{G}} \mathbf{HFG} \xrightarrow{\mathbf{H}\varepsilon} \mathbf{H}\}$$

$$\Psi: \{\tau: \mathbf{SG} \to \mathbf{H}\} \longrightarrow \{\mathbf{S} \xrightarrow{\mathbf{S}\eta} \mathbf{SGF} \xrightarrow{\tau\mathbf{F}} \mathbf{HF}\}.$$

The two composites can easily be checked to be the identity: in fact we have given $\sigma: \mathbf{S} \to \mathbf{H}$, the image of $\sigma$ by $\Psi\Phi$ is

$$\begin{aligned}\Psi\Phi(\sigma) &= (\mathbf{H}\varepsilon\mathbf{F}) \cdot (\sigma\mathbf{GF}) \cdot (\mathbf{S}\eta) \\ &= (\mathbf{H}\varepsilon\mathbf{F})(\mathbf{HF}\eta)(\sigma) \quad \text{by naturality of } \eta \\ &= \mathbf{H}(\varepsilon\mathbf{F} \cdot \mathbf{F}\eta)\sigma \\ &= \mathbf{H}(\mathbf{Id_F})\sigma \qquad \text{by adjunction properties} \\ &= \sigma.\end{aligned}$$

The other composite is also the identity. The calculation is similar. Now consider the case $\mathbf{H} = \mathbf{Id_A}$, $\mathbf{C} = \mathbf{A}$:

$$\mathrm{Nat}(\mathbf{S}, \mathbf{F}) \cong \mathrm{Nat}(\mathbf{SG}, \mathbf{Id_A});$$

hence $\mathbf{F} = \mathbf{Ran_G}(\mathbf{Id_A})$ and the counit of this Kan extension is

$$\Phi(\mathbf{Id_F}) = \varepsilon.$$

Moreover for any $\mathbf{H}$, we have

$$\mathrm{Nat}(\mathbf{S}, \mathbf{H}\, \mathbf{Ran_G}(\mathbf{Id_A})) \cong \mathrm{Nat}(\mathbf{SG}, \mathbf{H})$$

so this Kan extension is preserved by all functors.

**Remark**

The converse of this is also true: see MacLane [71], p. 244.

Given $\mathbf{G}: \mathbf{A} \to \mathbf{B}$, with left adjoint $\mathbf{F}: \mathbf{B} \to \mathbf{A}$, we know that $(\mathbf{G}, \mathbf{F}, \eta, \varepsilon)$ determines a monad on $\mathbf{B}$. (See section 2.2, example (b).) In fact we obtain an endofunctor

$\mathbf{T}: \mathbf{B} \to \mathbf{B}$, defined to be $\mathbf{GF}$, a unit $\eta: \mathbf{Id} \to \mathbf{T} = \mathbf{GF}$ and $\mu = \mathbf{G}\varepsilon\mathbf{F}$, a multiplication, $\mu: \mathbf{T}^2 \to \mathbf{T}$.

We now find

$$\begin{aligned} \mathbf{GF} &= \mathbf{G} \circ \mathbf{Ran_G\, Id_A} \quad \text{(by Proposition 2)} \\ &= \mathbf{Ran_G\, G} \end{aligned}$$

since $\mathbf{G}$ preserves all Kan extensions (and anyway $\mathbf{Ran_G\, Id_A}$ is an absolute Kan extension).

Thus if $\mathbf{G}$ has a left adjoint, the monad induced by the corresponding adjoint pair can be calculated directly from $\mathbf{Ran_G\, G}$, together with $\varepsilon$, $\eta$ and $\mu$ given by (on writing $\mathbf{R} = \mathbf{Ran_G\, G}$):

$$\text{if} \quad \phi_{\mathbf{S}}: \mathrm{Nat}(\mathbf{S}, \mathbf{R}) \cong \mathrm{Nat}(\mathbf{SG}, \mathbf{G})$$

$$\begin{aligned} \varepsilon &= \phi(Id_{\mathbf{R}}) \\ \eta &= \phi^{-1}(\mathrm{Id}_{\mathbf{G}}) \\ \mu &= \phi^{-1}(\varepsilon \cdot \mathbf{R}\varepsilon). \end{aligned}$$

More generally one has the following:

**Proposition 3**

*Suppose that* $\mathbf{G}: \mathbf{A} \to \mathbf{B}$ *is such that* $\mathbf{Ran_G\, G}$ *exists and is given for* $B \in |\mathbf{B}|$ *by the formula*

$$(\mathbf{Ran_G\, G})(B) = \mathrm{Lim}(B \downarrow \mathbf{G} \xrightarrow{\delta_{\mathbf{B}}} \mathbf{A} \xrightarrow{\mathbf{G}} \mathbf{B}).$$

*Then if*

$$\phi: \mathrm{Nat}(\mathbf{S}, \mathbf{Ran_G\, G}) \cong \mathrm{Nat}(\mathbf{SG}, \mathbf{G})$$

*denotes the natural isomorphisms giving universality of* $(\mathbf{Ran_G\, G}, \varepsilon)$ *with* $\varepsilon = \phi(\mathrm{Id}_{\mathbf{R}})$ *then defining*

$$\eta = \phi^{-1}(\mathrm{Id}_{\mathbf{G}}), \qquad \mu = \phi^{-1}(\varepsilon \cdot \mathbf{R}\varepsilon),$$

$(\mathbf{Ran_G\, G}, \eta, \mu)$ *is a monad on* $\mathbf{B}$.

Proof

For brevity we will write $\mathbf{R} = \mathbf{Ran_G\, G}$.

As $\mathbf{R}B = \mathrm{Lim}(B \downarrow \mathbf{G} \xrightarrow{\delta_{\mathbf{B}}} \mathbf{A} \xrightarrow{\mathbf{G}} \mathbf{B})$, for each $(f: B \to \mathbf{G}P, P)$ in $|B \downarrow \mathbf{G}|$ there is a projection, $p(f, p): \mathbf{R}B \to \mathbf{G}P$. Using these projections we can very simply describe the action of $\mathbf{R}$ on morphisms:

if $g: B_1 \to B_2$ is a morphism in $\mathbf{B}$,

$$\mathbf{R}B_1 \xrightarrow{\mathbf{R}g} \mathbf{R}B_2 \xrightarrow{p(f,P)} \mathbf{G}P = \mathbf{R}B_1 \xrightarrow{p(fg,P)} \mathbf{G}P.$$

Also it is clear that if $P \in |\mathbf{A}|$,

$$\varepsilon(P): \mathbf{RG}P \to \mathbf{G}P$$

is exactly $p(\mathrm{Id}_{\mathbf{G}P}, P)$.

Conversely we can retrieve $p(f, P)$ from $\mathbf{R}$ and $\varepsilon$ since, given $(f, P) \in |B \downarrow \mathbf{G}|$,

$$\begin{aligned} \varepsilon(P)\mathbf{R}(f) &= p(\mathrm{Id}_{\mathbf{G}P}, P) \cdot \mathbf{R}(f) \\ &= p(f, P). \end{aligned}$$

It is also useful to have an explicit description of $\phi$ in terms of $\varepsilon$: Given $\alpha\colon \mathbf{S} \to \mathbf{R}$, $\phi(\alpha) = \varepsilon \cdot \alpha\mathbf{G}\colon \mathbf{SG} \to \mathbf{G}$. Thus given $P \in |\mathbf{A}|$, and $\alpha \in \mathrm{Nat}(\mathbf{S}, \mathbf{R})$, $\phi(\alpha)(P)\colon \mathbf{SG}P \to \mathbf{G}$ is the composite

$$\mathbf{SG}P \xrightarrow{\alpha\mathbf{G}P} \mathbf{RG}P \xrightarrow{p(\mathrm{Id}_{\varepsilon P}, P)} \mathbf{G}P.$$

We can now give $\eta$ and $\mu$ in terms of their projections: if $B \in |\mathbf{B}|$, consider $\eta(B)\colon B \to \mathbf{R}B$. We calculate $p(f, p)\eta(B)$:

$$\begin{aligned} p(f, P)\eta(B) &= \varepsilon(P)\mathbf{R}(f)\eta(B) \\ &= \varepsilon(P)\eta(\mathbf{G}P)f \\ &= \phi(\eta)(P)f \\ &= f. \end{aligned}$$

Similarly consider $\mu(B)\colon \mathbf{R}^2B \to \mathbf{R}B$ and $p(f, P)\mu(B)$;

$$\begin{aligned} p(f, P)\mu(B) &= \varepsilon(P)\mathbf{R}(f)\mu(B) \\ &= \varepsilon(P)\mu(\mathbf{G}P)\mathbf{R}^2 f \\ &= \varepsilon(P)\cdot \mathbf{R}\varepsilon(P)\mathbf{R}^2(f) \\ &= \varepsilon(P)\mathbf{R}(\varepsilon(P)\mathbf{R}(f)) \\ &= \varepsilon(P)\mathbf{R}(p(f, P)) \\ &= p(p(f, P), P). \end{aligned}$$

It is now simple to check the monad axioms:

(i) $\mu \cdot \mu\mathbf{R} = \mu \cdot \mathbf{R}\mu$.
In fact for $B \in |\mathbf{B}|$, $f\colon B \to \mathbf{K}P$,

$$\begin{aligned} p(f, P)\mu(B)\cdot \mu\mathbf{R}(B) &= p(p(p(f, P), P), P) \\ &= p(f, P)\mu(B)\mathbf{R}\mu(B). \end{aligned}$$

(ii) $\mu \cdot \eta\mathbf{R} = \mathrm{Id} = \mu \cdot \mathbf{R}\eta$.
In fact for $B \in |\mathbf{B}|$, $f\colon B \to \mathbf{K}P$,

$$\begin{aligned} p(f, P)\mu(B)\cdot \eta\mathbf{R}(B) &= p(p(f, P), P)\eta(\mathbf{R}B) \\ &= p(f, P). \end{aligned}$$

Similarly,

$$p(f, P)\mu(B)\mathbf{R}(\eta B) = p(f, P) \quad \text{as required.}$$

This completes the proof.

We next calculate the Kleisli category of $(\mathbf{R}, \eta, \mu)$. We will denote it by $\mathbf{Kl\,R}$.

—$|\mathbf{Kl\,R}| = |\mathbf{B}|$
—$\mathbf{Kl\,R}(X, Y) = \mathbf{B}(X, \mathbf{R}(Y))$

$$= \mathrm{Nat}(\mathbf{B}(Y, -), \mathbf{B}(X, \mathbf{R}-))$$

(by the Yoneda Lemma)

$$= \mathrm{Nat}(\mathbf{B}(Y, -), \mathbf{B}(X, -)\mathbf{R}).$$

But $\mathbf{R}$ is a pointwise Kan extension, as it was calculated by the limit formula (i.e. it is preserved by representable functors), so

$$\begin{aligned}\mathbf{B}(X, -)\mathbf{R} &= \mathbf{Ran_G}(\mathbf{B}(X, -)\mathbf{G}) \\ &= \mathbf{Ran_G}(\mathbf{B}(X, \mathbf{G}-))\end{aligned}$$

and

$$\begin{aligned}\mathbf{Kl\,R}(X, Y) &= \mathrm{Nat}(\mathbf{B}(Y, -), \mathbf{Ran_G}(\mathbf{B}(X, \mathbf{G}-))) \\ &= \mathrm{Nat}(\mathbf{B}(Y, \mathbf{G}-), \mathbf{B}(X, \mathbf{G}-)) \\ &= \mathbf{S_G}(X, Y).\end{aligned}$$

Thus we have:

**Proposition 4**

*Suppose that* $\mathbf{G}: \mathbf{A} \to \mathbf{B}$ *is such that* $\mathbf{Ran_G\,G}$ *exists and is given for* $B \in |\mathbf{B}|$ *by the formula*

$$\mathbf{Ran_G\,G}(\mathbf{B}) = \mathrm{Lim}(B\downarrow\mathbf{G} \xrightarrow{\delta_B} \mathbf{A} \xrightarrow{\mathbf{G}} \mathbf{B}).$$

*Then there is a natural isomorphism between* $\mathbf{Kl\,R}$ *for* $\mathbf{R}$, *the monad corresponding to* $\mathbf{G}$ *(as in Proposition* 3) *and* $\mathbf{S_G}$, *the Holsztyński shape category of* $\mathbf{G}$.

**Definition**

A functor $\mathbf{G}: \mathbf{A} \to \mathbf{B}$ *admits* a *codensity monad* if for all $B \in |\mathbf{B}|$

$$\mathrm{Lim}(B\downarrow\mathbf{G} \xrightarrow{\delta_B} \mathbf{A} \xrightarrow{\mathbf{G}} \mathbf{B})$$

exists. A choice of limit together with projections $p(f, P)$ then determines the *codensity monad*, $\mathbf{R}$, generated by $\mathbf{G}$ (i.e. the monad described in Proposition 3).

Thus so far we have that if $\mathbf{G}$ generates a codensity monad, $\mathbf{R}$, then $\mathbf{Kl\,R}$ and $\mathbf{S_G}$ are ismorphic. This, however, does not give us that ($\mathbf{Kl\,R}$, $\mathbf{F}$) with $\mathbf{F}$ the 'free' functor, $\mathbf{B} \xrightarrow{\mathbf{F}} \mathbf{Kl\,R}$, is a shape theory. To check this, it suffices, of course, to check that (S2) is satisfied.

The natural function

$$\mathbf{B}(X, \mathbf{G}Q) \to \mathbf{S}(X, \mathbf{G}Q) \cong \mathbf{Kl\,R}(X, \mathbf{G}Q)$$

is given by

$$f \to \mathbf{R}(f)\eta_X.$$

We must show it to be a bijection. There is a function

$$\mathbf{Kl\,R}(X, \mathbf{G}Q) \to \mathbf{B}(X, \mathbf{G}Q)$$

given by

$$\begin{aligned}\mathbf{Kl\,R}(X, \mathbf{G}Q) = \mathbf{B}(X, \mathbf{RG}Q) &\to \mathbf{B}(X, \mathbf{G}Q) \\ s &\to \varepsilon(Q)s.\end{aligned}$$

Thus to verify (S2) it suffices to check

$$\mathbf{R}(\varepsilon\mathbf{G}(Q))\eta(\mathbf{G}(Q)) = \mathrm{Id}_{\mathbf{RG}(Q)}.$$

We check this by composing with $p(f, P)$ for $f\colon \mathbf{G}Q \to \mathbf{G}P$.

$$\begin{aligned} p(f, P)\mathbf{R}(\varepsilon\mathbf{G}Q)\eta(\mathbf{G}Q) &= p(f\varepsilon(Q), P)\eta(\mathbf{G}Q) \\ &= f\varepsilon(Q) \\ &= \varepsilon(P)\mathbf{R}(f) \\ &= p(f, P) \end{aligned}$$

as required.

## HISTORICAL NOTE

The first categorical approach to shape theory was that of Holsztyński [58], who continued with a study [59] of the continuity of the shape functor. The categorical characterisation of shape thus obtained led Bacon [5] to consider the notion of continuity in the sense of Kan and thus to axiomatise a shape theory for a category relative to a full subcategory. Another categorical approach was studied by Le Van [64].

The connection between categorical shape theory and the ANR-systems approach was published about the same time by Mardešić [72] and Weber [102]; it is also implicitly considered in Dold [29] in which one already finds the link with Kan extensions. Deleanu and Hilton [26, 27] and Frei [42, 43] developed these ideas.

A first approach to the notion of a Čech extension of a general functor is made in a remark by Eilenberg and Steenrod [38] and was studied in depth by Lee and Raymond [65]. The results describing the shape category of a functor with left adjoint were noted in Deleanu and Hilton [27] and Frei [43].

The link between categorical shape theory and the theory of procategories and proadjoints, although implicit in the work of Deleanu and Hilton in [26], was explicitly recognised only in the very much more recent work of Mardešić [74], Giuli [48] and Stramaccia [97]. The original results on procategories by Grothendieck and Verdier were available long before shape theory appeared on the scene, yet many of their results have been reproved independently since. The treatment in SGA 4 [51] deserves to be noted as the best starting point for any deep study of procategories and related concepts.

The results on codensity monads are noted by Frei [43] for the case when $\mathbf{K}$ is full. The original work on codensity monads is by Linton [67, 68], and by Applegate and Tierney [3] in their interesting dual form of 'model-induced triples'. Codensity monads are related to the problem of monadic or equational completions: given a functor $\mathbf{G}\colon \mathbf{A} \to \mathbf{B}$, factorise it minimally, say $\mathbf{A} \xrightarrow{\mathbf{G}_1} \mathbf{C} \xrightarrow{\mathbf{G}_2} \mathbf{B}$, such that $\mathbf{G}_2$ is monadic, i.e. has a left adjoint making $\mathbf{C}$ into a category of algebras for the resulting monad on $\mathbf{B}$. Linton [67] gives two ways of constructing $\mathbf{C}$; the first involves what is known as the full clone of operations on $\mathbf{G}$ (which is, in fact, the dual of $\mathbf{S}_{\mathbf{G}}$) and the other is the codensity monad on $\mathbf{B}$ giving $\mathbf{C}$ directly: both constructions are thus closely related to shape theory. (It must be admitted, however, that the idea of geometrical applications of an equational theory interpretation of shape theory seems rather strange.)

# 3

# Shape Theory for Topological Spaces

Let **H** be the homotopy category of topological spaces and **W** the full subcategory of **H** having as objects the spaces which have the homotopy type of a CW-complex; one has that if $X$ has the homotopy type of a CW-complex then it has the homotopy type of an ANR or alternatively the homotopy type of a simplicial complex with either the weak or the strong (metric) topology. If $f$ is a map from $X$ to $Y$, we will denote by $[f]$ the homotopy class of $f$.

We will start this chapter by recalling certain properties of simplicial complexes which will be needed in the study of the shape of topological spaces.

## 3.1 SIMPLICIAL COMPLEXES AND NUMERABLE COVERINGS

**Definition**

A *simplicial complex* $K$ is a set of objects, $V(K)$, called *vertices* and a set of finite subsets called *simplexes*. The simplexes satisfy the condition that each subset of a simplex is also a simplex. We will write '$s \in K$' to mean that $s$ is a simplex of $K$.

To each complex $K$ is associated a topological space called the *polyhedron of* $K$, denoted $|K|$, which is formed from the set of all functions from $V(K)$ to $[0, 1]$ such that

(1) if $\alpha \in |K|$, the set

$$\{v \in V(K) | \alpha(v) \neq 0\}$$

is a simplex of $K$

(2) $\sum_{v \in V(K)} \alpha(v) = 1;$

if $s \in K$, we denote by $|s|$ the set

$$|s| = \{\alpha \in |K| \,|\, \alpha(v) \neq 0 \text{ implies } v \in s\}$$

and

$$\langle s\rangle = \{\alpha\in|K|\,|\,\alpha(v)\neq 0 \text{ if and only if } v\in s\}.$$

$\alpha(v)$ is called the $v$th *barycentric coordinate of* $\alpha$ and the mapping from $|K|$ to $[0, 1]$ defined by $p_v(\alpha)=\alpha(v)$ is called the $v$th *barycentric projection.*

We can put a metric $\phi$ on $|K|$ defined by

$$\phi(\alpha, \beta)=\left(\sum_{v\in V(K)} (p_v(\alpha)-p_v(\beta))^2\right)^{1/2};$$

the topology so determined is the initial topology defined by the barycentric projections. The *weak topology* of $|K|$ consists of the subsets $U$ of $|K|$ such that $U\cap|s|$ is open in the metric topology on $|s|$ for each simplex $s$ of $K$.

A *simplicial mapping* $\phi$ from $K$ to another simplicial complex $L$ is a mapping from $V(K)$ to $V(L)$ such that if $s$ is a simplex of $K$, $\phi(s)$ is a simplex of $L$. $\phi$ induces a continuous mapping

$$|\phi|: |K|\to|L|$$

for either the strong (metric) topology or the weak topology. Two simplicial maps $\phi$ and $\phi'$ are said to be *contiguous* if for each $s\in K$, $\phi(s)\cup\phi'(s)$ is a simplex of $L$; thus if $\phi$ and $\phi'$ are contiguous, $|\phi|\simeq|\phi'|$. (Later on we will need to look at this in some detail, but for the moment it suffices that $|\phi|$ and $|\phi'|$ are homotopic.)

More generally, let $X$ be a topological space and let $f$ and $g$ be two continuous maps from $X$ to $|K|$; $f$ and $g$ are said to be *contiguous* if for each $x\in X$, there is some $s$ in $K$ such that $f(x), g(x)\in|s|$; this implies $f\simeq g$.

**Definition**

Let $K$ be a simplicial complex and $v$ a vertex of $K$; we denote by st $v$ the set of $\alpha\in|K|$ such that $\alpha(v)\neq 0$; this is called the *star of* $v$. We will denote by $\mathscr{U}^*$ the covering of $|K|$ formed by stars of vertices.

As st $v=p_v^{-1}((0, 1])$, $\mathscr{U}^*$ is an open covering in either weak or strong topologies.

**Definition**

Let $X$ be a topological space. A covering $\mathscr{U}$ of $X$ is said to be *numerable* if there is a partition of unity $r=(r_U)_{U\in\mathscr{U}}$ (for each $x\in X$, $\sum_{U\in\mathscr{U}} r_U(x)=1$) such that $r_U^{-1}((0, 1])\subset U$ for each $U\in\mathscr{U}$. We say $r$ is a *numeration* if the covering formed by the $r_U^{-1}((0, 1])$ is pointwise finite, that is each $x\in X$ is in only finitely many of the $r_U^{-1}((0, 1])$. In particular if $\mathscr{U}$ is numerable, $\mathscr{U}$ admits a locally finite numeration [29].

We will denote by cov($X$) the set of all open numerable coverings of $X$.

**Remark**

If $X$ is a normal space, then all locally finite covers are numerable [2].

**Definition**

Let $X$ be a topological space, $\mathscr{U}$ an open covering of $X$. The nerve of $\mathscr{U}$, denoted $\mathrm{N}(X, \mathscr{U})$ or $\mathrm{N}(\mathscr{U})$ if no confusion is possible, is the simplicial complex having as vertices the open sets in $\mathscr{U}$ and for simplexes those finite families of open sets in $\mathscr{U}$

whose intersection is non-empty. A continuous map $p$ from $X$ to $|N(X, \mathscr{U})|$ is said to be *canonical* if for each $U$ in $\mathscr{U}$, $p^{-1}(\text{st } U) \subset U$.

If $\mathscr{U}$ is numerable with numeration $r = (r_U)_{U \in \mathscr{U}}$, there is a canonial map $p_{\mathscr{U}}: X \to |N(X, \mathscr{U})|$ defined by

$$p_{\mathscr{U}}(x) = (r_U(x))_{U \in \mathscr{U}} = \sum_{U \in \mathscr{U}} r_U(x)U.$$

Any other choice of canonical map from $X$ to $|N(X, \mathscr{U})|$ is then contiguous to $p_{\mathscr{U}}$, thus homotopic to it.

**Definition**

Let $\mathscr{U}$ be a covering of a topological space $X$, $f$ and $g$ two continuous maps from $Y$ to $X$; then $f$ and $g$ are said to be $\mathscr{U}$-near if for each $y \in Y$, $f(y)$ and $g(y)$ are contained in some open set in $\mathscr{U}$.

**Lemma 1**

*Let $(r_i)_{i \in I}$ be a partition of unity for a space $X$ and let $\varepsilon > 0$; then for every $x \in X$ there is a neighbourhood $V$ of $x$ such that only a finite number of the $r_i$ have values greater than or equal to $\varepsilon$ in $V$.*

*Proof*

Let $x \in X$ and $\varepsilon > 0$. As $\sum_{i \in I} r_i(x) = 1$, there is a finite subset $F' \subset I$ such that for all finite subsets $F \subset I$, with $F \supset F'$,

$$\left| \sum_{i \in F} r_i(x) - 1 \right| < \varepsilon.$$

$\sum_{i \in F} r_i$ being continuous from $X$ to $[0, 1]$, $V = [\sum_{i \in F} r_i]^{-1}(1 - \varepsilon, 1)$ is a neighbourhood of $x$ and for all $x' \in V$ and $i \in I - F$, $r_i(x) < \varepsilon$.

**Lemma 2**

*Let $(r_i)_{i \in I}$ be a partition of unity for a topological space $X$; then*

$$m(x) = \sup_{i \in I} (r_i(x)) = \max_{i \in I} (r_i(x))$$

*is a continuous map from $X$ to $[0, 1]$.*

*Proof*

Let $x \in X$. As $\sum_{i \in I} r_i(x) = 1$, there is an index, $i \in I$, such that $r_i(x) > 0$. Let $\varepsilon > 0$ be such that $m(x) > \varepsilon$. From the preceding lemma, we have that there is a finite set $F \subset I$ such that

$$m(x') = \sup_{i \in I} r_i(x') = \max_{i \in F} r_i(x')$$

for all $x'$ in some neighbourhood $V$ of $x$. Clearly, $m$ is thus a continuous function from $X$ to $[0, 1]$.

**Proposition 1**

*Let $K$ be a simplicial complex and $f_1$ and $f_2$ two continuous maps from $X$ to $|K|$. If $f_1$ and $f_2$ are $\mathscr{U}^*$-near then they are homotopic.*

*Proof*

Let $x \in X$; then there is a vertex $v \in V(K)$ such that $f_1(x)$ and $f_2(x)$ are in st $v$. Let $s_1$ and $s_2$ be simplexes of $K$ with $f_i(x) \in \langle s_i \rangle$, $i = 1, 2$.

For each $v \in V(K)$, we have denoted by $p_v\colon |K| \to [0, 1]$ the $v$th barycentric projection of $|K|$. Now put $f_v^i = p_v \circ f_i$ for $i = 1, 2$. Let $m = \sup_{v \in V(K)} g_v$ where $g_v = \min(f_v^1, f_v^2)$. As $m(x) > 0$, let $\varepsilon = 1/(4m(x))$, $V$ a neighbourhood of $x$ and $V_x$ a finite subset of $V(K)$ be such that $f_v^1(x') < \varepsilon$ for all $x' \in V$ and $v \in V(K) - V_x$. Lemma 1 gives us that as $(f_v^i)_{v \in V(K)}$ is a partition of unity, one has for each $x' \in V$ and $v \in V(K)$, $g_v(x') < \varepsilon$. As there is a $v' \in V(K)$ with $g_{v'}(x) \geqslant \varepsilon$, there is a neighbourhood $V'$ of $x$, $V' \subset V$ such that

$$m(x') = \max_{v \in V_x} (g_v(x'))$$

for all $x' \in V'$; consequently $m$ is a continuous function, $m\colon X \to [0, 1]$.

Now let $s_v(x) = \max(0, 2g_v(x) - m(x))$; $s_v$ is a continuous function from $X$ to $[0, 1]$ for each $v \in V(K)$ such that $(s_v^{-1}((0, 1])_{v \in V(K)})$ is locally finite. Thus if $x \in X$, $M = m^{-1}((2\varepsilon, 1])$ is a neighbourhood of $x$, and for all $x' \in V' \cap M$ and $v \in V(K) - V_x$, $s_v(x') = 0$.

Let $t(x) = \sum_{v \in V(K)} s_v(x)$. $t$ is continuous and $t(x) > 0$ for all $x$. Let $h_v(x) = s_v(x)/t(x)$ then $h$ is continuous, $h\colon X \to [0, 1]$ and $(h_v)_{v \in V(K)}$ is a locally finite partition of unity. Let $h\colon X \to |K|$ be given by $h(x) = (h_v(x))_{v \in V(K)}$. $h$ is continuous and as $g_v(x) = 0$ implies $h_v(x) = 0$, if $s$ is a simplex of $K$ such that $h(x) \in \langle s \rangle$, then $s \subset s_1 \cap s_2$. Consequently $h$ is contiguous to both $f_1$ and $f_2$ and so

$$f_1 \simeq h \simeq f_2.$$

**Definition**

Let $X$ be a topological space, $\mathscr{U} = (U_i)_{i \in I}$ an open covering of $X$, and $S$ a mapping from $I$ to the set of regular open coverings of $[0, 1]$, that is, the set of finite open coverings $(V_j)_{j=0}^n$ such that each $V_j$ is connected, $0 \in V_0$, $0 \notin V_1$, $1 \in V_n$, $1 \notin V_{n-1}$, $V_j \cap V_{j+1} \neq \varnothing$ for $j = 0, \ldots, n-1$ and $V_j \cap V_k = \varnothing$ for $j < k - 1$. $S$ will be called a *stacking function*. We denote by $\mathscr{U} \times S$ the open covering of $X \times [0, 1]$ formed of the $U_i \times V$ for $i \in I$, $V \in S(i)$. $\mathscr{U} \times S$ is called a *stacked cover* of $X \times [0, 1]$, '*stacked over* $\mathscr{U}$' if we need to be more precise. We will denote by $V_0^i$ and $V_1^i$ the uniquely determined open sets of $S(i)$ containing 0 and 1 respectively.

**Proposition 2**

*Let $b_0$ and $b_1$ be the simplicial maps from $|\mathrm{N}(\mathscr{U})|$ to $|(\mathrm{N}(\mathscr{U} \times S)|$ defined on vertices by*

$$b_\varepsilon(U_i) = U_i \times V_\varepsilon^i \quad \text{for } \varepsilon = 0, 1.$$

*Then $b_0$ and $b_1$ are homotopic.*

*Proof*
Let $\mathscr{U}^*$ be the star covering of $|N(\mathscr{U})|$; $\mathscr{U}^* \in \text{cov}(|N(\mathscr{U})|$, the numeration being defined by the barycentric projections. Note, moreover, that $\mathscr{U}^* \times S$ is a numerable covering of $|N(\mathscr{U})| \times [0, 1]$. Thus we have a canonical map

$$p_{\mathscr{U}^* \times S}\colon |N(\mathscr{U})| \times [0, 1] \to |N(\mathscr{U}^* \times S)|.$$

Let $e\colon |N(\mathscr{U}^* \times S)| \to |N(\mathscr{U} \times S)|$ be the simplicial homeomorphism defined by

$$e(\text{st } U_i \times V) = U_i \times V \quad \text{for } V \in S(i)$$

and let $h_\varepsilon\colon |N(\mathscr{U})| \to |N(\mathscr{U})| \times [0, 1]$ be defined by

$$h_\varepsilon(\alpha) = (\alpha, \varepsilon)\,. \qquad \varepsilon = 0, 1.$$

As, for $\alpha \in |N(\mathscr{U})|$,

$$p_{\mathscr{U}^* \times S} h_\varepsilon(\alpha) \in \text{st}(\text{st}(U_i) \times V) \quad \text{for some } U_i \in U,\ V \in S(i)$$

and as

$$p_{\mathscr{U}^* \times S}(\text{st}(\text{st}(U_i) \times V)) \subset \text{st}(U_i) \times V,$$

we have $\alpha \in \text{st}(U_1)$ and $V = V'_\varepsilon$. Consequently

$$e p_{\mathscr{U}^* \times S} h_\varepsilon(\alpha) \in \text{st}(U_i \times V^i_\varepsilon).$$

Thus $\alpha \in \text{st}(U_i)$ implies that

$$b_\varepsilon(\alpha) \in \text{st}(U_i \times V'_\varepsilon)$$

and $e p_{\mathscr{U}^* \times S} h_\varepsilon$ and $b_\varepsilon$ are $(\mathscr{U} \times S)^*$-near, thus homotopic. As $h_0 \simeq h_1$, we have

$$e p_{\mathscr{U}^* \times S} h_0 \simeq b_0 \simeq e p_{\mathscr{U}^* \times S} h_1 \simeq b_1.$$

**Remark**
Let $\mathscr{U}' \in \text{cov}(X \times [0, 1])$, then there is a $\mathscr{U} \in \text{cov}(X)$ and $S$ a stacking function such that $\mathscr{U} \times S$ is finer than $\mathscr{U}'$. $\mathscr{U} \times S$ is, in particular, an element of $\text{cov}(X \times [0, 1])$—see [29].

## 3.2 SHAPE FOR TOPOLOGICAL SPACES

Let **T** be the category of topological spaces, **P** the full subcategory of **T** with objects the polyhedra, **H** the homotopy category of topological spaces and **W** the full subcategory of **H** having as objects those spaces which have the homotopy type of a CW-complex. If $P \in |\mathbf{W}|$, we can thus take $P$ to be a polyhedron. We will denote by $[\ ]\colon \mathbf{T} \to \mathbf{H}$ the canonical 'homotopy' functor.

Let $X \in |\mathbf{T}|$, and $\text{cov}(X)$, as before, be the set of numerable covers of $X$. If $\mathscr{U}, \mathscr{V}$ are two coverings of $X$, we write $\mathscr{U} \leqslant \mathscr{V}$ if $\mathscr{V}$ is finer than $\mathscr{U}$. With this ordering, $\text{cov}(X)$ is then a directed set. We will denote by $p^{\mathscr{V}}_{\mathscr{U}}$ the simplicial map from $|N(X, \mathscr{V})|$ to $|N(X, \mathscr{U})|$ defined by

$$p^{\mathscr{V}}_{\mathscr{U}}(V) = U$$

where $V \subset U$ for $V \in \mathscr{V}$, $U \in \mathscr{U}$. This, of course, depends on a choice of $U \in \mathscr{U}$, but as any other choice defines a simplicial map contiguous to $p^{\mathscr{V}}_{\mathscr{U}}$, $p^{\mathscr{V}}_{\mathscr{U}}$ defines a unique homotopy class, $[p_{\mathscr{U}}^{\mathscr{V}}]$, from $|\mathrm{N}(X, \mathscr{V})|$ to $|\mathrm{N}(X, \mathscr{U})|$. Also the canonical map

$$p_{\mathscr{U}}: X \to |\mathrm{N}(X, \mathscr{U})|$$

defined in section 3.1 defines a unique homotopy class, $[p_{\mathscr{U}}]$, from $X$ to $|\mathrm{N}(X, \mathscr{U})|$.

We will say that an inverse system $\underset{\sim}{X} = (X_\alpha, p_\alpha^{\alpha'}, A)$ is (**H**, **W**)-*associated* to $X$ if it is **K**-associated to $X$ for **K** the inclusion **W** into **H** (section 2.4).

**Definition**

Let (**S**, $S$) be the Holsztyński shape theory for the inclusion, **K**, of **W** into **H**. Two topological spaces $X$ and $Y$ are said to *have the same shape* if there is an invertible morphism from $X$ to $Y$ in **S**.

**Theorem 1**

*For a topological space X, the inverse system*

$$\underset{\sim}{X} = (|\mathrm{N}(X, \mathscr{U})|, [p_{\mathscr{U}}^{\mathscr{V}}], \mathrm{cov}(X))$$

*is* (**H**, **W**)-*associated to X with pseudoprojection* $[p_{\mathscr{U}}]$.

*Proof*

Let $P \in |\mathbf{W}|$—we may suppose that $P$ is a polyhedron, so $P = |K|$ for some simplicial complex, $K$. Let $\mathscr{K}^*$ be the star covering of $P$ and $[f] \in \mathbf{H}(X, P)$. As $f^{-1}(\mathscr{K}^*) = \mathscr{U} \in \mathrm{cov}(X)$, there is a simplicial map

$$f_{\mathscr{U}}: |\mathrm{N}(X, \mathscr{U})| \to P$$

defined by

$$f_{\mathscr{U}}(f^{-1}(\mathrm{st}\ v)) = v$$

for $v \in V(K)$ and $f_{\mathscr{U}} p_{\mathscr{U}}$ and $f$ are $\mathscr{K}^*$-near since, if $x \in f^{-1}(\mathrm{st}(v))$, $p_{\mathscr{U}}(x) \in \mathrm{st}(f^{-1}(\mathrm{st}(v)))$ and $f_{\mathscr{U}} p_{\mathscr{U}}(x) \in \mathrm{st}(f_{\mathscr{U}}(f^{-1}(\mathrm{st}(v)))) = \mathrm{st}(v)$, consequently $[f] = [f_{\mathscr{U}}][p_{\mathscr{U}}]$.

Let $P \in |\mathbf{W}|$ and $f_0$ and $f_1$ be two continuous maps from $|\mathrm{N}(X, \mathscr{U})|$ to $P$ such that

$$[f_0][p_{\mathscr{U}}] = [f_1][p_{\mathscr{U}}].$$

Let $H: X \times [0, 1] \to \mathrm{P}$ be a homotopy such that $H_\varepsilon = f_\varepsilon P_{\mathscr{U}}$ for $\varepsilon = 0, 1$, where $H_\varepsilon = H|X \times \{\varepsilon\}$ and let $K''$ be the second barycentric subdivision of $K$; thus $P = |K| = |K''|$ and $\mathscr{K}''^*$, the star covering of $|K''|$, is a star refinement of $\mathscr{K}^*$, i.e. for each vertex $v_1''$ of $K''$, there is a vertex $v$ of $K$ such that

$$\bigcup \{\mathrm{st}(v'') | \mathrm{st}(v'') \cap \mathrm{st}(v_1'') \neq \varnothing\} \subset \mathrm{st}(v).$$

As $\mathscr{K}''^*$ is a numerable covering of $P$, $H^{-1}(\mathscr{K}''^*)$ is a numerable covering of $X \times [0, 1]$ so there is a $\mathscr{U}_1 \in \mathrm{cov}(X)$ and a stacking function $S$ such that $\mathscr{U}_1 \times S$ is finer than $H^{-1}(\mathscr{K}''^*)$.

As $f_\varepsilon^{-1}(\mathscr{K}''^*)$ for $\varepsilon = 0, 1$ is a numerable covering of $|\mathrm{N}(X, \mathscr{U})|$, one can find a subdivision, $Y$, of $|\mathrm{N}(X, \mathscr{U})|$ finer than $f_\varepsilon^{-1}(\mathscr{K}''^*)$ for $\varepsilon = 0, 1$, i.e. $Y$ is a simplicial complex such that $|Y| = |\mathrm{N}(X, \mathscr{U})|$ and $f_\varepsilon^{-1}(\mathscr{K}''^*) \leqslant \mathscr{Y}^*$. Now let $\mathscr{U}_2 = p_U^{-1}(\mathscr{Y}^*) \in \mathrm{cov}(X)$ and pick $\mathscr{U}' \in \mathrm{cov}(X)$ such that $\mathscr{U}_1 \leqslant \mathscr{U}'$ and $\mathscr{U}_2 \leqslant \mathscr{U}'$.

As $\mathscr{U}_1 \leqslant \mathscr{U}'$, there is a stacking function $S'$ (induced from $S$) such that

$$H^{-1}(\mathscr{K}''^*) \leqslant \mathscr{U}_1 \times S \leqslant \mathscr{U}' \times S'.$$

Let $d$ be the simplicial map

$$d\colon |\mathrm{N}(X \times [0,1], \mathscr{U}' \times S')| \to P = |K''|$$

defined on vertices by:

if $\mathscr{U}' = (U_i)_{i \in I}$

$$\mathrm{d}(U_i \times V) = V'' \quad \text{where } H(U_i \times V) \subset \mathrm{st}_{K''}(V'').$$

Now since $\mathscr{U}_2 \leqslant \mathscr{U}''$, there is a simplicial map

$$C\colon |\mathrm{N}(X, \mathscr{U}')| \to |\mathrm{N}(X, \mathscr{U})| = |Y|$$

defined by

$$C(U_i) = V_Y \quad \text{where } p_U(U_i) \subset \mathrm{st}(V_Y).$$

As $\mathscr{U} \leqslant \mathscr{U}'$ and $Y$ is a subdivision of $|\mathrm{N}(X, \mathscr{U})|$, $p^{\mathscr{U}}_{\mathscr{U}'}$ and $C$ are contiguous and thus homotopic.

Let $b_\varepsilon$ be the simplicial maps

$$b_\varepsilon\colon |\mathrm{N}(X, \mathscr{U}')| \to |\mathrm{N}(X \times [0,1], \mathscr{U}' \times S')|$$

defined on vertices by

$$b_\varepsilon(U_i) = U_i \times V^i_\varepsilon, \qquad \varepsilon = 0, 1.$$

We have

—$f_\varepsilon p_U(U_i) = H(U_i \times \{\varepsilon\}) \subset H(U_i \times V^i_\varepsilon) \subset \mathrm{st}_{K''}(db_\varepsilon(U_i))$
—$f_\varepsilon p_U(U_i) \subset f_\varepsilon(\mathrm{st}_Y(C(U_i)))$

and so $f_\varepsilon(\mathrm{st}_Y(C(U_i))) \cap \mathrm{st}_{K''}(db_\varepsilon(U_i)) \neq \varnothing$. As $f^{-1}(\mathscr{K}''^*) \leqslant \mathscr{Y}^*$, there is a vertex $v''$, of $K''$, such that

$$f_\varepsilon(\mathrm{st}_Y(C(U_i))) \subset \mathrm{st}_{K''}(v'').$$

Again, as $\mathscr{K}''^*$ is a star refinement of $\mathscr{K}^*$, there is a vertex $v$ of $K$ such that

$$\mathrm{st}_{K''}(v'') \cup \mathrm{st}_{K''}(db_\varepsilon(U_i)) \subset \mathrm{st}\, v.$$

Finally as $(f_\varepsilon C)(\mathrm{st}\, U_i) \subset f_\varepsilon(\mathrm{st}_Y(C(U_i))) \subset \mathrm{st}_{K''}(v'')$, $f_\varepsilon C$ and $db_\varepsilon$ are $K^*$-near and hence homotopic. Thus one has

$$f_0\, p^{\mathscr{U}}_{\mathscr{U}'} \simeq f_0 C \simeq db_0 \simeq db_1 \simeq f_1 C \simeq f_1\, p^{\mathscr{U}}_{\mathscr{U}'}$$

since $b_0$ and $b_1$ are homotopic by Proposition 2 of section 3.1.

**Theorem 2**

*If two topological spaces X and Y have the same shape, they have isomorphic Čech homology and cohomology groups.*

*Proof*

Let $\mathbf{F}\colon \mathbf{P} \to \mathbf{G}$ be a functor (resp. a cofunctor, i.e. a functor from $\mathbf{P}^{\mathrm{op}}$ to $\mathbf{G}$) and assume $\mathbf{G}$ is complete (resp. cocomplete) and that $\mathbf{F}$ is homotopy invariant. Let $\mathbf{F}'$ be the functor (resp. cofunctor)

$$\mathbf{F}'\colon \mathbf{W} \to \mathbf{G}$$

such that $\mathbf{F}'[\ ] = \mathbf{F}$. The Čech extension of $\mathbf{F}$ from $\mathbf{T}$ to $\mathbf{G}$ is the functor defined by

$$\check{\mathbf{F}}X = \mathrm{Lim}(\mathbf{F}(|\mathrm{N}(X, \mathscr{U})|), \mathbf{F}p_{\mathscr{U}}^{\mathscr{V}}, \mathrm{cov}(X))$$

(resp. the cofunctor defined by

$$\check{\mathbf{F}}X = \mathrm{Colim}(\mathbf{F}(|\mathrm{N}(X, \mathscr{U})|), \mathbf{F}p_{\mathscr{U}}^{\mathscr{V}}, \mathrm{cov}(X))) \quad \text{(see [29])}.$$

$\check{\mathbf{F}}$ is a homotopy invariant functor, the factorisation via the functor $[\ ]: \mathbf{T} \to \mathbf{H}$, being $\check{\mathbf{F}} = \check{\mathbf{F}}'[\ ]$ where $\check{\mathbf{F}}': \mathbf{H} \to \mathbf{G}$ is as in Chapter 2 since $\mathbf{F}p_{\mathscr{U}}^{\mathscr{V}} = \mathbf{F}'[p_{\mathscr{U}}^{\mathscr{V}}]$.

As a result, if $\mathbf{G}$ is the category of Abelian groups, $\mathbf{H}_n(-, G)$ (resp. $\mathbf{H}^n(-, G)$) the $n$th homology (resp. cohomology) functor with coefficients in an Abelian group, $G$, considered as a functor from $\mathbf{P}$ to $\mathbf{G}$, then, as $\mathbf{H}_n(-, \mathbf{G})$ (resp. $\mathbf{H}^n(-, \mathbf{G})$) is homotopy invariant, if $X$ and $Y$ have the same shape, $\check{\mathbf{H}}_n(X, G)$ and $\check{\mathbf{H}}_n(Y, G)$ (resp. $\check{\mathbf{H}}_n(X, G)$ and $\check{\mathbf{H}}_n(Y, G)$) are isomorphic by Proposition 5 of section 2.3.

## 3.3 COMPARISON OF THE SHAPE THEORIES OF BORSUK, FOX, MARDEŠIĆ–SEGAL AND MARDEŠIĆ

**Definition**

Let $\underline{X} = (X_\alpha, p_\alpha^{\alpha'}, A)$ be an inverse system of topological spaces and $X = \mathrm{Lim}\, \underline{X}$ with projections, $p_\alpha: X \to X_\alpha, \alpha \in A$. Let $\mathscr{U}' \in \mathrm{cov}(X_{\alpha'})$; $\mathscr{U}'$ is said to be *proper* if $|\mathrm{N}(X, p_{\alpha'}^{-1}(\mathscr{U}'))|$ is homeomorphic to $|\mathrm{N}(X_{\alpha'}, \mathscr{U}')|$ via $p_{\alpha'}$.

If for each $\alpha \in A$, $\mathscr{U} \in \mathrm{cov}(X)$, $\mathscr{V} \in \mathrm{cov}(X_\alpha)$, there is an $\alpha' \geqslant \alpha$ and a proper covering $\mathscr{V}'$ of $X_{\alpha'}$ such that $\mathscr{U} \leqslant p_{\alpha'}^{-1}(\mathscr{V}')$ and $p_\alpha^{\alpha'-1}(\mathscr{V}) \leqslant \mathscr{V}'$, $\underline{X}$ is said to be a *proper inverse system*.

**Proposition 1**

*Let $\underline{X} = (X_\alpha, p_\alpha^{\alpha'}, A)$ be an inverse system of topological spaces, $X = \mathrm{Lim}\, \underline{X}$ and suppose that $\underline{X}$ is proper. Then the inverse system $(X_\alpha, [p_\alpha^{\alpha'}], A)$ is an inverse system $(\mathbf{H}, \mathbf{W})$-associated to $X$ with pseudoprojection, $[p_\alpha]$.*

*Proof*

Using Theorem 1 of section 3.2, we have that $(|\mathrm{N}(X, \mathscr{U})|, [p_{\mathscr{U}}^{\mathscr{V}}], \mathrm{cov}(X))$ is $(\mathbf{H}, \mathbf{W})$-associated to $X$ with pseudoprojections $[p_{\mathscr{U}}]$. The condition that $\underline{X}$ is proper clearly should give us an isomorphism in $\mathbf{pro(W)}$ between this canonical inverse system and $(X_\alpha, [p_\alpha^{\alpha'}], A)$, compatible with the pseudoprojections.

Thus to prove the theorem we have merely to check this in detail; we do this explicitly, i.e. we will use Theorem 1 of section 3.2 to verify the condition on $(X_\alpha, [p_\alpha^{\alpha'}], A)$ directly.

Let $\phi \in \mathbf{H}(X, P)$ with $P \in |\mathbf{W}|$; then there is a $\mathscr{U} \in \mathrm{cov}(X)$, $f_{\mathscr{U}}: |\mathrm{N}(X, \mathscr{U})| \to P$ such that $[f_{\mathscr{U}}][p_{\mathscr{U}}] = \phi$.

Now let $\alpha \in A$ and $\mathscr{V} \in \mathrm{cov}(X_\alpha)$; there is an $\alpha' \geqslant \alpha$, and a proper $\mathscr{V}' \in \mathrm{cov}(X_\alpha)$, such that $\mathscr{U} \leqslant p_{\alpha'}^{-1}(\mathscr{V}')$ and $(p_\alpha^{\alpha'})^{-1}(\mathscr{V}) \leqslant \mathscr{V}'$. Let $q_{\mathscr{V}'}$ be a canonical map from $X_{\alpha'}$ to $|\mathrm{N}(X_{\alpha'}, \mathscr{V}')|$ so that on identifying $|\mathrm{N}(X_{\alpha'}, \mathscr{V}')|$ and $|\mathrm{N}(X, p_{\alpha'}^{-1}(\mathscr{V}')|$ by the given homeomorphism, $q_{\mathscr{V}'}p_{\alpha'}$ is a canonical map from $X$ to $|\mathrm{N}(X, p_{\alpha'}^{-1}(\mathscr{V}')|$; consequently $q_{\mathscr{V}'}p_{\alpha'} \simeq p_{\mathscr{U}'}$ where $\mathscr{U}' = p_{\alpha'}^{-1}(\mathscr{V}')$. Let $f_{\alpha'} = f_{\mathscr{U}} p_{\mathscr{U}}^{\mathscr{U}'} q_{\mathscr{V}'}: X_{\alpha'} \to P$, then

$$[f_{\alpha'}][p_{\alpha'}] = [f_{\mathscr{U}}][p_{\mathscr{U}}^{\mathscr{U}'}][q_{\mathscr{V}'}][p_{\alpha'}] = [f_{\mathscr{U}}][p_{\mathscr{U}}] = \phi.$$

Next suppose $f_1, f_2\colon X_\alpha \to P$ are such that $f_1 p_\alpha \simeq f_2 p_\alpha$; let $\mathscr{K}^*$ be the star covering of $P = |K|$; then there is some $\alpha' \geqslant \alpha$, $\mathscr{V}'$ a proper covering of $X_{\alpha'}$ such that $(f_\varepsilon p_\alpha)^{-1}(\mathscr{K}^*) \leqslant p_{\alpha'}^{-1}(\mathscr{V}')$ and $(f_\varepsilon p_\alpha^{\alpha'})^{-1}(\mathscr{K}^*) \leqslant \mathscr{V}'$ for $\varepsilon = 0, 1$. Let $f'_\varepsilon\colon |\mathrm{N}(X_{\alpha'}, \mathscr{V}')| \to P$ be the simplicial map defined on vertices by

$$f'_\varepsilon(V') = V \quad \text{where } V' \subset (f_\varepsilon p_\alpha^{\alpha'})^{-1}(\mathrm{st}\, V) \quad \text{for } V' \in \mathscr{V}'.$$

As $f'_\varepsilon q_{\mathscr{V}'}$ and $f_\varepsilon p_\alpha^{\alpha'}$ are $\mathscr{K}^*$-near,

$$f'_\varepsilon q_{\mathscr{V}'} \simeq f_\varepsilon p_\alpha^{\alpha'}$$

where $q_{\mathscr{V}'}\colon X_{\alpha'} \to |\mathrm{N}(X_{\alpha'}, \mathscr{V}')|$ is a canonical map.

As $q_{\mathscr{V}'} p_{\alpha'} \simeq p_{\mathscr{U}'}$ where as before, $\mathscr{U}' = p_{\alpha'}^{-1}(\mathscr{V}')$, we have

$$f'_1 p_{\mathscr{U}'} \simeq f'_2 p_{\mathscr{U}'}.$$

Thus there is a $\mathscr{U}_1 \geqslant \mathscr{U}'$ such that

$$f'_1 p_{\mathscr{U}'}^{\mathscr{U}_1} \simeq f'_2 p_{\mathscr{U}'}^{\mathscr{U}_1},$$

by Theorem 1 of section 3.2. Moreover there is an $\alpha'' \geqslant \alpha'$ and $\mathscr{V}''$ a proper covering of $X_{\alpha''}$ such that

$$\mathscr{U}_1 \leqslant p_{\alpha''}^{-1}(\mathscr{V}'') = \mathscr{U}'' \quad \text{and} \quad (p_{\alpha'}^{\alpha''})^{-1}(\mathscr{V}') \leqslant \mathscr{V}''.$$

One thus has

$$f'_1 p_{\mathscr{U}_1}^{\mathscr{U}''} \simeq f'_2 p_{\mathscr{U}_1}^{\mathscr{U}''}$$

and

$$q_{\mathscr{V}'} p_{\alpha'}^{\alpha''} \simeq p_{\mathscr{U}'}^{\mathscr{U}''} q_{\mathscr{V}''},$$

where $q_{\mathscr{V}''}\colon X_{\alpha''} \to |\mathrm{N}(X_{\alpha''}, \mathscr{V}'')|$ is a canonical map. Putting this lot together we therefore obtain

$$\begin{aligned} f_1 p_\alpha^{\alpha''} &= f_1 p_\alpha^{\alpha'} p_{\alpha'}^{\alpha''} \simeq f'_1 q_{\mathscr{V}'} p_{\alpha'}^{\alpha''} \simeq f'_1 p_{\mathscr{U}'}^{\mathscr{U}''} q_{\mathscr{V}''} \simeq f'_1 p_{\mathscr{U}'}^{\mathscr{U}_1} p_{\mathscr{U}_1}^{\mathscr{U}''} q_{\mathscr{V}''} \\ &\simeq f'_2 p_{\mathscr{U}'}^{\mathscr{U}_1} p_{\mathscr{U}_1}^{\mathscr{U}''} q_{\mathscr{V}''} \simeq f'_2 q_{\mathscr{V}'} p_{\alpha'}^{\alpha''} \simeq f_2 p_\alpha^{\alpha''}, \end{aligned}$$

as required.

We will now consider two examples of proper inverse systems: those which occur in the definitions of shape given by Fox and by Mardešić and Segal.

Let $X$ be a metric space; then [9] there is a metrisable space $P$ which is an ANR for metric spaces and which contains $X$ as a closed subspace. Let $\mathrm{N}(X)$ be the set of open neighbourhoods of $X$ in $P$, ordered by $U \leqslant V$ if $V \subset U$; then $\underline{X} = (U, p_U^V, \mathrm{N}(X))$ is an inverse system (with $p_U^V$ the inclusion) such that $X = \mathrm{Lim}\, \underline{X}$.

**Proposition 2**

*Let $X$ be a metric space and $\underline{X} = (U, p_V^U, \mathrm{N}(X))$ the inverse system introduced above; then $\underline{X}$ is a proper inverse system.*

*Proof*

Let $\mathscr{U} \in \mathrm{cov}(X)$ and $\mathscr{V} \in \mathrm{cov}(U)$ for $U \in \mathrm{N}(X)$. If $A$ is an open set of $X$, we will denote by $\phi(A)$ the set of $y$ in $P$ such that

$$d(y, A) < d(y, X - A);$$

$\phi(A)$ is an open set of $P$ such that $\phi(A)\cap X = A$ and if $A_i$, for $i=1,\ldots,n$, are open sets of $X$ such that $\bigcap_{i=1}^{n} A_i=\varnothing$; then $\bigcap_{i=1}^{n}\phi(A_i)=\varnothing$ (see [63]).

Let $\mathscr{V}\cap X$ be the induced covering of $X$; then there is a locally finite open covering $\mathscr{U}'$ of $X$ such that $\mathscr{U}\leqslant\mathscr{U}'$ and $\mathscr{V}\cap X\leqslant\mathscr{U}'$. For each $A'\in\mathscr{U}'$, let $V_{A'}\in\mathscr{V}$ be such that

$$A'\subset V_{A'}\cap X$$

and let

$$\phi_0(A')=\phi(A)\cap V_{A'}.$$

As $\mathscr{U}'$ is locally finite, let $V_x$ be an open neighbourhood of $x$ in $X$ which intersects non-trivially only finite many elements of $U'$ and let

$$V_0=\bigcup_{x\in X}\phi(V_x) \quad\text{and}\quad V=\bigcup_{A'\in\mathscr{U}'}(\phi_0(A')\cap V_0).$$

One has $X\subset V\subset U$ and the covering $\mathscr{V}'=(\phi_0(A')\cap V_0)_{A'\in\mathscr{U}'}$ is locally finite—in fact, let $y\in V$ then as $V\subset V_0$, there is some $\phi(V_x)$ such that $y\in\phi(V_x)\cap V$. Let $A_1',\ldots,A_n'$ be those open sets in $U'$ such that $V_x\cap A_i'\neq\varnothing$, $i=1,\ldots,n$. As

$$\phi(V_x)\cap(\phi_0(A')\cap V_0)\neq\varnothing$$

implies that

$$\phi(V_x)\cap\phi(A')\neq\varnothing,$$

$V_x\cap A'\neq\varnothing$. Consequently there is an $i\in\{1,\ldots,n\}$ such that $A'=A_i'$. As $\mathscr{V}$ is locally finite, $\mathscr{V}'\cap V\leqslant\mathscr{V}'$, $\mathscr{V}'\cap X=\mathscr{U}'\geqslant\mathscr{U}$ and $|\mathrm{N}(X,\mathscr{U}')|$ is homeomorphic to $|\mathrm{N}(V,\mathscr{V}')|$.

**Proposition 3**

*Let $\underline{X}=(X_\alpha, p_\alpha^{\alpha'}, A)$ be an inverse system of compact metric spaces such that $X=\mathrm{Lim}\,\underline{X}$; then $\underline{X}$ is a proper inverse system.*

*Proof*

We will start by recalling a property of finite coverings of normal spaces which will be needed later in the proof [38].

If $X$ is a normal space, $A$ a closed subset of $X$ and $\mathscr{U}$ a finite open covering of $X$, there is a finite open covering $\mathscr{V}$ of $X$, finer than $\mathscr{U}$ and regular relative to $A$, i.e. such that

(1) $V\cap A=\varnothing$ implies $\bar{V}\cap A=\varnothing$ for $V\in\mathscr{V}$

(2) $V_i\cap A\neq\varnothing$ for $i=1,\ldots,n$ and $\bigcap_{i=1}^{n} V_i\neq\varnothing$ imply that $\left(\bigcap_{i=1}^{n} V_i\right)\cap A\neq\varnothing$.

Now let $\mathscr{U}\in\mathrm{cov}(X)$, $\mathscr{V}\in\mathrm{cov}(X_\alpha)$ for some $\alpha\in A$; then there is an $\alpha'\geqslant\alpha$ and $\mathscr{V}'$ a finite covering of $X_{\alpha'}$ such that $(p_\alpha^{\alpha'})^{-1}(\mathscr{V})\leqslant\mathscr{V}'$ and $p_{\alpha'}^{-1}(\mathscr{V}')\geqslant\mathscr{U}$. As $p_{\alpha'}(X)$ is closed in $X_{\alpha'}$, there is a finite open covering $\mathscr{W}$ of $X_{\alpha'}$ finer than $\mathscr{V}'$ and regular relative to $p_{\alpha'}(X)$. Let $W_1,\ldots,W_n$ be the open sets in the covering $\mathscr{W}$ satisfying $W_i\cap p_{\alpha'}(X)\neq\varnothing$ for $i=1,\ldots,n$ and $p_{\alpha'}(X)\subset W=\bigcup_{i=1}^{n} W_i$. Then there is some $\alpha''\geqslant\alpha'$ such that $p_{\alpha'}^{\alpha''}(X_{\alpha''})\subset W$. Let $\mathscr{W}''=\{W_1'',\ldots,W_n''\}$ be the finite open covering of $X_{\alpha''}$ given by

$$W_i''=(p_{\alpha'}^{\alpha''})^{-1}(W_i),\qquad i=1,\ldots,n.$$

$\mathscr{W}''$ is a proper open cover and $(p_\alpha^{\alpha''})^{-1}(\mathscr{V})\leqslant\mathscr{W}''$, $p_{\alpha''}^{-1}(\mathscr{W}'')\geqslant\mathscr{U}$.

**Remarks**

(1) Let $\underline{X}' = (X'_\beta, q_\beta^{\beta'}, B)$ be an inverse system in a category; $\underline{X}'$ is said to be of finite closure if for each $\beta \in B$, $\beta$ has only a finite number of predecessors. We have seen (Proposition 4 of section 2.3) that any inverse system is isomorphic to one which is of finite closure; in fact we proved more: if $X = \mathrm{Lim}\,\underline{X}$ exists, where $\underline{X} = (X_\alpha, p_\alpha^{\alpha'}, A)$ we gave an explicit construction of an $\underline{X}' = (X'_\beta, q_\beta^{\beta'}, B)$ which was of finite closure with the same limit, $X$.

(2) Let **C** be the category of compact spaces, **HC** the homotopy category of **C** and **WC** the full subcategory of **HC** having as objects those compact spaces which have the homotopy type of a CW-complex.

Let $\underline{X} = (X_\alpha, [p_\alpha^{\alpha'}], A)$ and $\underline{Y} = (Y_\beta, [q_\beta^{\beta'}], B)$ be objects of **pro(WC)** with both $(X_\alpha, p_\alpha^{\alpha'}, A)$ and $(Y_\beta, q_\beta^{\beta'}, B)$ being of finite closure. Let $[\underline{h}] \in$ **pro(WC)**$(\underline{X}, \underline{Y})$. Then there is a morphism $\phi$ of inverse systems (in the sense of section 1.3) from $(X_\alpha, p_\alpha^{\alpha'}, A)$ to $(Y_\beta, q_\beta^{\beta'}, B)$ such that $[\underline{\phi}] = [\underline{h}]$.

**Proposition 4**

*Let* **B** *be the Borsuk shape category, and $B$ the corresponding functor, for compact metric spaces; then* (**B**, $B$) *is a shape theory in the sense of Chapter* 2.

*Proof*

(We will, for this proof, always take the Hilbert cube as the AR in which our compact metric spaces are embedded.)

It follows from Proposition 3 of section 1.2 and the definition of **B**, that (S1) and (S2) are satisfied. It therefore remains to prove (S3), i.e. that $B$ is **K**-continuous.

Let $Y$ be a compact metric space embedded in $I^\infty$. $Y = \mathrm{Lim}(Y_n, j_n^{n+1}, \mathbb{N})$ where $(Y_n)_{n\in\mathbb{N}}$ is a decreasing sequence of compact ANR neighbourhoods of $Y$ in $I^\infty$, $j_n^{n+1}$ being the inclusion of $Y_{n+1}$ in $Y_n$. By Proposition 3, $\underline{Y} = (Y_n, [j_n^{n+1}], \mathbb{N})$ is an inverse system associated to $Y$ with pseudoprojections $[j_n]$, where $j_n$ is the inclusion of $Y$ in $Y_n$; thus if $Y = \mathrm{Lim}\,B(\underline{Y})$ where $B(\underline{Y}) = (Y_n, [\underline{j}_n^{n+1*}], \mathbb{N})$ with projection $[\underline{j}_n^*]$, using Theorem 1 of section 2.4 we will have that $B$ is **K**-continuous for **K** the inclusion of **WMC** into **HMC**, **WMC** being the subcategory of compact metric spaces of the homotopy type of a CW-complex.

So let $\underline{g}_n$ for each $n$ be a fundamental sequence from $X$ to $Y_n$ such that $\underline{j}_n^{n+1*}\underline{g}_{n+1} \simeq \underline{g}_n$. As $Y_n$ is an ANR, there is an $h_n\colon X \to Y_n$, unique up to homotopy, such that $[\underline{h}_n^*] = [\underline{g}_n]$; this implies that

$$[\underline{j}_n^{n+1*}][\underline{h}_{n+1}^*] = [(\underline{j_n^{n+1}h_{n+1}})^*] = [\underline{h}_n^*],$$

so

$$j_n^{n+1}h_{n+1} \simeq h_n$$

as maps from $X$ to $Y_n$.

Let $X = \mathrm{Lim}(X_n, i_n^{n+1}, \mathbb{N})$ where $X = \bigcap_{n\in\mathbb{N}} X_n$, $(X_n)_{n\in\mathbb{N}}$ being a decreasing sequence of compact ANR neighbourhoods of $X$, with $i_n$ the inclusion of $X$ into $X_n$. As $\underline{X} = (X_n, [i_n^{n+1}], \mathbb{N})$ is an inverse system associated to $X$, there is a map of systems $\underline{\psi} = ((\psi_n)_{n\in\mathbb{N}}, \gamma)$ from $\underline{X}$ to $\underline{Y}$ (section 2.4) such that for each $n$, $\psi_n i_{\gamma(n)} \simeq j_n\colon X \to Y_n$.

By the remark preceding this proposition, there is a morphism of inverse systems

$$\underline{\psi}' = ((\psi)_{n\in\mathbb{N}}, \theta')$$

in the sense of section 1.3, such that $[\underline{\psi}'] = [\underline{\psi}]$. Let $\underline{\phi} = ((\phi_n)_{n \in \mathbb{N}}, \theta)$ be a regular morphism (see section 1.3) from $\underline{X}$ to $\underline{Y}$ homotopic to $\underline{\psi}'$ and thus also to $\underline{\psi}$, i.e. for each $n \in \mathbb{N}$, there is an index $\alpha$, $\alpha \geqslant \theta(n)$, $\alpha \geqslant \gamma(n)$, such that

$$\phi_n i^{\alpha}_{\theta(n)} \simeq \psi_n i^{\alpha}_{\gamma(n)}.$$

As $\underline{\psi}$ is regular, there is a fundamental sequence $\underline{h} = (h_n)_{n \in \mathbb{N}}$ (by Proposition 3 of section 1.3) such that $h_{\theta(n)} | X_{\theta(n)} = \phi_n$ and, for each $m$, $m' \geqslant \theta(n)$,

$$h_m | X_{\theta(n)} \simeq h_{m'} | X_{\theta(n)} \quad \text{within } Y_n.$$

We set $V_n = X_{\theta(n)}$ and $f_n = h_{\theta(n)}$. As $\theta$ is strictly increasing, $\underline{f} = (f_n)_{n \in \mathbb{N}}$ is a fundamental sequence from $X$ to $Y$ such that

$$f_n | V_n = \phi_n$$

and, for each $m \geqslant n$,

$$f_m | V_n \simeq f_n | V_n \quad \text{within } Y_n.$$

$[\underline{f}]$ is the unique morphism of **B** from $X$ to $Y$ such that

$$[\underline{j}^*_n][\underline{f}] = [\underline{g}_n] \quad \text{for each } n$$

as we now show.

Let $\mathrm{Id}^*_{Y_n}$ be an extension of $\mathrm{Id}_{Y_n}$ over $I^\infty$ and $\underline{l} = (l_m)_{m \in \mathbb{N}}$ where $l_m = \mathrm{Id}^*_{Y_n}$ for each $m$; $\underline{l}$ is a fundamental sequence from $Y$ to $Y_n$ and $[\underline{j}^*_n] = [\underline{l}]$ by Proposition 1 of section 1.2. Let $\underline{h}^*_n = (h^n_m)_{m \in \mathbb{N}}$ where $h^n_m = h^*_n$ for each $m$ and let $V$ be an open neighbourhood of $Y_n$ in $I^\infty$. For each $m \geqslant n$,

$$l_m f_m | V_n \simeq l_m f_n | V_n = \phi_n \quad \text{within } Y_n.$$

As $\phi_n | X = \phi_n i_{\theta(n)} = \phi_n i^{\alpha}_{\theta(n)} i_\alpha \simeq \psi_n i^{\alpha}_{\gamma(n)} i_\alpha = \psi_n i_{\gamma(n)} \simeq h_n$ from $X$ to $Y_n$, and thus within $V$, open in $I^\infty$, and as $\phi_n$ and $h^*_n | V$ are maps from $V_n$ to $I^\infty$, it follows from Proposition 3 of section 1.1 that there is a neighbourhood, $W_n$, of $X$ within $V_n$, and thus within $I^\infty$, such that

$$\phi_n | W_n \simeq h^*_n | W_n \quad \text{within } V.$$

Consequently for all $m \geqslant n$, we have

$$l_m f_m | W_n \simeq l_m f_n | W_n = \phi_n | W_n \simeq h^*_n | W_n = h^n_m | W_n$$

within $V$ as required.

**Remark**

As $B$ satisfies (S1) and (S2), we could equally have shown that $(\mathbf{B}, B)$ is a shape theory by showing that the factorisation of $B$ via the Holsztyński shape functor for the inclusion of **WMC** into **HMC** given by Proposition 2 of section 2.1 is an isomorphism.

Let **H** (resp. **HC**, **HM**, **HMC**) be the homotopy category of topological spaces (resp. compact spaces, metric spaces and compact metric spaces), **W** (resp. **WC**, **WM**, **WMC**) being the corresponding subcategories of spaces having the homotopy type of a CW-complex.

In [80], Morita considers for $(\mathbf{H}, \mathbf{W})$ the category $\mathbf{S}_\mathbf{M}$ where $\mathbf{S}_\mathbf{M}(X, Y) = \mathbf{Pro(W)}(\underline{X}, \underline{Y})$, $\underline{X}$ and $\underline{Y}$ being inverse systems $(\mathbf{H}, \mathbf{W})$-associated to $X$ and $Y$ respectively.

In [76, 77] Mardešić and Segal consider for (**HC**, **WC**) the category $\mathbf{S}_{MS}$ where $\mathbf{S}_{MS}(X, Y)$ is the set of homotopy classes of morphisms of inverse systems from $\underline{X}$ to $\underline{Y}$, $\underline{X}$ and $\underline{Y}$ being two closure finite inverse systems such that $X = \text{Lim } \underline{X}$, $Y = \text{Lim } \underline{Y}$, any compact space being obtainable as an inverse system of compact polyhedra (see [38] and the remark preceding Proposition 4 of section 3.3).

In [40, 41], Fox considers the category $\mathbf{S}_F$ for (**HM**, **WM**), having as morphisms from $X$ to $Y$, the promorphisms from $\underline{X}$ to $\underline{Y}$, where $\underline{X} = (U, [p_U^V], \mathrm{N}(X))$, $X$ being embedded in $P$, a metric ANR, and $\mathrm{N}(X)$ being the set of open neighbourhoods of $X$ in $P$.

In [72], Mardešić considers for (**HC**, **WC**) the Holsztyński shape theory for the corresponding inclusion.

In general, if $S: \mathbf{H}' \to \mathbf{S}$, $S': \mathbf{H}' \to \mathbf{S}'$ are two functors, we will say $(\mathbf{S}, S)$ and $(\mathbf{S}', S')$ are *identical on* $\mathbf{H}'$ if there is a unique isomorphism from $\mathbf{S}$ to $\mathbf{S}'$ commuting with $S$ and $S'$. By the corollary to Proposition 2 of section 2.1, if $(\mathbf{S}, S)$ and $(\mathbf{S}', S')$ are two shape theories for a pair $(\mathbf{H}', \mathbf{W}')$ where $\mathbf{W}'$ is a full subcategory of $\mathbf{H}'$, then they are identical on $\mathbf{H}'$.

**Proposition 5**

*$(\mathbf{S}_M, S_M)$ is identical to $(\mathbf{S}_{MS}, S_{MS})$ on compact spaces, and with $(\mathbf{S}_F, S_F)$ on metric spaces. $(\mathbf{B}, B)$ is identical to $(\mathbf{S}_F, B_F)$ and to $(\mathbf{S}_{MS}, S_{MS})$ on compact metric spaces.*

*Proof*

By Theorem 2 of section 2.4, $(\mathbf{S}_M, S_M)$ is identical to the Holsztyński shape theory for the inclusion of **W** into **H** and, *a fortiori*, is identical to Mardešić's theory on compact spaces. Thus $(\mathbf{S}_M, S_M)$ is clearly a shape theory.

Let $(\mathbf{S}_M, S_M)$ be considered restricted to (**HC**, **WC**); following the remark after Proposition 3 of section 3.3 and that proposition itself, we can take as an inverse system associated to $X$, a closure inverse system $\underline{X}$ satisfying $X = \text{Lim } \underline{X}$. $(\mathbf{S}_M, S_M)$ is thus trivially seen to be the same as $(\mathbf{S}_{MS}, S_{MS})$ on compact spaces.

Proposition 2 of section 3.3 likewise gives us the identity of $(\mathbf{S}_M, S_M)$ and $(\mathbf{S}_F, S_F)$ on (**HM**, **WM**). As $(\mathbf{S}_{MS}, S_{MS})$ and $(\mathbf{S}_F, S_F)$ are shape theories on (**HC**, **WC**), they are shape theories, by restriction, on the inclusion corresponding to the pair, (**HMC**, **WMC**), and hence they are identical to $(\mathbf{B}, B)$ on **HMC**.

**Example 1**

The Warsaw circle is the subset $S_W$ of $\mathbb{R}^2$ consisting of the set

$$\left\{\left(x, \sin\frac{1}{x}\right)\middle| 0 < x \leqslant \frac{1}{2\pi}\right\} \cup \{(0, y) | -1 \leqslant y \leqslant 1\} \cup C$$

where $C$ is an arc in $\mathbb{R}^2$ joining $(0, 0)$ and $[1/(2\pi), 0]$, disjoint from the other two subsets specified above except at its endpoints. Clearly $S_W$ has not the same homotopy type as $S^1$, but from the classification of connected compacta in $\mathbb{R}^2$, treated in Chapter 1, $S_W$ has the same shape as $S^1$. One can obtain this result more directly in the following way:

$$S_W = \text{Lim } \underline{X}$$

where $\underline{X} = (X_n, p_n^{n+1}, \mathbb{N})$ where, for each $n$, $X_n$ has the same homotopy type as $S^1$, and $p_n^{n+1}$ is a homotopy equivalence, $X_n$ being defined by

$$\left\{\left(x, \sin\frac{1}{x}\right)\Big|\frac{1}{2n\pi} \leqslant x \leqslant \frac{1}{2\pi}\right\} \cup \left\{(x, 0)|0 \leqslant x \leqslant \frac{1}{2n\pi}\right\} \cup \{(0, y)| -1 \leqslant y \leqslant 1\} \cup C.$$

The projection $p_n: S_W \to X_n$ is given by

$$p_n\left(x, \sin\frac{1}{x}\right) = (x, 0) \quad \text{if } 0 < x \leqslant \frac{1}{2n\pi}$$

$$p_n(x, y) = (x, y) \quad \text{otherwise.}$$

The bonding maps $p_n^{n+1}: X_{n+1} \to X_n$ are given in the obvious way and are clearly homotopy equivalences. Following remark 2 of section 2.2, $S_{MS}(p_n)$ is invertible in $\mathbf{S}_{MS}$ for each $n$, and so $S_W$ and $S^1$ have the same shape.

**Example 2**

Let $P = (r_1, \ldots, r_n, \ldots)$ be a sequence of prime numbers. We shall give the name '$P$-adic solenoid' to the space

$$S_P = \text{Lim}(X_n, p_n^{n+1}, \mathbb{N})$$

where for each $n$, $X_n = S^1$ and (considering $S^1$ as the unit circle in $\mathbb{C}$, i.e. $S^1 = \{z \in \mathbb{C} \| z| = 1\}$), $p_n^{n+1}(z) = z^{r_n}$. $S_P$ is a topological group which is compact and connected. (A purely geometric description of this space is given in [13].)

Let $Q = (S_1, \ldots, S_n, \ldots)$ be another sequence of prime numbers; we will write $P \sim Q$ if on omitting a finite number of elements from each sequence the resulting sequences $P'$ and $Q'$ are such that each prime appears with the same frequency in each sequence.

If $S_P$ and $S_Q$ have the same shape, $\check{\mathbf{H}}^1(S_P, \mathbb{Z}) \cong \check{\mathbf{H}}^1(S_Q, \mathbb{Z})$ by Theorem 2 of section 3.2. Continuity of Čech cohomology gives

$$\check{\mathbf{H}}^1(S_P, \mathbb{Z}) \cong \text{Lim}(\check{\mathbf{H}}^1(X_n, \mathbb{Z}), p_n^{n+1*}, \mathbb{N})$$

where for each $n \in \mathbb{N}$,

$$\check{\mathbf{H}}^1(X_n, \mathbb{Z}) \cong \mathbb{Z} = A_n, \quad \text{say}$$

and

$$p_n^{n+1*}(c) = r_n c.$$

Let $G_P$ be the Abelian group formed of elements $u/(r_1 \cdots r_i)$ for $u \in \mathbb{Z}, i = 1, \ldots, m, \ldots$; then $G_P \simeq \check{\mathbf{H}}^1(S_P; \mathbb{Z})$ since there is for each $n$ a homomorphism

$$V_n: A_n \to G_P$$

defined by

$$V_n(t) = \frac{t}{r_1 \cdots r_n}.$$

$G_P \cong G_Q$ implies $P \sim Q$—in fact if $\phi$ is an isomorphism,

$$\phi: G_P \to G_Q,$$

let

$$\phi(1)=\frac{v_0}{s_1\cdots s_{j_0}} \quad \text{and} \quad \phi^{-1}(1)=\frac{u_0}{r_1\cdots r_{i_0}}.$$

Let $P'$ be the sequence obtained from $P$ in the following manner:

If $p\in\{r_1,\ldots,r_{i_0}\}\cup\{s_1,\ldots,s_{j_0}\}$, $p\in\{P\}$ and $p$ appears $n$ times, then let $r_{i_1}=\cdots=r_{i_n}=p$ and we omit $r_{i_1},\ldots,r_{i_n}$; similarly one obtains $Q'$. Then if $p\in\{P'\}$ and $p$ appears an infinite number of times, one has $p\in\{Q'\}$ and $p$ also appears an infinite number of times in $Q'$.

If $p\in P'$ and $p$ appears $n$ times in $P$, then $p\in Q'$ and it appears there $n$ times.

Conversely $P\sim Q$ implies $G_P\cong G_Q$·

—if $\{P\}=\{Q\}$, the elements appearing the same number of times in $P$ and in $Q$ then clearly $G_P\cong G_Q$;

—if $Q$ is obtained from $P$ by omitting an element $r_n$ then there is an isomorphism from $G_P$ to $G_Q$ defined by

—if $i<n$, $\phi\left(\dfrac{u}{r_1\cdots r_i}\right)=\dfrac{ur_n}{r_1\cdots r_i}$;

—if $j\geqslant n$, $\phi\left(\dfrac{u}{r_1\cdots r_n\cdots r_j}\right)=\dfrac{ur_{j+1}}{r_1\cdots r_{n-1}r_{n+1}\cdots r_jr_{j+1}}$.

Let $P'$ and $Q'$ be the sequences defined by $P\sim Q$, on omitting from $P$ the elements $r_{i_1},\ldots,r_{i_n}$ and from $Q$ the elements $s_{j_1},\ldots,s_{j_m}$. Let

$$Q_1=Q-\{s_{j_1}\},\qquad Q_2=Q_1-\{s_{j_2}\},\ldots,Q'=Q_{m-1}-\{s_{j_m}\};$$

similarly $P_1=P-\{r_{i_1}\}$ etc. One then has

$$G_Q\cong G_{Q_1}\cong\cdots\cong G_{Q'}\cong G_{P'}\cong\cdots\cong G_{P_1}\cong G_P.$$

This isomorphism is, moreover, compatible with the natural limit topologies on $G_P$ and $G_Q$ (i.e. $G_P \underset{\text{Top}}{\cong} G_Q$).

By remark 1 of section 2.2, $\mathbf{H}(S_P, S^1)$ is in one–one correspondence with Colim $\mathbf{H}(X_n, S^1)$ and $H(X_n, S^1)\cong\mathbb{Z}$. As $S^1$ is arcwise connected, if $f\colon S_P\to S^1$ is continuous, there is a unique continuous homomorphism, $h$, from $S_P$ to $S^1$ such that $f\simeq h$ (W. Scheffer, *Proc. Amer. Math. Soc.* **33**, 562–567 (1972)). $G_P \underset{\text{Top}}{\cong} \mathrm{Char}(S_P)$ where $\mathrm{Char}(S_P)$ is the topological group of characters of $S_P$, consequently $G_P \underset{\text{Top}}{\cong} G_Q$ implies that $S_P$ and $S_Q$ are homeomorphic. In general, let $S^m$ be the $m$-sphere, $\underline{X}=(X_n, p_n^{n+1}, \mathbb{N})$ an inverse system such that $X_n=S^m$, $p_n^{n+1}$ being of degree $r_n$ for each $n$ and let $S_P^m=\mathrm{Lim}\,\underline{X}$, then $S_P^m$ and $S_Q^m$ have the same shape if and only if $P\sim Q$ [76]. One can thus deduce that given any '$S^m$-like' compact connected metric space $X$ (which is thus obtainable as an inverse limit of such a system $(X_n, p_n^{n+1}, \mathbb{N})$ with $X_n=S^m$ for all $n$), $X$ has the shape of a point, of $S^m$ or of a $S_P^m$ for $P$ a sequence of prime numbers. (A space $X$ is $S^m$-like if for any $\varepsilon>0$, there is a continuous map

$$f_\varepsilon\colon X\to S^m$$

such that for any $y \in S^m$,

$$\operatorname{diam}(f_\varepsilon^{-1}(y)) < \varepsilon.)$$

## HISTORICAL NOTE

In reply to a problem posed by his notion of overlay—a sort of generalised covering space—Fox [40, 41] introduced the notion of shape independently of Borsuk's work. The connection between the notion of shape based on the Čech construction of nerves of finite open coverings and the ANR systems approach was studied for compact metric spaces by Porter [85]. This was extended to numerable open covers giving a comparison with the generalised shape theories of Holsztyński and Mardešić by Le Van [64], Morita [80, 81] and Bacon [6].

# 4

# Distributors and Shape Theory

We have seen several times the link between the shape theory $(\mathbf{S_K}, S_\mathbf{K})$ associated with a functor $\mathbf{K}: \mathbf{A} \to \mathbf{B}$ and certain Kleisli categories associated to $\mathbf{K}$ via constructions such as the codensity monad introduced in section 2.6. The results obtained depended on the existence of certain adjoints, sometimes to $\mathbf{K}$ itself, sometimes via the notion of Kan extension to a functor induced by $\mathbf{K}$. To obtain the most natural and general results, we pass to a situation in which all functors have adjoints. Such a context was introduced by Bénabou [8] and is obtained from the 2-category of (small) categories, **Cat**, by formally adjoining adjoints to all functors. This construction leads not to a 2-category but to a bicategory, that is a generalised 2-category in which equalities are replaced by 'coherent' isomorphisms in the formulation of the rules relating to associativity of composition and to the existence of identities. In this bicategory, the categorical description of shape theory simplifies enormously.

## 4.1 THE BICATEGORY Dist

Rather than start formally with a definition of the notion of a bicategory, it would seem better to let that wait until we can outline the axioms as being abstractions of the relevant properties of our main example of a bicategory, namely the bicategory of distributors.

**Definition**
Let $\mathbf{A}$, $\mathbf{B}$ be two categories; a *distributor* $\phi$ from $\mathbf{A}$ to $\mathbf{B}$ is a functor

$$\phi: \mathbf{B}^{op} \times \mathbf{A} \to \mathbf{Sets}.$$

The notation we will use is

$$\phi: \mathbf{A} \dashrightarrow \mathbf{B}.$$

**Examples**

If $\mathbf{F}: \mathbf{A} \to \mathbf{B}$ is a functor, then we may associate two distributors to $\mathbf{F}$:

(i) $\phi_{\mathbf{F}}: \mathbf{A} \dashrightarrow \mathbf{B}$ defined by $\phi_{\mathbf{F}}(B, A) = \mathbf{B}(B, \mathbf{F}A)$;
(ii) $\phi^{\mathbf{F}}: \mathbf{B} \dashrightarrow \mathbf{B}$ defined by $\phi^{\mathbf{F}}(A, B) = \mathbf{B}(\mathbf{F}A, B)$.

A *natural transformation* or *2-cell* between distributors is exactly a natural transformation between the corresponding functors, for instance if $t: \mathbf{F} \to \mathbf{G}$ is a natural transformation of functors from $\mathbf{A}$ to $\mathbf{B}$ then $t$ induces a 2-cell

$$\phi_t: \phi_{\mathbf{F}} \Rightarrow \phi_{\mathbf{G}}$$

of distributors. Composition of natural transformations makes the set $\mathbf{Dist}(\mathbf{A}, \mathbf{B})$ of all distributors from $\mathbf{A}$ to $\mathbf{B}$ into a category. In fact

$$\mathbf{Dist}(\mathbf{A}, \mathbf{B}) = \mathbf{Cat}(\mathbf{B}^{\mathrm{op}} \times \mathbf{A}, \mathbf{Sets}).$$

Diagrammatially we will represent a 2-cell, $t$, from $\phi$ to $\phi'$ by

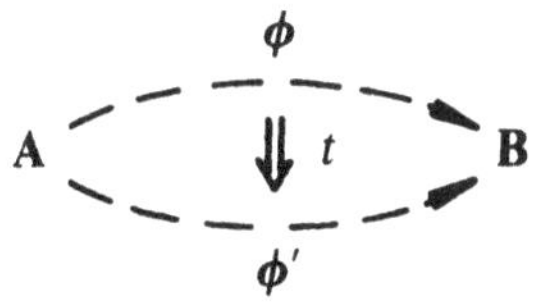


In order to complete the definition of **Dist**, we need to define composition. To see a feasible way to define a composition it helps to view a distributor.

$$\phi: \mathbf{A} \dashrightarrow \mathbf{B},$$

as a bi-indexed collection of sets $\{\phi(B, A) \mid B \in |\mathbf{B}|, A \in |\mathbf{A}|\}$ together with actions of $\mathbf{B}$ on the left and $\mathbf{A}$ on the right:

—if $\beta: B \to B'$, then $\beta$ acts on $x \in \phi(B', A)$ to give $\phi(\beta, A)(x) \in \phi(B, A)$ and similarly for the action of $\alpha: A' \to A$.

If $\mathbf{A}$ and $\mathbf{B}$ were small additive categories with $|\mathbf{A}| = |\mathbf{B}| =$ a one-point set (i.e. if $\mathbf{A}$ and $\mathbf{B}$ were rings) and $\phi$ were an additive functor

$$\phi: \mathbf{B}^{\mathrm{op}} \times \mathbf{A} \to \mathbf{Ab}$$

then $\phi$ would be precisely a left $\mathbf{B}$-, right $\mathbf{A}$-bimodule. Continuing with this idea given a second left $\mathbf{C}$-, right $\mathbf{B}$-bimodule $\psi$, one can form a new left $\mathbf{C}$-, right $\mathbf{A}$-bimodule by constructing the tensor product $\psi \underset{\mathbf{B}}{\otimes} \phi$ and in fact it is difficult to imagine another sensible way of arriving at such a bimodule.

We do not need additivity to make this work and so proceed as follows:

Let $\phi: \mathbf{A} \dashrightarrow \mathbf{B}$ and $\psi: \mathbf{B} \dashrightarrow \mathbf{C}$ be two distributors. The composite $\psi \otimes \phi: \mathbf{A} \dashrightarrow \mathbf{C}$ is defined by

$$(\psi \otimes \phi)(C, A) = \left( \coprod_{B \in |\mathbf{B}|} \psi(C, B) \times \phi(B, A) \right) \Big/ R.$$

where $R$ is the equivalence relation generated by the relation

$$(\psi(C, b)(\beta), \alpha) \sim (\beta, \phi(b, A)(\alpha))$$

for

$$\alpha \in \phi(B, A), \qquad \beta \in \psi(C, B') \quad \text{and} \quad b \in \mathbf{B}(B', B).$$

We will denote by $y \otimes x$ the equivalence class determined by $(y, x)$, $x \in \phi(B, A)$, $y \in \psi(C, B)$. The above thus implies

$$\psi(C, b)(\beta) \otimes \alpha = \beta \otimes \phi(b, A)(\alpha).$$

Just as in the well known case of tensor product of bimodules, it is clear that this composition will not be associative, but that there will be a natural isomorphism between multiple composites in the usual way.

To see how this composition behaves with regard to functors we note the following properties:

(1) If

$$\mathbf{A} \xrightarrow{\mathbf{G}} \mathbf{B} \xrightarrow{\mathbf{F}} \mathbf{C}$$

are functors then

$$\phi_{\mathbf{F}} \otimes \phi_{\mathbf{G}} \cong \phi_{\mathbf{FG}}.$$

In fact

$$(\phi_{\mathbf{F}} \otimes \phi_{\mathbf{G}})(C, A) = \left( \coprod_{B \in |\mathbf{B}|} \mathbf{C}(C, \mathbf{F}B) \times \mathbf{B}(B, \mathbf{G}A) \right) \Big/ R.$$

Suppose $x \in \mathbf{B}(B, \mathbf{G}A)$, $y \in \mathbf{C}(C, \mathbf{F}B)$ then

$$\begin{aligned} y \otimes x &= y \otimes \mathbf{B}(x, \mathbf{G}A)(\mathrm{Id}_{\mathbf{G}A}) \\ &= \mathbf{C}(C, \mathbf{F}(x))(y) \otimes \mathrm{Id}_{\mathbf{G}A} \\ &= \mathbf{F}(x)y \otimes \mathrm{Id}_{\mathbf{G}A}. \end{aligned}$$

Thus any $y \otimes x$ has a representative in $\mathbf{C}(C, \mathbf{FG}A) \times \{\mathrm{Id}_{\mathbf{G}A}\}$. Define the transformation

$$\phi_{\mathbf{F}} \otimes \phi_{\mathbf{G}} \to \phi_{\mathbf{FG}}$$

by

$$y \otimes x \to \mathbf{F}(x)y.$$

This is well defined, natural in both arguments and defines the desired equivalence.

(2) If one has

$$\mathbf{A} \xrightarrow{\mathbf{F}} \mathbf{B} \overset{\phi}{\dashrightarrow} \mathbf{C}$$

then

$$(\phi \otimes \phi_{\mathbf{F}})(C, A) \cong \phi(C, \mathbf{F}A)$$

and if

$$\mathbf{B} \overset{\phi}{\dashrightarrow} \mathbf{C} \xleftarrow{\mathbf{F}} \mathbf{A}, \qquad \mathbf{A} \overset{\phi^{\mathbf{F}}}{\dashrightarrow} \mathbf{C}$$

then

$$(\phi^{\mathbf{F}} \otimes \phi)(A, B) \cong \phi(\mathbf{F}A, B).$$

The proof of these is nearly the same as that of (1).

As a corollary of (2), we have that if $\mathbf{I}: \mathbf{A} \to \mathbf{A}$ is the identity functor then $\phi_{\mathbf{I}} = \phi^{\mathbf{I}}$ is the left and right identity for the composition in **Dist**, but that these identities are identities only up to natural isomorphisms

$$\phi \otimes \phi_{\mathbf{I}} \cong \phi \cong \phi^{\mathbf{I}} \otimes \phi.$$

We note finally the easily verified fact that in the situation

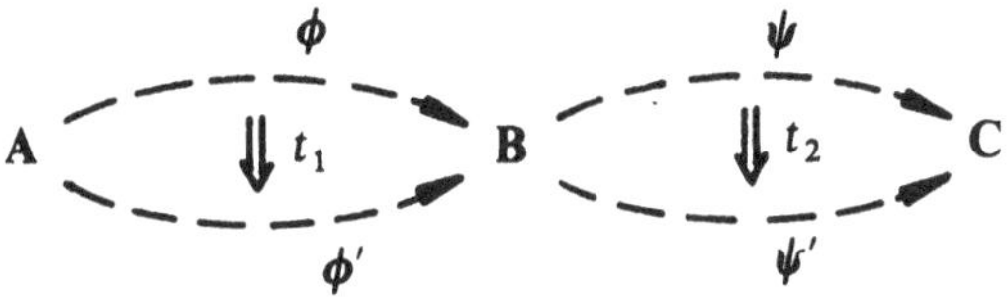


one always has a formula

$$t_2 \otimes t_1 \overset{\text{defn}}{=} (t_2 \otimes \phi')\cdot(\psi \otimes t_1) = (\psi' \otimes t_1)\cdot(t_2 \otimes \phi)$$

where $\cdot$ denotes composition of 2-cells.

With these facts now available, we may now give a formal definition of a bicategory and will then develop the formal results on them that will be needed later.

**Definition**

A *bicategory* $\mathbf{A}$ is determined by the following data:

(i) a class $|\mathbf{A}|$ whose elements, denoted $A, B, C$, etc., are called the *objects* of $\mathbf{A}$

(ii) for each pair, $(A, B)$, of objects, a category $\mathbf{A}(A, B)$; an object $f$ of $\mathbf{A}(A, B)$ is called an *arrow* and we will write $A \xrightarrow{f} B$; a morphism $\alpha$ from the arrow $f$ to the arrow $f'$ is called a *2-cell* and one writes

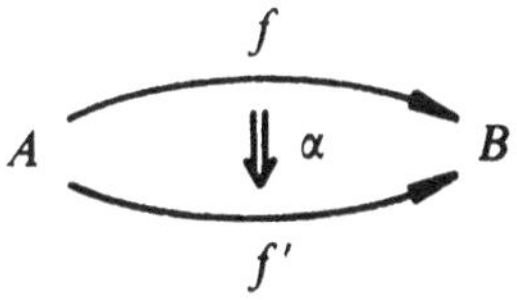


(iii) for each triplet $(A, B, C)$, a *composition functor*,

$$C(A, B, C): \mathbf{A}(B, C) \times \mathbf{A}(A, B) \to \mathbf{A}(A, C).$$

(We will denote this composition by $\otimes$ to tie it in with our principal example. Note that (iii) implies that in the situation

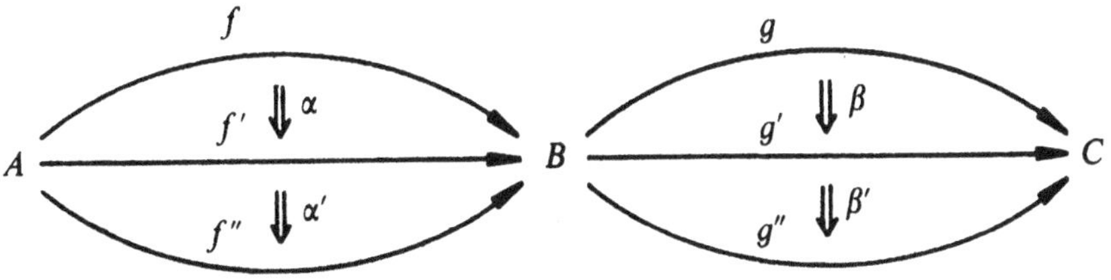

we will have a Godement interchange law:

$$(\beta' \otimes \alpha') \circ (\beta \otimes \alpha) = (\beta' \circ \beta) \otimes (\alpha' \circ \alpha).)$$

(iv) for each object, $A$, an arrow, denoted $1_A$, in $\mathbf{A}(A, A)$, called the *identity arrow on* $A$. The identity on $1_A$ will be denoted $i_A$

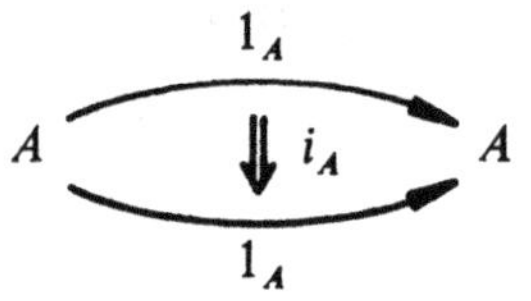


(v) for each quadruple $(A, B, C, D)$ a natural '*associativity isomorphism*', denoted

$$a(A, B, C, D): C(A, B, D) \circ [C(B, C, D) \times \mathrm{Id}] \xrightarrow{\cong} C(A, C, D) \circ [\mathrm{Id} \times C(A, B, C)].$$

For $f: A \to B$, $g: B \to C$ and $h: C \to D$, it will be usual to write $a$ or $a(f, g, h)$ instead of $a(A, B, C, D)(f, g, h)$

(vi) for each pair $(A, B)$, two natural isomorphisms

$$r(A, B): C(A, A, B) \circ (\mathrm{Id} \times 1_A) \to \rho$$

$$l(A, B): C(A, B, B) \circ (1_B \times \mathrm{Id}) \to \gamma$$

where for the moment we have written $\rho, \gamma$ for the canonical functors

$$\rho: \mathbf{A}(A, B) \times 1 \xrightarrow{\cong} \mathbf{A}(A, B),$$

$$\gamma: 1 \times \mathbf{A}(A, B) \xrightarrow{\cong} \mathbf{A}(A, B).$$

The 2-cell $r(A, B)(f): f \otimes 1_A \xrightarrow{\cong} f$ will also be denoted by $r(f)$ or $r$ and $l(A, B)(f): 1_B \otimes f \xrightarrow{\cong} f$ by $l(f)$ or simply $l$ is no confusion will so arise.

**Examples**

Apart from **Dist**, the bicategory of distributors introduced earlier, one has that any 2-category is a bicategory, a particularly nice form of bicategory in which the natural isomorphism of (v) and (vi) are the respective identities. In fact the notion of bicategory is best thought of as being a 'lax' generalisation of that of 2-category. Thus in particular **Cat**, the 2-category of small categories, is a bicategory.

Another example mentioned briefly is that of bimodules over all rings with composition given by a tensor product.

**Remark**

Because of the analogy with bimodules, some authors call distributors '*bimodules*' and because they stand in place of functors some use the term '*profunctors*'. We shall use Bénabou's original term, as both the other terms have associations which tend to distract from the exact meaning: the first gives some indication of additivity; the second gives a hint of a connection with pro-categories which, although not too distant, are best kept out of the discussion for the moment.

We next turn to morphisms between bicategories.

**Definition**

Let $\mathbf{A}$, $\mathbf{A}'$ be two bicategories. A *morphism* $(\mathbf{F}, \phi)$ from $\mathbf{A}$ to $\mathbf{A}'$ is given by the following:

(i) a function $\mathbf{F}\colon |\mathbf{A}| \to |\mathbf{A}'|$

(ii) a family of functors

$$\mathbf{F}_{A,B}\colon \mathbf{A}(A, B) \to \mathbf{A}'(FA, FB)$$

(iii) for each $A \in |\mathbf{A}|$ a 2-cell in $\mathbf{A}'(FA, FA)$

$$\phi_A\colon 1_{\mathbf{F}A} \to \mathbf{F}_{AA}(1_A)$$

(iv) for each triple $(A, B, C)$ of objects of $\mathbf{A}$, a natural transformation

$$\phi_{A,B,C}\colon C'(\mathbf{F}A, \mathbf{F}B, \mathbf{F}C) \circ (\mathbf{F}_{B,C} \times \mathbf{F}_{A,B}) \to \mathbf{F}_{A,C} \circ C(A, B, C).$$

We write $\phi(g, f)$ for $\phi_{A,B,C}(g, f)$ for simplicity.

We suppose that these fit together coherently in the following way:

(M1) for each triple $(f, g, h)$ of composable arrows the following diagram is commutative:

$$\begin{array}{ccc}
(\mathbf{F}h \otimes' \mathbf{F}g) \otimes' \mathbf{F}f & \xrightarrow{a'(\mathbf{F}f,\, \mathbf{F}g,\, \mathbf{F}h)} & \mathbf{F}h \otimes' (\mathbf{F}g \otimes' \mathbf{F}f) \\
{\scriptstyle \phi(h, g) \otimes' \mathrm{Id}} \downarrow & & \downarrow {\scriptstyle \mathrm{Id} \otimes' \phi(g, h)} \\
\mathbf{F}(h \otimes g) \otimes' \mathbf{F}(f) & & \mathbf{F}h \otimes' \mathbf{F}(g \otimes f) \\
{\scriptstyle \phi(h \otimes g, f)} \downarrow & & \downarrow {\scriptstyle \phi(h, g \otimes f)} \\
\mathbf{F}((h \otimes g) \otimes f) & \xrightarrow{\mathbf{F}(a(f, g, h))} & \mathbf{F}(h \otimes (g \otimes f))
\end{array}$$

(M2) for all $f$ in $\mathbf{A}(A, B)$, the following two diagrams are commutative:

$$\begin{array}{ccc}
\mathbf{F}(1_B \otimes f) & \xrightarrow{\mathbf{F}_{A,B}(l)} & \mathbf{F}f \\
{\scriptstyle \phi(1_B, f)} \uparrow & & \uparrow {\scriptstyle l'} \\
\mathbf{F}(1_B) \otimes' Ff & \xleftarrow{\phi_B \otimes' \mathrm{Id}} & 1_{FB} \otimes' Ff
\end{array}$$

$$\begin{array}{ccc}
\mathbf{F}(f \otimes 1_A) & \xrightarrow{\mathbf{F}_{AB}(r)} & \mathbf{F}(f) \\
{\scriptstyle \phi(f, 1_A)} \uparrow & & \uparrow {\scriptstyle r'} \\
\mathbf{F}(f) \otimes' \mathbf{F}(1_A) & \xleftarrow{\mathrm{Id} \otimes' \phi_A} & \mathbf{F}f \otimes' 1_{\mathbf{F}A}
\end{array}$$

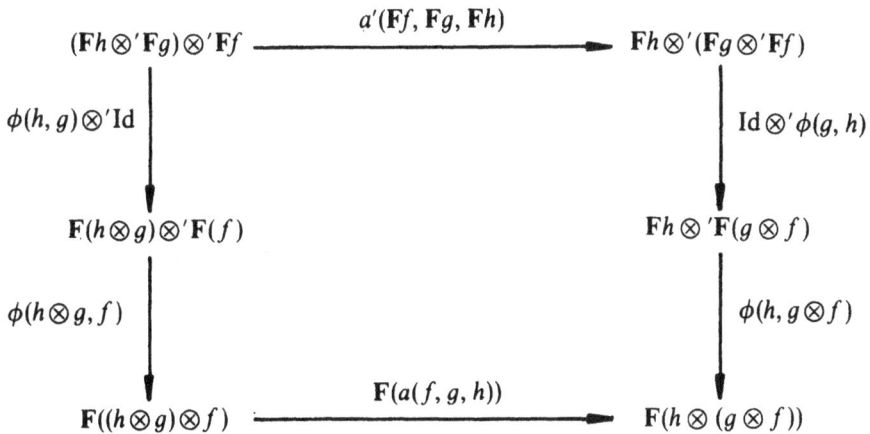

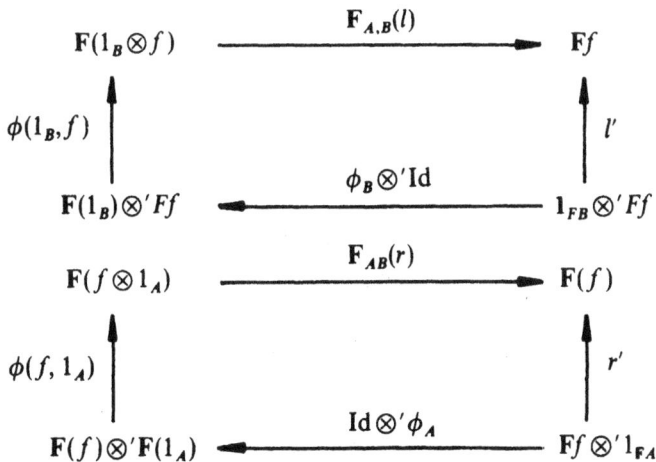

**Examples of morphisms**

(1) 2-functors between 2-categories considered as bicategories are morphisms in which the natural transformation in (iii) and (iv) are identities.

(2) Let **1** denote the bicategory with a single object * such that $\mathbf{1}(*, *)$ is the final category. A morphism **F** from **1** to the bicategory **Cat** consists of a category $\mathbf{A} = \mathbf{F}(*)$, an endofunctor $\mathbf{T} = \mathbf{F}_{*,*}(*)$ on **A**, i.e. $\mathbf{F}_{**}(*) \in \mathbf{Cat}(\mathbf{F}(*), \mathbf{F}(*))$, a 2-cell, i.e. a natural transformation

$$\varepsilon = \phi_*: \mathbf{1} \to \mathbf{T},$$

and a natural transformation

$$\mu = \phi_{*,*,*}: \mathbf{T}^2 \to \mathbf{T}$$

such that the diagrams

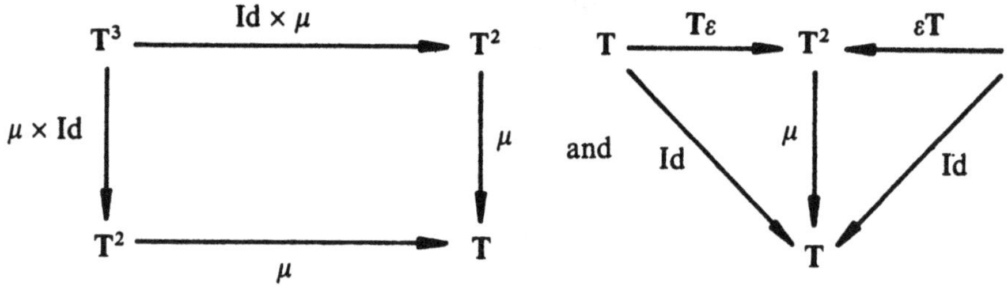


commute.

Thus **F** from **1** to **Cat** corresponds exactly to a monad on the category **A**. Because of this we use the terminology '*monad in the bicategory* **A**' for a morphism from **1** to **A**.

(3) The assignment to each functor $\mathbf{F}: \mathbf{A} \to \mathbf{B}$ in **Cat** of the corresponding distributor $\phi_{\mathbf{F}}: \mathbf{A} \dashrightarrow \mathbf{B}$ gives part of the specification of a bicategory morphism from **Cat** to **Dist**. At a 2-cell level this morphism is given by:

—to the transformation $\alpha: \mathbf{F} \Rightarrow \mathbf{F}'$, one associates the 2-cell

$$\phi_\alpha: \mathbf{B}(-, \mathbf{F}-) \Rightarrow \mathbf{B}(-, \mathbf{F}'-)$$

such that $\phi_\alpha(B, A) = \mathbf{B}(\mathrm{B}, \alpha(\mathrm{A}))$. In fact, for each pair $(A, B)$ of categories the functor

$$\mathbf{Cat}(A, B) \to \mathbf{Dist}(A, B) = \mathbf{Cat}(\mathbf{B}^{\mathrm{op}} \times \mathbf{A}, \mathbf{Sets})$$

is an embedding (fully faithful and injective on objects). The proof results from the Yoneda embedding.

We mentioned in the introduction to this chapter that in **Dist** all functors have adjoints. To make sense of this in detail, we need to have a notion of adjointness of arrows in a bicategory such that in the example of **Cat**, we obtain exactly the notion of an adjoint and in **Dist** each $\phi_{\mathbf{F}}$ has an adjoint.

**Definition**

Let $f: A \to B$, $u: B \to A$ be two arrows of a bicategory **A**. We shall say that $f$ is *left adjoint to* $u$ (written $f \dashv u$) if there are two cells

$$\eta: 1_A \Rightarrow u \otimes f$$

$$\varepsilon: f \otimes u \Rightarrow 1_B$$

such that the following diagrams commute:

$$\begin{array}{ccccc}
u & \xrightarrow{\;\cong\;} & 1_A \otimes u & \xrightarrow{\;\eta \otimes u\;} & (u \otimes f) \otimes u \\
\Big\| & & & & \Big\downarrow \cong \\
u & \xleftarrow{\;\cong\;} & u \otimes 1_B & \xleftarrow{\;u \otimes \varepsilon\;} & u \otimes (f \otimes u)
\end{array}$$

$$\begin{array}{ccccc}
f & \xrightarrow{\;\cong\;} & f \otimes 1_A & \xrightarrow{\;f \otimes \eta\;} & f \otimes (u \otimes f) \\
\Big\| & & & & \Big\downarrow \cong \\
f & \xleftarrow{\;\cong\;} & 1_B \otimes f & \xleftarrow{\;\varepsilon \otimes f\;} & (f \otimes u) \otimes f
\end{array}$$

**Remark**

It is clear that in the bicategory **Cat**, $f$—a functor from $A$ to $B$—is left adjoint to $u$ in this sense if and only if it is the usual sense since $f \otimes u$ is merely $fu$, $f \otimes (u \otimes f) = (f \otimes u) \otimes f$ etc.

**Proposition 1**

*For each functor* $\mathbf{F}\colon \mathbf{A} \to \mathbf{B}$, *the distributor* $\boldsymbol{\phi}_{\mathbf{F}}$ *is left adjoint to* $\boldsymbol{\phi}^{\mathbf{F}}$.

*Proof*

We define

$$\eta_{A,A'}\colon \mathbf{A}(A, A') \to (\boldsymbol{\phi}^{\mathbf{F}} \otimes \boldsymbol{\phi}_{\mathbf{F}})(A, A')$$

by

$$\eta_{A,A'}(\alpha) = \mathbf{F}(\alpha) \otimes \mathrm{Id}_{\mathbf{F}(A')}$$

and

$$\varepsilon_{B,B'}\colon (\boldsymbol{\phi}_{\mathbf{F}} \otimes \boldsymbol{\phi}^{\mathbf{F}})(B, B') \to \mathbf{B}(B, B')$$

by

$$\varepsilon_{B,B'}(\psi \otimes \phi) = \phi\psi$$

where $\phi \in \mathbf{B}(\mathbf{F}A, B')$, $\psi \in \mathbf{B}(B, \mathbf{F}A)$. This is easily seen to be independent of the choice of representative; in fact if $\phi = \phi'\mathbf{F}(\beta)$, $\beta\colon A' \to A$ then

$$\psi \otimes \phi'\mathbf{F}(\beta) = \mathbf{F}(\beta) \cdot \psi \otimes \phi'$$

and

$$\phi\psi = (\phi \cdot \mathbf{F}(\beta))\psi = \phi' \cdot (\mathbf{F}(\beta) \cdot \psi).$$

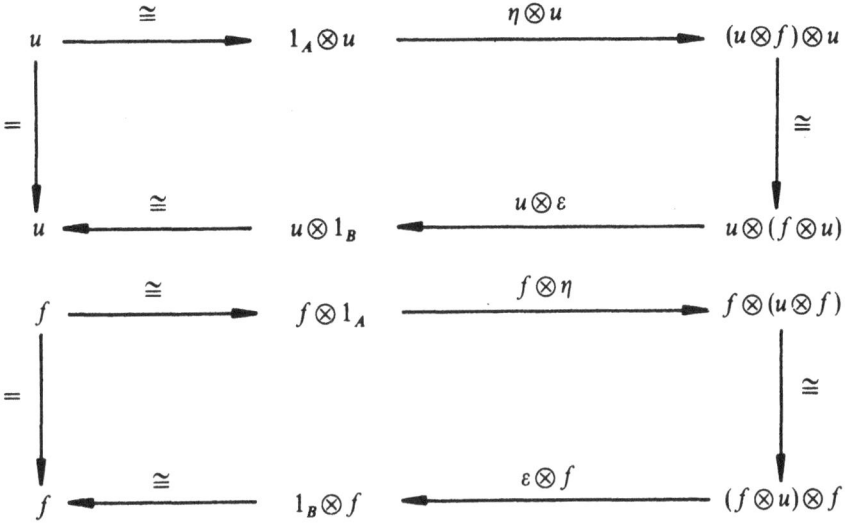

We verify that the first diagram commutes, by evaluating it on the pair $(A, B)$. We assume $\alpha \in \boldsymbol{\phi}^{\mathbf{F}}(A, B) = \mathbf{B}(\mathbf{F}A, B)$ and follow it around the diagram

$$\begin{array}{ccccc}
\boldsymbol{\phi}^{\mathbf{F}} & \xrightarrow{\;\cong\;} & 1_{\mathbf{A}} \otimes \boldsymbol{\phi}^{\mathbf{F}} & \xrightarrow{\;\eta \otimes \boldsymbol{\phi}^{\mathbf{F}}\;} & (\boldsymbol{\phi}^{\mathbf{F}} \otimes \boldsymbol{\phi}_{\mathbf{F}}) \otimes \boldsymbol{\phi}^{\mathbf{F}} \\
{\scriptstyle =}\big\downarrow & & & & \big\downarrow{\scriptstyle \cong} \\
\boldsymbol{\phi}^{\mathbf{F}} & \xleftarrow{\;\cong\;} & \boldsymbol{\phi}^{\mathbf{F}} \otimes 1_{\mathbf{B}} & \xleftarrow{\;\boldsymbol{\phi}^{\mathbf{F}} \otimes \varepsilon\;} & \boldsymbol{\phi}^{\mathbf{F}} \otimes (\boldsymbol{\phi}_{\mathbf{F}} \otimes \boldsymbol{\phi}^{\mathbf{F}})
\end{array}$$

i.e.

$$\begin{array}{c}
\alpha \mapsto \mathrm{Id}_A \otimes \alpha \mapsto (\mathrm{Id}_{F(A)} \otimes \mathrm{Id}_{F(A)}) \otimes \alpha \\
\downarrow \\
\alpha \leftarrow\!\shortmid\; \mathrm{Id}_{\mathbf{F}(A)} \otimes \alpha \leftarrow\!\shortmid\; \mathrm{Id}_{\mathbf{F}(A)} \otimes (\mathrm{Id}_{\mathbf{F}(A)} \otimes \alpha).
\end{array}$$

Similarly for the diagram

$$\begin{array}{ccccc}
\boldsymbol{\phi}_{\mathbf{F}} & \xrightarrow{\;\cong\;} & \boldsymbol{\phi}_{\mathbf{F}} \otimes 1_{\mathbf{A}} & \xrightarrow{\;\boldsymbol{\phi}_{\mathbf{F}} \otimes \eta\;} & \boldsymbol{\phi}_{\mathbf{F}} \otimes (\boldsymbol{\phi}^{\mathbf{F}} \otimes \boldsymbol{\phi}_{\mathbf{F}}) \\
{\scriptstyle =}\big\downarrow & & & & \big\downarrow \\
\boldsymbol{\phi}^{\mathbf{F}} & \xleftarrow{\;\cong\;} & 1_{\mathbf{B}} \otimes \boldsymbol{\phi}_{\mathbf{F}} & \xleftarrow{\;\varepsilon \otimes \boldsymbol{\phi}_{\mathbf{F}}\;} & (\boldsymbol{\phi}_{\mathbf{F}} \otimes \boldsymbol{\phi}^{\mathbf{F}}) \otimes \boldsymbol{\phi}_{\mathbf{F}}
\end{array}$$

Thus $\boldsymbol{\phi}_{\mathbf{F}} \dashv \boldsymbol{\phi}^{\mathbf{F}}$.

**Remark**

Given this result it is natural to ask the question: what if $\mathbf{F}$ already had a right adjoint functor $\mathbf{G}: \mathbf{B} \to \mathbf{A}$? What is the connection between $\boldsymbol{\phi}_{\mathbf{G}}$ and $\boldsymbol{\phi}^{\mathbf{F}}$?

The answer is simple: there is, for any $A, B$, a natural isomorphism

$$\mathbf{B}(\mathbf{F}A, B) \cong \mathbf{A}(A, \mathbf{G}B)$$

by the fact that $\mathbf{F} \dashv \mathbf{G}$; this, of course, says there is a natural isomorphism

$$\boldsymbol{\phi}^{\mathbf{F}}(A, B) \cong \boldsymbol{\phi}_{\mathbf{G}}(A, B),$$

i.e. $\boldsymbol{\phi}^{\mathbf{F}} \cong \boldsymbol{\phi}_{\mathbf{G}}$.

Just as one can extend the definition of adjointness from **Cat** to an arbitrary bicategory, so also for the notion of Kan extension.

**Definition**

Let **A** be a bicategory. Suppose we are given a diagram

$$\begin{array}{ccc}
A & \xrightarrow{\;f\;} & B \\
 & {\scriptstyle g}\searrow & \\
 & & C
\end{array}$$

in **A**; then a *right extension of $g$ along $f$* is an arrow $h: B \to C$ such that for any $k: B \to C$, one has a natural bijection

$$\delta(k): \mathbf{A}(B, C)(k, h) \cong \mathbf{A}(A, C)(k \otimes f, g).$$

In other words, the functor $(-) \otimes f: \mathbf{A}(B, C) \to \mathbf{A}(A, C)$ has a right adjoint. The 2-cell $\delta(h)(1_h): h \otimes f \Rightarrow g$ is called the *counit of the extension.*

**Remark**

Bénabou [7] uses the term *right closed* for a bicategory in which all the right extensions exist. In this terminology the following result means that **Dist** is right closed.

**Proposition 2**

*Given a diagram*

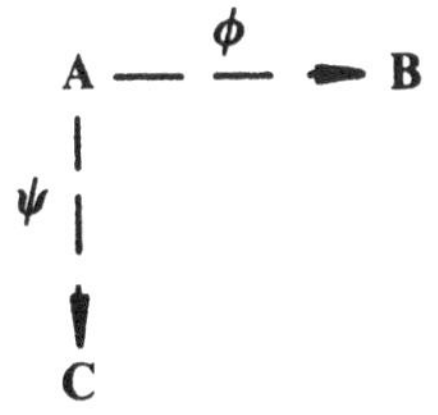

*the distributor*

$$\tilde{\psi}_\phi: \mathbf{B} \dashrightarrow \mathbf{C}$$

*given by*

$$\tilde{\psi}_\phi(C, B) = \mathrm{Nat}(\phi(B, -), \psi(C, -))$$

*is a right extension of $\psi$ along $\phi$.*

*Proof*

(This proof is the non-additive version of the well known proof of the adjunction between $\otimes$ and Hom.)

We have to prove that there is a natural isomorphism

$$\mathbf{Dist}(\mathbf{A}, \mathbf{C})(\theta \otimes \phi, \psi) \overset{\cong}{\Rightarrow} \mathbf{Dist}(\mathbf{B}, \mathbf{C})(\theta, \tilde{\psi}_\phi)$$

for $\theta: \mathbf{B} \dashrightarrow \mathbf{C}$.

Suppose

$$\alpha: \theta \otimes \phi \to \psi$$

is a 2-cell, and define

$$\tilde{\alpha}: \theta \to \tilde{\psi}_\phi$$

as follows:

Let $B \in |\mathbf{B}|$, $C \in |\mathbf{C}|$;

$$\tilde{\alpha}_{C,B}: \theta(C, B) \to \mathrm{Nat}(\phi(B, -), \psi(C, -))$$

sends $x \in \theta(C, B)$ to $\bar{\alpha}(x)\colon \phi(B, -) \to \psi(C, -)$ where

$$\bar{\alpha}(x)_A(y) = \alpha_{C,A}(x \otimes y) \quad \text{for } y \in \phi(B, A).$$

It is clear that $\bar{\alpha}(x)$ is a natural transformation, for if $f\colon A \to A'$ then

$$\begin{aligned}\alpha_{C,A'}(x \otimes \phi(B, f)(y)) &= \alpha_{C,A'}(\theta \otimes \phi(C, f)(x \otimes y))\\ &= (\theta \otimes \phi)(C, f)\alpha_{C,A}(x \otimes y)\\ &= (\theta \otimes \phi)(C, f)\bar{\alpha}(x)_A(y)\end{aligned}$$

as required.

In the reverse direction, suppose

$$\beta\colon \theta \to \tilde{\psi}_\phi$$

is a 2-cell, and define

$$\underline{\beta}\colon \theta \otimes \phi \to \psi$$

as follows:

Let $A \in |\mathbf{A}|$, $B \in |\mathbf{B}|$, $C \in |\mathbf{C}|$ as before and $x \in \theta(C, B)$, $y \in \phi(B, A)$;

$$\underline{\beta}_{C,A}\colon (\theta \otimes \phi)(C, A) \to \psi(C, A)$$

is given by

$$\underline{\beta}_{C,A}(x \otimes y) = (\beta_{C,B}(x))_A(y).$$

Again it is simple to check that the naturality of $\beta$ implies independence of the definition of $\underline{\beta}$ from the choice of representative $(x, y)$ for $x \otimes y$ and also the naturality of $\underline{\beta}$.

Finally, it is obvious that we have thus set up a natural isomorphism between $\mathbf{Dist}(\mathbf{A}, \mathbf{C})(\theta \otimes \phi, \psi)$ and $\mathbf{Dist}(\mathbf{B}, \mathbf{C})(\theta, \tilde{\psi}_\phi)$.

**Remark**

This formula for $\tilde{\psi}_\phi$ should clearly be suggesting to the attentive reader the definition of the hom-sets in $\mathbf{S_K}$. More precisely for $\mathbf{K}\colon \mathbf{A} \to \mathbf{B}$

$$\begin{aligned}\mathbf{S_K}(X, Y) &= \mathrm{Nat}(\mathbf{B}(Y, \mathbf{K}-), \mathbf{B}(X, \mathbf{K}-))\\ &= \mathrm{Nat}(\phi_{\mathbf{K}}(Y, -), \phi_{\mathbf{K}}(X, -))\\ &= (\widetilde{\phi_{\mathbf{K}}})_{\phi_{\mathbf{K}}}(X, Y).\end{aligned}$$

We shall explore this link with shape theory later; before this we will examine briefly the subject of codensity monad distributors since the existence of these right extensions shows that such codensity monads will always exist.

Again we must pause to ask about $\tilde{\psi}_\phi$ when $\phi$ and $\psi$ come from functors $\phi = \phi_{\mathbf{F}}$, $\psi = \phi_{\mathbf{G}}$ and $\mathbf{R} = \mathbf{Ran_F}\, \mathbf{G}$, the right Kan extension of $\mathbf{G}$ along $\mathbf{F}$ exists.

**Proposition 3**
*Suppose*

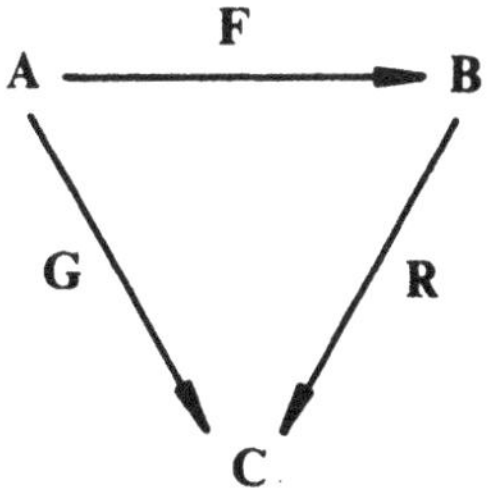


*is a diagram in* **Cat**; *then* $\phi_{\mathbf{R}}$ *is a right extension of* $\phi_{\mathbf{G}}$ *along* $\phi_{\mathbf{F}}$ *if and only if* $\mathbf{R} = \mathbf{Ran_G\,F}$, *the right Kan extension of* **G** *along* **F** *and this extension if pointwise (i.e. is preserved by all representable functors—see section 2.6).*

*Proof*
Suppose $\mathbf{R} = \mathbf{Ran_G\,F}$ and this extension is pointwise. Thus writing $\mathbf{h}^C$ for the functor

$$\mathbf{h}^C\colon \mathbf{C} \to \mathbf{Sets}$$
$$\mathbf{h}^C(C') = \mathbf{C}(C, C'),$$

we have

$$\mathbf{h}^C\mathbf{R} = \mathbf{Ran_F}\,\mathbf{h}^C\mathbf{G}.$$

Consequently

$$\begin{aligned}\phi_{\mathbf{R}}(C, B) &= \mathbf{C}(C, \mathbf{R}B)\\ &\cong \mathrm{Nat}(\mathbf{h}^B, \mathbf{h}^C\mathbf{R})\\ &\cong \mathrm{Nat}(\mathbf{h}^B\mathbf{F}, \mathbf{h}^C\mathbf{G})\\ &\cong \mathrm{Nat}(\phi_{\mathbf{F}}(B, -), \phi_{\mathbf{G}}(C, -))\end{aligned}$$

as required.

Conversely, let **H** be a functor from **B** to **C**. We have

$$\begin{aligned}\mathrm{Nat}(\mathbf{H}, \mathbf{R}) &\cong \mathrm{Nat}(\mathbf{C}(-, \mathbf{H}-), \mathbf{C}(-, \mathbf{R}-))\\ &= \mathrm{Nat}(\phi_{\mathbf{H}}, \phi_{\mathbf{R}})\\ &\cong \mathrm{Nat}(\phi_{\mathbf{HF}}, \phi_{\mathbf{G}})\\ &\cong \mathrm{Nat}(\mathbf{HF}, \mathbf{G}),\end{aligned}$$

i.e. $\mathbf{R} = \mathbf{Ran_F}\,\mathbf{G}$.

Now let $C$ be an object of **C**. The functor $\mathbf{h}^C\mathbf{G}\colon \mathbf{A} \to \mathbf{Sets}$ has as right Kan extension, the functor $\mathbf{R}'$ given by

$$\begin{aligned}\mathbf{R}'B &= \mathrm{Nat}(\mathbf{B}(B, \mathbf{F}-), \mathbf{C}(C, \mathbf{G}-))\\ &= \mathrm{Nat}(\phi_{\mathbf{F}}(B, -), \phi_{\mathbf{G}}(C, -)),\end{aligned}$$

but thus

$$\mathbf{R}'B \cong \phi_{\mathbf{R}}(C, B)$$
$$= (\mathbf{h}^C\mathbf{R})B;$$

consequently $\mathbf{h}^C\mathbf{R} = \mathbf{Ran_F}\ \mathbf{h}^C\mathbf{G}$.

**Remark**

Before turning to shape theory again, it is perhaps worth noting a useful way of interpreting the defintion of an extension:

Given

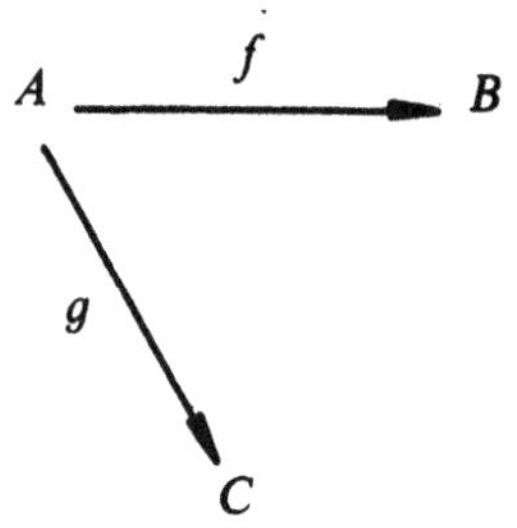


$h: B \to C$ is an extension if there is a counit $\varepsilon: h \otimes f \Rightarrow g$ such that given any other diagram

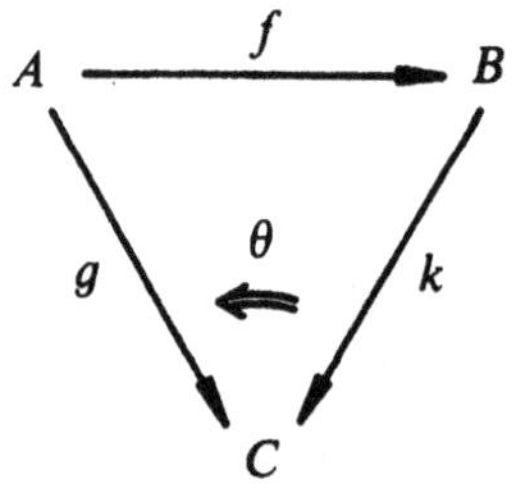


(i.e. $\theta: k \otimes f \Rightarrow g$) there is a unique $\psi: k \Rightarrow h$ such that

$$\theta = \varepsilon(\psi \otimes f).$$

## 4.2 CODENSITY MONADS AND KLEISLI CATEGORIES IN Dist

Earlier we gave a general definition of a monad in a bicategory, **A**, as being a bicategory morphism

$$\mathbf{F}: \mathbf{1} \to \mathbf{A}$$

and checked that for the 2-category **Cat**, this gave us the usual notion. It is a very simple matter to check that if **A** is **Dist** then a monad in **Dist** corresponds to the following data:

—an endodistributor $\mathbf{T}: \mathbf{A} \dashrightarrow \mathbf{A}$
—2-cells

$$\eta: \mathbf{1_A} \to \mathbf{T}$$

and

$$\mu : \mathbf{T} \otimes \mathbf{T} \to \mathbf{T}$$

such that

$$\mu \cdot (\mathbf{T} \otimes \eta) = (\eta \otimes \mathbf{T}) \cdot \mu = \mathbf{T}$$

and

$$\mu \cdot (\mathbf{T} \otimes \mu) = \mu \cdot (\mu \otimes \mathbf{T}).$$

Clearly any monad in **Cat** gives one in **Dist** in the obvious way.

**Examples**

(1) In **Cat** one first encounters monads as being generated by adjoint pairs; hence when every $\phi_{\mathbf{F}}$ has an adjoint, it is not surprising that one obtains a monad in **Dist** for each functor:

—If $\mathbf{F}: \mathbf{A} \to \mathbf{B}$, we denote by ${}_{\mathbf{F}}\mathbf{T}$, the monad in **Dist** on **A** determined by the adjunction $\phi_{\mathbf{F}} \dashv \phi^{\mathbf{F}}$, thus

$$_{\mathbf{F}}\mathbf{T} = \phi^{\mathbf{F}} \otimes \phi_{\mathbf{F}}$$

$$_{\mathbf{F}}\eta : \mathbf{1_A} \to \phi^{\mathbf{F}} \otimes \phi_{\mathbf{F}}$$

${}_{\mathbf{F}}\mu = \phi^{\mathbf{F}} \otimes \varepsilon \otimes \phi_{\mathbf{F}}$ with $\varepsilon$ as in Proposition 1 of section 4.1.

The verification that this gives a monad is entirely analogous to the classical case.

(2) In section 2.6, we discussed the codensity monad of a functor $\mathbf{F}: \mathbf{A} \to \mathbf{B}$ when $\mathbf{Ran_F}\, \mathbf{F}$ existed and was pointwise.

We saw earlier (Proposition 2 of section 4.1) that any diagram

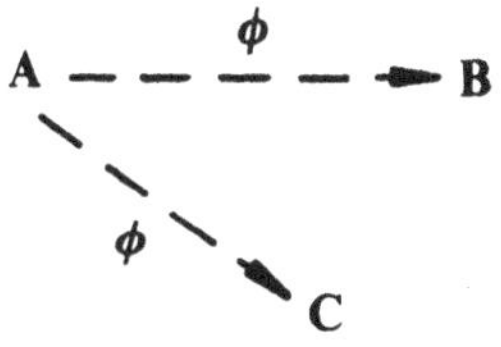

could be filled by a right extension $\tilde{\psi}_{\phi}: \mathbf{B} \dashrightarrow \mathbf{C}$; this means in particular that given any functor $\mathbf{F}: \mathbf{A} \to \mathbf{B}$, we have a diagram

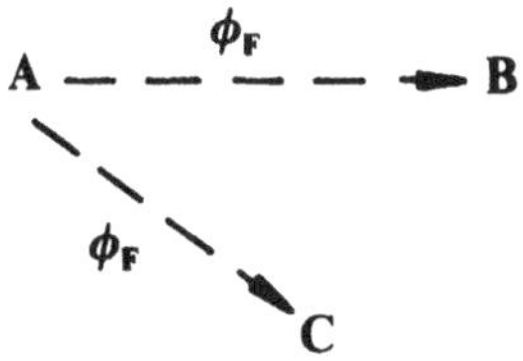

and hence an endodistributor $(\tilde{\phi}_{\mathbf{F}})_{\phi_{\mathbf{F}}}: \mathbf{B} \dashrightarrow \mathbf{B}$. (We will abbreviate $(\tilde{\phi}_{\mathbf{F}})_{\phi_{\mathbf{F}}}$ to $\tilde{\phi}_{\mathbf{F}}$ for ease of printing.)

We know the following about $\tilde{\phi}_{\mathbf{F}}$:

$$—\tilde{\phi}_{\mathbf{F}}(B, B') = \mathrm{Nat}(\phi_{\mathbf{F}}(B', -), \phi_{\mathbf{F}}(B, -))$$
$$= \mathrm{Nat}(\mathbf{B}(B', \mathbf{F}-), \mathbf{B}(B, \mathbf{F}-))$$

—for any $\psi$: **B** ----→ **B**, there is a natural isomorphism

$$\delta_{\mathbf{B}}(\psi): \mathbf{Dist}(\mathbf{B}, \mathbf{B})(\psi, \bar{\phi}_{\mathbf{F}}) \cong \mathbf{Dist}(\mathbf{A}, \mathbf{B})(\psi \otimes \phi_{\mathbf{F}}, \phi_{\mathbf{F}}).$$

By analogy with the case (in **Cat**) that we examined in section 2.6, we put

$$\eta = \delta_{\mathbf{B}}(1_B)^{-1}(1_{\phi_{\mathbf{F}}})$$

$$\mu = \delta_{\mathbf{B}}(\bar{\phi}_{\mathbf{F}} \otimes \bar{\phi}_{\mathbf{F}})^{-1}(\theta \cdot (\bar{\phi}_{\mathbf{F}} \otimes \theta))$$

where $\theta = \delta_{\mathbf{B}}(\bar{\phi}_{\mathbf{F}})(1_{\bar{\phi}_{\mathbf{F}}})$ is the counit of the extension.

Checking that this gives a monad in **Dist** is formally similar to the verification of the **Cat** version given in section 2.6 and so will be omitted. We denote this codensity monad for **F** by $\mathbf{T}_{\mathbf{F}} = (\bar{\phi}_{\mathbf{F}}, \eta, \mu)$.

The identity of $\bar{\phi}_{\mathbf{F}}(B, B')$ (as given above) with $\mathbf{S}_{\mathbf{F}}(B, B')$ and our earlier results on Kleisli categories of monads should indicate the direction we shall now take.

Clearly we should try to form a Kleisli category for a monad in **Dist** and then in the particular case of $\bar{\phi}_{\mathbf{F}}$ above we should compare this with $\mathbf{S}_{\mathbf{F}}$.

The construction of the Kleisli category for a monad in **Dist** is simple. Suppose

$$\mathbf{T} = (\mathbf{T}, \eta, \mu)$$

is a monad in **Dist** on the category **A**, so that

$$\mathbf{T}: \mathbf{1} \to \mathbf{Dist} \text{ is a morphism}$$

$$\mathbf{T}(*) = \mathbf{A}.$$

We form a category **KlT** called the Kleisli category of **T** with the objects of **A** and with

$$\mathbf{KlT}(A, A') = \mathbf{T}(A, A').$$

The composition law is given by

$$\mathbf{T}(A, A') \times \mathbf{T}(A', A'') \to \mathbf{T}(A, A'')$$

$$(\alpha, \alpha') \mapsto \mu(\alpha' \otimes \alpha)$$

and identities: to $A$ in **A**, associate $\eta(A)(1_A) \in \mathbf{T}(A, A)$. There is, moreover, a functor

$$\mathbf{F}_{\mathbf{T}}: \mathbf{A} \to \mathbf{KlT}$$

bijective on objects, defined by

$$\mathbf{F}_{\mathbf{T}}(\alpha) = \eta(A, A')(\alpha) \quad \text{for } \alpha \in \mathbf{A}(A, A').$$

**Proposition 1**

*Given any functor* $\mathbf{F}: \mathbf{A} \to \mathbf{B}$ *such that the adjunction* $(\phi_{\mathbf{F}}, \phi^{\mathbf{F}})$ *determines the monad* **T** *on* **A**, *there is a unique functor* $\hat{\mathbf{F}}: \mathbf{KlT} \to \mathbf{B}$ *satisfying* $\mathbf{F} = \hat{\mathbf{F}} \cdot \mathbf{F}_{\mathbf{T}}$.

*Proof*

As $\mathbf{F}_{\mathbf{T}}$ is the identity on objects, we must have $\hat{\mathbf{F}}A = \mathbf{F}A$ for all objects $A$ of **A**.

If $\alpha \in \mathbf{KlT}(A, A') = \mathbf{T}(A, A')$ and since **T** is the same endodistributor as $\phi^{\mathbf{F}} \otimes \phi_{\mathbf{F}}$, we have

$$\alpha \in \phi^{\mathbf{F}} \otimes \phi_{\mathbf{F}}(A, A') = \mathbf{B}(\mathbf{F}A, \mathbf{F}A')$$

and the obvious definition of $\mathbf{F}(\alpha)$ is $\alpha$ itself. The uniqueness of $\hat{\mathbf{F}}$ such that $\mathbf{F} = \hat{\mathbf{F}} \cdot \mathbf{F}_{\mathbf{T}}$ is now obvious.

**Remarks**

(1) Thus the Kleisli category **KlT** of $\mathbf{T} = {}_{\mathbf{F}}\mathbf{T}$ is a category which provides a decomposition of **F** as composition of a functor which is bijective on objects with a fully faithful functor.

(2) The Kleisli category $\mathbf{KlT_F}$ as noted before has the same objects as **B** and has

$$\mathbf{KlT_F}(B, B) = \mathrm{Nat}(\mathbf{B}(B', \mathbf{F}-), \mathbf{B}(B, \mathbf{F}-))$$

and hence is isomorphic to the Holsztyński shape category of **F**. We will shortly examine from the point of view of **Dist**, the axioms necessary to make this a shape theory in the sense of section 2.1. In this we will need to compare the two monads ${}_{\mathbf{F}}\mathbf{T}$ and $\mathbf{T_F}$ associated with $\mathbf{F}: \mathbf{A} \to \mathbf{B}$, and to this end we give explicitly below a description of a morphism between two monads.

**Definition**

Let $\mathbf{T} = (\mathrm{T}, \eta, \mu)$ and $\mathbf{T}' = (\mathrm{T}, \eta', \mu')$ be two monads in **Dist** on the category **A**; a *morphism from* **T** *to* **T**′ is given by a 2-cell

$$\tau: \mathbf{T} \to \mathbf{T}'$$

such that

$$\tau: \eta = \eta'$$

and

$$\tau \cdot \mu = \mu' \cdot \tau \otimes \tau.$$

**Proposition 2**

*Let* **T** *be a monad in* **Dist** *on* **A** *and* **F** *a functor from* **A** *to* **B**, *then there is a bijection between the set of functors* $\mathbf{G}: \mathbf{KlT} \to \mathbf{B}$ *such that* $\mathbf{GF_T} = \mathbf{F}$ *and the set of morphisms from the monad* **T** *to* ${}_{\mathbf{F}}\mathbf{T}$.

*Proof*

Let $\mathbf{G}: \mathbf{KlT} \to \mathbf{B}$ be a functor such that $\mathbf{GF_T} = \mathbf{F}$; then $\mathbf{G}(A) = \mathbf{F}(A)$ for all $A$ in **A**.

If $\alpha \in \mathbf{T}(A, A')$, $\mathbf{G}(\alpha) \in \mathbf{B}(\mathbf{G}A, \mathbf{G}A') = \mathbf{B}(\mathbf{F}A, \mathbf{F}A')$ so

$$\mathbf{G}(\alpha) \in (\boldsymbol{\phi}^{\mathbf{F}} \otimes \boldsymbol{\phi}_{\mathbf{F}})(A, A') = {}_{\mathbf{F}}\mathbf{T}(A, A')$$

and

$$\tau_{A,A'}(\alpha) = \mathbf{G}(\alpha)$$

gives a 2-cell $\tau: \mathbf{T} \to {}_{\mathbf{F}}\mathbf{T}$.

Since $\mathbf{GF_T} = \mathbf{F}$, we have for $\alpha \in \mathbf{A}(A, A')$,

$$\mathbf{GF_T}(\alpha) = \tau_{A,A'}\eta(A, A')(\alpha) = \mathbf{F}(\alpha) = {}_{\mathbf{F}}\eta(A, A')(\alpha)$$

so

$$\tau \cdot \eta = {}_{\mathbf{F}}\eta.$$

Likewise **G** preserves composition so given $\alpha \in \mathbf{T}(A, A')$, $\alpha' \in \mathbf{T}(A', A'')$, we have

$$\begin{aligned} \tau_{A',A''}(\mu(\alpha \otimes \alpha)) &= \mathbf{G}(\mu(\alpha' \otimes \alpha)) \\ &= {}_{\mathbf{F}}\mu(\mathbf{G}(\alpha') \otimes \mathbf{G}(\alpha)) \\ &= {}_{\mathbf{F}}\mu(\tau_{A',A''}(\alpha') \otimes \tau_{A,A'}(\alpha)) \end{aligned}$$

and $\tau: \mathbf{T} \to \mathbf{T_F}$ gives the required morphism.

As this process is reversible, we obtain the desired bijection

Taking the case where $\mathbf{B} = \mathbf{KlT}'$, $\mathbf{F} = \mathbf{F}_{\mathbf{T}'}$ gives:

**Corollary 1**

*Let* $\mathbf{T}$ *and* $\mathbf{T}'$ *be monads in* **Dist** *on* $\mathbf{A}$, *then there is a bijection between the set of morphisms from* $\mathbf{T}$ *to* $\mathbf{T}'$ *and the set of functors* $\mathbf{G}: \mathbf{KlT} \to \mathbf{KlT}'$ *such that* $\mathbf{GF}_{\mathbf{T}} = \mathbf{F}_{\mathbf{T}'}$.

**Proposition 3**

*There is a bijection between isomorphism classes of monads in* **Dist** *on* $\mathbf{A}$ *and isomorphism classes of functors* $\mathbf{F}$ *with domain* $\mathbf{A}$ *which are bijective on objects.*

*Proof*

Let $\mathbf{T}$ be a monad in **Dist** on $\mathbf{A}$; then $\mathbf{F}_{\mathbf{T}}: \mathbf{A} \to \mathbf{KlT}$ is a functor bijective on objects. Conversely if $\mathbf{F}: \mathbf{A} \to \mathbf{B}$ is a functor (which is bijective on objects) then ${}_{\mathbf{F}}\mathbf{T}$ is a monad in **Dist** on $\mathbf{A}$.

Following through the two processes, one finds

$$\mathbf{T} \longrightarrow \mathbf{F} = \mathbf{F}_{\mathbf{T}} \longrightarrow {}_{\mathbf{F}}\mathbf{T}$$

and

$$\mathbf{F} \longrightarrow \mathbf{T} = {}_{\mathbf{F}}\mathbf{T} \longrightarrow \mathbf{F}_{\mathbf{T}}$$

yield isomorphic objects (by Propositions 1 and 2 above).

In fact given $\mathbf{T}$ and writing $\mathbf{F}$ for $\mathbf{F}_{\mathbf{T}}$ we find

$$\begin{aligned} {}_{\mathbf{F}}\mathbf{T}(A, A') &= (\phi^{\mathbf{F}} \otimes \phi_{\mathbf{F}})(A, A') \\ &= (\sqcup \mathbf{KlT}(A, A'') \times \mathbf{KlT}(A'', A'))/R \\ &\cong \mathbf{KlT}(A, A'), \end{aligned}$$

since $\mathbf{F}$ is bijective on objects (in fact $\mathbf{F}$ is the identity on objects!). Similarly ${}_{\mathbf{F}}\eta_{A,A'} \cong \eta_{A,A'}$ and likewise for the corresponding $\mu$s.

For the other calculation, we factor $\mathbf{F} = \hat{\mathbf{F}} \cdot \mathbf{F}_{\mathbf{T}}$. Both $\mathbf{F}$ and $\mathbf{F}_{\mathbf{T}}$ are essentially the identity on objects and $\hat{\mathbf{F}}$ (after the proof of Proposition 1 above) is fully faithful.

## 4.3 SHAPE THEORIES AND DISTRIBUTORS

(In this section we have, for simplicity of formulae, omitted the coherence isomorphisms occurring in associativity and identity expressions.)

We have seen that the Holsztyński shape category of a functor $\mathbf{K}: \mathbf{A} \to \mathbf{B}$ has a neat description as the Kleisli category of the codensity monad in **Dist** generated by $\mathbf{K}$. Now we will examine the interpretation of the axioms for a shape theory in terms of distributors.

Recall that we are given a functor $\mathbf{K}: \mathbf{A} \to \mathbf{B}$ and a pair $(\mathbf{S}, S)$ with $\mathbf{S}$ a category and $S$ a functor from $\mathbf{B}$ to $\mathbf{S}$.

**Axiom (S1):** $\mathbf{S}$ is bijective on objects

In the light of Proposition 3 of section 4.2 just proved the following is not that surprising.

**Proposition 1**
*(S, $S$) satisfies (S1) if and only if* **S** *is isomorphic to the Kleisli category of* ${}_S\mathbf{T}$ *(respecting the functors $S$ and $\mathbf{F_T}$).*

*Proof*
One way is trivial since if

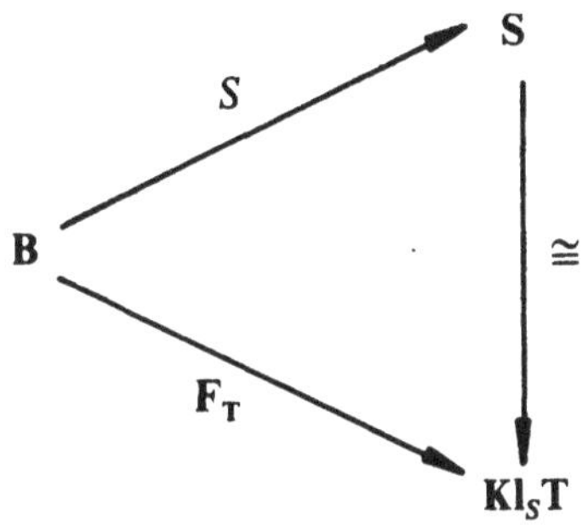


then as $\mathbf{F_T}$ is bijective on objects, so is $S$.

Now suppose that $S: \mathbf{B} \to \mathbf{S}$ is bijective on objects; then, as we have just seen,

$$\mathbf{Kl}_S\mathbf{T}(X, X') = \boldsymbol{\phi}^S \otimes \boldsymbol{\phi}_S(X, X') \cong \mathbf{S}(X, X')$$

and it is simple to use these natural isomorphisms to construct the required isomorphism between **S** and $\mathbf{Kl}_S\mathbf{T}$ (relative to **B**).

**Axiom (S2):** If $s \in \mathbf{S}(X, \mathbf{K}Q)$ *then there is a unique* $f \in \mathbf{B}(X, \mathbf{K}Q)$ *with* $Sf = s$.
This expresses that $S$ induces a bijection between $\mathbf{B}(X, \mathbf{K}Q)$ and $\mathbf{S}(SX, S\mathbf{K}Q)$ (being pedantic and writing $SX$ instead of just $X$ to emphasise the nature of the bijection).

**Proposition 2**
*(S, $S$) satisfies (S2) if and only if* $\eta' \otimes \boldsymbol{\phi}_\mathbf{K}$ *is invertible where* ${}_S\mathbf{T} = (\boldsymbol{\phi}^S \otimes \boldsymbol{\phi}_S, \eta', \mu')$ *is the monad in* **Dist** *on* **B** *generated by the adjunction* $\boldsymbol{\phi}_S \dashv \boldsymbol{\phi}^S$.

*Proof*
Let $\mathbf{K}: \mathbf{A} \to \mathbf{B}$ and $S: \mathbf{B} \to \mathbf{S}$ be functors such that for all $X \in |\mathbf{B}|$ and $P \in |\mathbf{A}|$, $S$ induces a bijection

$$\mathbf{B}(X, \mathbf{K}P) \xrightarrow{\cong} \mathbf{S}(SX. S\mathbf{K}P) \cong \boldsymbol{\phi}^S \otimes \boldsymbol{\phi}_{S\mathbf{K}}(X, P).$$

Thus there is an invertible 2-cell

$$\sigma: \boldsymbol{\phi}_\mathbf{K} \Rightarrow \boldsymbol{\phi}^S \otimes \boldsymbol{\phi}_S \otimes \boldsymbol{\phi}_\mathbf{K}$$

(where we have tacitly used the isomorphism $\boldsymbol{\phi}_S \otimes \boldsymbol{\phi}_\mathbf{K} \cong \boldsymbol{\phi}_{S\mathbf{K}}$).

It remains to identify this 2-cell $\sigma$.

Recall that the unit of the adjunction $\boldsymbol{\phi}_S \dashv \boldsymbol{\phi}^S$,

$$\eta': \mathrm{Id}_\mathbf{B} \to \boldsymbol{\phi}^S \otimes \boldsymbol{\phi}_S,$$

is given by

$$\eta'(X, X')(\alpha) = S(\alpha) \otimes \mathrm{Id}_{X'}.$$

where

$$\alpha \in \mathbf{B}(X, X').$$

Thus

$$\eta' \otimes \phi_{\mathbf{K}}(\alpha \otimes f) = S(\alpha) \otimes \mathrm{Id}_{X'} \otimes f$$

for $f \in \mathbf{B}(X, \mathbf{K}P)$, say, but

$$S(\alpha) \otimes \mathrm{Id}_{X'} \otimes f = S(f\alpha) \otimes \mathrm{Id}_{\mathbf{K}P} \otimes \mathrm{Id}_{\mathbf{K}P}.$$

Thus clearly $\sigma$ is isomorphic to $\eta' \otimes \phi_{\mathbf{K}}$ and consequently axiom (S2) states simply that $\eta' \otimes \phi_{\mathbf{K}}$ is invertible.

**Remark**

(S2) induces a comparison between ${}_S\mathbf{T}$ and $\mathbf{T}_{\mathbf{K}}$. In fact let $\mathbf{T}_{\mathbf{K}} = (\tilde{\phi}_{\mathbf{K}}, \eta, \mu)$ be as before the codensity monad of $\mathbf{K}$ in **Dist**. The 2-cell

$$(\eta' \otimes \phi_{\mathbf{K}})^{-1}\colon \phi^S \otimes \phi_S \otimes \phi_{\mathbf{K}} \Rightarrow \phi_{\mathbf{K}},$$

i.e.

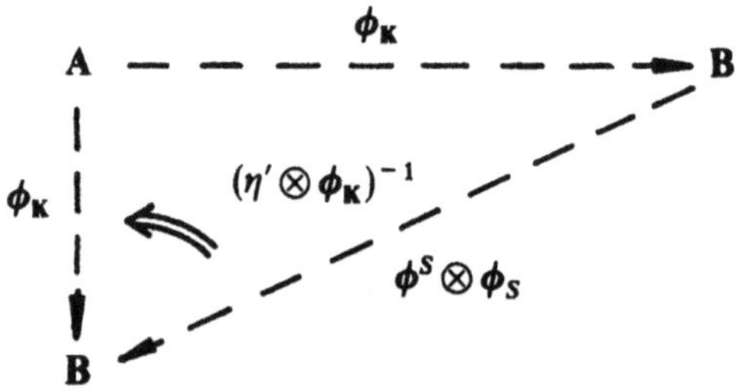


induces (by the universal property of $\tilde{\phi}_{\mathbf{K}}$) a 2-cell

$$\tau\colon \phi^S \otimes \phi_S \Rightarrow \tilde{\phi}_{\mathbf{K}}$$

such that

$$\theta \cdot (\tau \otimes \phi_{\mathbf{K}}) = (\eta' \otimes \phi_{\mathbf{K}})^{-1}$$

with $\theta$ the counit of the extension $\theta\colon \tilde{\phi}_{\mathbf{K}} \otimes \phi_{\mathbf{K}} \Rightarrow \phi_{\mathbf{K}}$.
One can check that $\tau$ induces a morphism of monads in **Dist** on **B**,

$$\tau\colon {}_S\mathbf{T} \to \mathbf{T}_{\mathbf{K}}.$$

**Axiom (S3):** *S is* **K**-*continuous.*

**Proposition 3**

(**S**, *S*) *satisfies* (*S*3) *if and only if* $\phi_S$ *is a right extension of* $\phi_S \otimes \phi_{\mathbf{K}}$ *along* $\phi_{\mathbf{K}}$ *or equivalently if S is a pointwise Kan extension of S***K** *along* **K**.

*Proof*

Let $S\colon \mathbf{B} \to \mathbf{S}$ be a **K**-continuous functor. We note that

$$\mathrm{Func}(X \downarrow \mathbf{K}, Z \downarrow S\mathbf{K}) \cong \mathrm{Nat}(\mathbf{B}(X, \mathbf{K}-), \mathbf{S}(Z, S\mathbf{K}-))$$

and this defines a distributor $\phi\colon \mathbf{B} \dashrightarrow \mathbf{S}$.

Moreover the formula defining $\phi$ implies via Proposition 2 of section 4.1 that the following diagram defines $\phi$ as a right extension

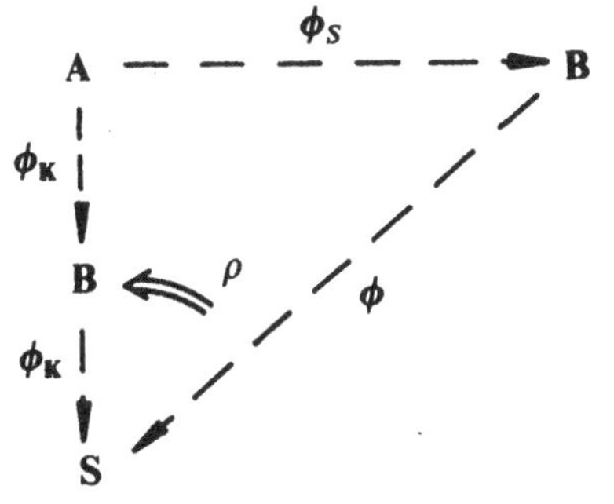


where $\rho: \phi \otimes \phi_{\mathbf{K}} \Rightarrow \phi_{S\mathbf{K}}$ is given by $\rho(Z, \mathbf{K}P)$ from $\phi(Z, \mathbf{K}P) \cong \phi \otimes \phi_{\mathbf{K}}(Z, P)$ to $\phi_{S\mathbf{K}}(Z, P) \cong \mathbf{S}(Z, S\mathbf{K}P)$ associates to $t$ the element $t(\mathrm{Id}_{\mathbf{K}P})$. As $\phi_S$ is a distributor from **B** to **S**, there is a unique 2-cell

$$a: \phi_S \Rightarrow \phi$$

such that

$$\rho \cdot (a \otimes \phi_{\mathbf{K}}) = \mathrm{Id}(\phi_S \otimes \phi_{\mathbf{K}}).$$

$a(Z, X)$ from $\phi_S(Z, X) = \mathbf{S}(Z, SX)$ to $\phi(Z, X)$ is the mapping which to $s \in \mathbf{S}(Z, SX)$ associates

$$s^* S^* \in \mathrm{Func}(X \downarrow \mathbf{K}, Z \downarrow S\mathbf{K})$$

where

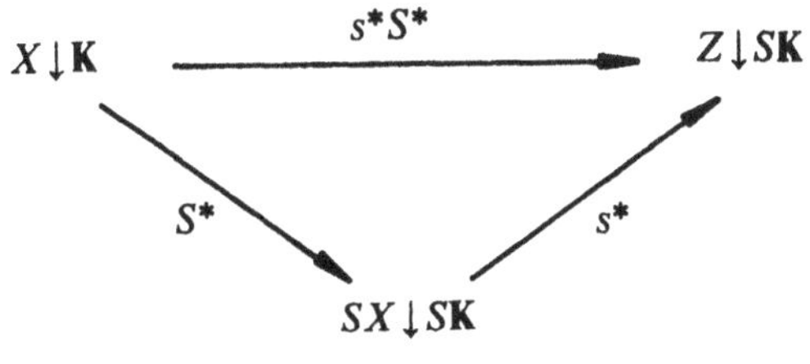


$S$ is **K**-continuous if and only if

$$a(Z, X): \mathbf{S}(Z, SX) \rightarrow \mathrm{Func}(X \downarrow \mathbf{K}, Z \downarrow S\mathbf{K})$$

is bijective so if and only if $\phi_S$ is a right extension of $\phi_S \otimes \phi_{\mathbf{K}}$ along $\phi_{\mathbf{K}}$. Proposition 3 of section 4.1 now shows that this is equivalent to saying that $S$ is the pointwise right Kan extension of $S\mathbf{K}$ along **K**.

**Remark**

(S3) implies there is a comparison between $\mathbf{T}_{\mathbf{K}}$ and ${}_S\mathbf{T}$. In fact since $\phi^S$ is a right adjoint in **Dist** of $\phi_S$, it preserves right extensions. Thus if (S3) is satisfied, $\phi^S \otimes \phi_S$ is a right extension of $\phi^S \otimes \phi_S \otimes \phi_{\mathbf{K}}$ along $\phi_{\mathbf{K}}$ with counit $\phi^S \otimes \rho$.

Thus there is an invertible 2-cell

$$\phi^S \otimes a: \phi^S \otimes \phi_S \Rightarrow \phi^S \otimes \phi$$

and a diagram

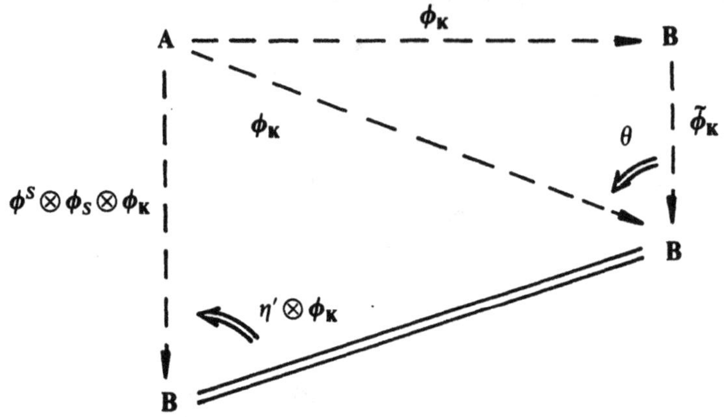


which induces a comparison morphism between $\tilde{\phi}_{\mathbf{K}}$ and $\phi^S \otimes \phi$ and thus with $\phi^S \otimes \phi_S$. This comparison morphism is the required morphism of monads from $\mathbf{T}_{\mathbf{K}}$ to ${}_S\mathbf{T}$.

In fact this morphism, which will be denoted by $\widetilde{(\eta' \otimes \phi_{\mathbf{K}})}$, is the unique 2-cell satisfying

$$(\phi^S \otimes \rho) \cdot (\widetilde{(\eta' \otimes \phi_{\mathbf{K}})} \otimes \phi_{\mathbf{K}}) = (\eta' \otimes \phi_{\mathbf{K}}) \cdot \theta$$

(remember $\theta$ is the counit of $\tilde{\phi}_{\mathbf{K}}$).

Now suppose that (S1) is satisfied and that $\phi^S \otimes a$ is invertible; then, in fact, $a$ was already invertible. To see this, note that for each pair $(X, X')$, the morphism

$$(\phi^S \otimes a)(X, X') \colon (\phi^S \otimes \phi_S)(X, X') \to (\phi^S \otimes \phi)(X, X')$$

can be very simply described explicitly; it is (up to isomorphism)

$$a \cdot (S^{\mathrm{op}} \times \mathrm{Id}_{\mathbf{B}})(X, X') \colon \phi^S(S^{\mathrm{op}} \times \mathrm{Id}_{\mathbf{B}})(X, X') \to \phi(S^{\mathrm{op}} \times \mathrm{Id}_{\mathbf{B}})(X, X').$$

Thus the 2-cell $\phi^S \otimes a$, considered as a natural transformation (thus in **Cat**), is given by

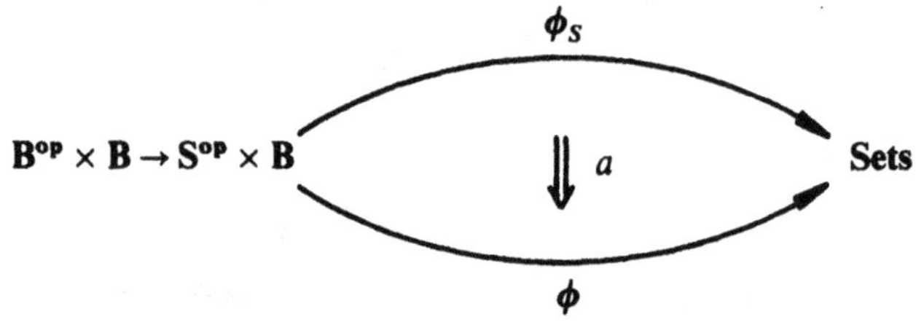


As (S1) is assumed to be satisfied, $S$, and hence also $S^{\mathrm{op}} \times \mathrm{Id}_{\mathbf{B}}$, is bijective on objects. Now for all $X, X'$, $a \cdot (S^{\mathrm{op}} \times \mathrm{Id}_{\mathbf{B}})(X, X')$ is invertible, but this clearly implies that each $a(X, X')$ is invertible and so $a$ itself is invertible as required.

Putting the results on the three axioms together we obtain the following theorem.

**Theorem 1**

*Let $\mathbf{K} \colon \mathbf{A} \to \mathbf{B}$ be a functor; then $(\mathbf{S}, S)$ is a shape theory if and only if the counit $\theta \colon \tilde{\phi}_{\mathbf{K}} \otimes \phi_{\mathbf{K}} \Rightarrow \phi_{\mathbf{K}}$ is invertible and $\mathbf{S}$ is a Kleisli category for the codensity monad (in* **Dist**) *of* $\mathbf{K}$, $\mathbf{T}_{\mathbf{K}} = (\tilde{\phi}_{\mathbf{K}}, \eta, \mu)$.

*Proof*

We start by pointing out that '$\mathbf{S}$ is a Kleisli category for $\mathbf{T}_\mathbf{K}$' means $\mathbf{S}$ is isomorphic to $\mathbf{KlT}_\mathbf{K}$ in such a way that $\mathbf{S}$ corresponds to $\mathbf{F}_\mathbf{T}$. In fact this is equivalent to saying that ${}_S\mathbf{T} \cong \mathbf{T}_\mathbf{K}$ as monads.

Now let $(\mathbf{S}, S)$ be a shape theory. We have monad morphisms

$$\tau\colon {}_S\mathbf{T} \to \mathbf{T}_\mathbf{K}$$
$$(\phi^S \otimes a)^{-1}\widetilde{(\eta' \otimes \phi_\mathbf{K})}\colon \mathbf{T}_\mathbf{K} \to {}_S\mathbf{T}.$$

Clearly one hopes that these are mutually inverse (perhaps modulo coherence isomorphisms—which will anyway be omitted!).

We have defined $\widetilde{(\eta' \otimes \phi_\mathbf{K})}$ by the diagram

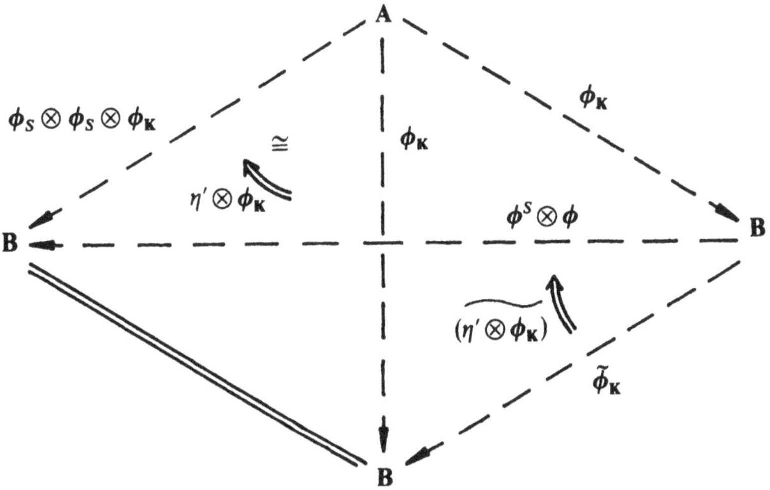


but as $\eta' \otimes \phi_\mathbf{K}$ is invertible, clearly $\widetilde{(\eta' \otimes \phi_\mathbf{K})}$ is as well and it remains to show that $\tau = \widetilde{(\eta' \otimes \phi_\mathbf{K})}^{-1}(\phi^S \otimes a)$.

We have the following defining universal properties for $\tau$, $a$ and $\widetilde{(\eta' \otimes \phi_\mathbf{K})}$

(1) $\theta \cdot (\tau \otimes \phi_\mathbf{K}) = (\eta' \otimes \phi_\mathbf{K})^{-1}$

(2) $\rho \cdot (a \otimes \phi_\mathbf{K}) = \mathrm{Id}_{\phi_S \otimes \phi_\mathbf{K}}$

(3)$(\phi^S \otimes \rho)(\widetilde{(\eta' \otimes \phi_\mathbf{K})} \otimes \phi_\mathbf{K}) = (\eta' \otimes \phi_\mathbf{K}) \cdot \theta$

(2) implies that

$$(\phi^S \otimes \rho)(\phi^S \otimes a \otimes \phi_\mathbf{K}) = \mathrm{Id}_{\phi^S \otimes \phi_S \otimes \phi_\mathbf{K}}$$

so

$$(\phi^S \otimes a \otimes \phi_\mathbf{K})^{-1} = (\phi^S \otimes \rho).$$

Now

$$
\begin{aligned}
&((\phi^S \otimes a)^{-1} \cdot \widetilde{(\eta' \otimes \phi_{\mathbf{K}})} \cdot \tau) \otimes \phi_{\mathbf{K}} \\
&= (\phi^S \otimes a \otimes \phi_{\mathbf{K}})^{-1} \cdot (\widetilde{(\eta' \otimes \phi_{\mathbf{K}})} \otimes \phi_{\mathbf{K}}) \cdot (\tau \otimes \phi_{\mathbf{K}}) \\
&= (\phi^S \otimes \rho) \cdot (\widetilde{(\eta' \otimes \phi_{\mathbf{K}})} \otimes \phi_{\mathbf{K}}) \cdot (\tau \otimes \phi_{\mathbf{K}}) \\
&= (\eta' \otimes \phi_{\mathbf{K}}) \cdot \theta \cdot (\tau \otimes \phi_{\mathbf{K}}) \\
&= (\eta' \otimes \phi_{\mathbf{K}}) \cdot (\eta' \otimes \phi_{\mathbf{K}})^{-1} \\
&= \mathrm{Id}_{\phi^S \otimes \phi_S \otimes \phi_{\mathbf{K}}}
\end{aligned}
$$

as required. Hence $\tau$ is an isomorphism of monads.

Now we have $\theta \cdot (\eta \otimes \phi_{\mathbf{K}}) = \mathrm{Id}_{\phi_{\mathbf{K}}}$ and the isomorphism between ${}_S\mathbf{T}$ and $\mathbf{T}_{\mathbf{K}}$ together with the invertibility of $\eta' \otimes \phi_{\mathbf{K}}$ gives us $\eta \otimes \phi_{\mathbf{K}}$ is invertible, but then $\theta$ is invertible as required.

Conversely, suppose ${}_S\mathbf{T}$ and $\mathbf{T}_{\mathbf{K}}$ are isomorphic; then $\mathbf{Kl}_S\mathbf{T}$ and $\mathbf{KlT}_{\mathbf{K}}$ are isomorphic, and we may as well assume ${}_S\mathbf{T} = \mathbf{T}_{\mathbf{K}}$ so $\tilde{\phi}_{\mathbf{K}} = \phi^S \otimes \phi_S$, $\eta = \eta'$, etc. We shall also be assuming that $\theta \colon \tilde{\phi}_{\mathbf{K}} \otimes \phi_{\mathbf{K}} \Rightarrow \phi_{\mathbf{K}}$ is invertible.

$\theta \cdot (\eta \otimes \phi_{\mathbf{K}}) = \mathrm{Id}_{\phi_{\mathbf{K}}}$ implies that $\eta \otimes \phi_{\mathbf{K}}$ and hence $\eta' \otimes \phi_{\mathbf{K}}$ is invertible, i.e. that (S2) is satisfied (Proposition 2).

The construction of $\tau$ now goes through as before, but, of course, $(\eta' \otimes \phi_{\mathbf{K}})^{-1} = \theta$ so $\tau = \mathrm{Id}_{\tilde{\phi}_{\mathbf{K}}}$. Again we have $\tau = \widetilde{(\eta' \otimes \phi_{\mathbf{K}})^{-1}}(\phi^S \otimes a)$, but we have to check this slightly differently as we are seeking to prove that $\phi^S \otimes a$ is invertible (i.e. (S3) holds). Here we calculate

$$
\begin{aligned}
&\theta \cdot (\widetilde{(\eta' \otimes \phi_{\mathbf{K}})^{-1}} \cdot (\phi^S \otimes a)) \otimes \phi_{\mathbf{K}} \\
&= \theta \cdot \widetilde{(\eta' \otimes \phi_{\mathbf{K}})^{-1}} \otimes \phi_{\mathbf{K}} \cdot (\phi^S \otimes a) \otimes \phi_{\mathbf{K}} \\
&= (\eta' \otimes \phi_{\mathbf{K}})^{-1} \cdot (\phi^S \otimes \rho) \cdot (\phi^S \otimes a) \otimes \phi_{\mathbf{K}} \\
&= (\eta' \otimes \phi_{\mathbf{K}})^{-1} \cdot \phi^S \otimes (\rho \cdot (a \otimes \phi_{\mathbf{K}})) \\
&= (\eta' \otimes \phi_{\mathbf{K}})^{-1}
\end{aligned}
$$

but $\tau$ is the unique 2-cell satisfying

$$\theta \cdot (\tau \otimes \phi_{\mathbf{K}}) = (\eta' \otimes \phi_{\mathbf{K}})^{-1}$$

so

$$\tau = (\eta' \otimes \phi_{\mathbf{K}})^{-1} \cdot (\phi^S \otimes a),$$

i.e. $\phi^S \otimes a = (\eta' \otimes \phi_{\mathbf{K}})$ and hence is invertible, i.e. by Proposition 3, and the remark that follows it, $a$ is invertible and (S3) is satisfied.

We have just seen that a functor $\mathbf{K}$ has a shape theory if and only if $\theta \colon \tilde{\phi}_{\mathbf{K}} \otimes \phi_{\mathbf{K}} \Rightarrow \phi_{\mathbf{K}}$ is invertible. We will here translate this invertibility into more usual functorial language.

We have

$$\theta(X, P) \colon \tilde{\phi}_{\mathbf{K}} \otimes \phi_{\mathbf{K}}(X, P) \cong \tilde{\phi}_{\mathbf{K}}(X, \mathbf{K}P) \to \phi_{\mathbf{K}}(X, P)$$

so, as

$$\tilde{\phi}_{\mathbf{K}}(X, \mathbf{K}P) = \mathrm{Nat}(\mathbf{B}(\mathbf{K}P, \mathbf{K}-), \mathbf{B}(X, \mathbf{K}-)),$$

$\theta(X, P)$ associates $s(\mathrm{Id}_{\mathbf{K}P})$ to $s \in \mathrm{Nat}(\mathbf{B}(\mathbf{K}P, \mathbf{K}-), \mathbf{B}(X, \mathbf{K}-))$. $\theta$ always admits $\eta \otimes \phi_{\mathbf{K}}$ as a section ($\theta \cdot (\eta \otimes \phi_{\mathbf{K}}) = \mathrm{Id}_{\phi_{\mathbf{K}}}$) and $(\eta \otimes \phi_{\mathbf{K}})(X, P)$ sends $f \in \mathbf{B}(X, \mathbf{K}P)$ to $S(f)$, the natural transformation induced from $f$ by composition. Thus to say $\theta$ is invertible is equivalent to saying

$$s = S(s(\mathrm{Id}_{\mathbf{K}P}))$$

for all $s \in \mathrm{Nat}(\mathbf{B}(\mathbf{K}P, \mathbf{K}-), \mathbf{B}(X, \mathbf{K}-))$ or alternatively, for all $f: X \to \mathbf{K}P$ in $\mathbf{B}$, and all $s \in \mathrm{Nat}(\mathbf{B}(X, \mathbf{K}-), \mathbf{B}(Y, \mathbf{K}-))$, we have the commutation formula

$$s \cdot S(f) = S(s(f)).$$

**Definition**

A functor $\mathbf{K}$ will be said to be *formal* if $\theta: \tilde{\phi}_{\mathbf{K}} \otimes \phi_{\mathbf{K}} \Rightarrow \phi_{\mathbf{K}}$ is invertible.

Several stronger conditions on $\mathbf{K}$ have been studied in papers on categorical shape theory. We list some below:

**Definition**

(a) A functor $\mathbf{K}: \mathbf{A} \to \mathbf{B}$ is *very rich* if for all $P, Q \in |\mathbf{A}|$ and any $g: \mathbf{K}P \to \mathbf{K}Q$, there is a diagram

$$P \xleftarrow{\;r\;} V \xrightarrow{\;r'\;} Q$$

in $\mathbf{A}$ with $\mathbf{K}(r)$ invertible in $\mathbf{B}$ and $g = \mathbf{K}(r')\mathbf{K}(r^{-1})$.

(*b*) *A functor* $\mathbf{K}: \mathbf{A} \to \mathbf{B}$ is *rich* if for all $P$, $Q \in |\mathbf{A}|$ and any $g: \mathbf{K}P \to \mathbf{K}Q$, there is a diagram (zigzag)

$$P \xrightarrow{\;r_1\;} V_1 \xleftarrow{\;r_2\;} V_2 \xrightarrow{\;r_3\;} \cdots \xrightarrow{\;r_{2k-1}\;} V_{2k-1} \xleftarrow{\;r_{2k}\;} Q$$

in $\mathbf{A}$ with $\mathbf{K}(r_{2i})$ invertible in $\mathbf{B}$ for each $i$, $1 \leqslant i \leqslant k$, and

$$g = \mathbf{K}(r_{2k})^{-1}\mathbf{K}(r_{2k-1}) \cdots \mathbf{K}(r_2)^{-1}\mathbf{K}(r_1).$$

(c) A functor $\mathbf{K}$: $\mathbf{A} \to \mathbf{B}$ is *opaque* if for all $P$, $Q \in |\mathbf{A}|$ and any $g$: $\mathbf{K}P \to \mathbf{K}Q$, there is a zigzag

$$P \xrightarrow{\;r_1\;} V_1 \xleftarrow{\;r_2\;} V_2 \xrightarrow{\;r_3\;} \cdots \xrightarrow{\;r_{2k-1}\;} V_{2k-1} \xleftarrow{\;r_{2k}\;} Q$$

in $\mathbf{A}$ whose image by $\mathbf{K}$ can be incorporated in a commutative 'lantern' diagram

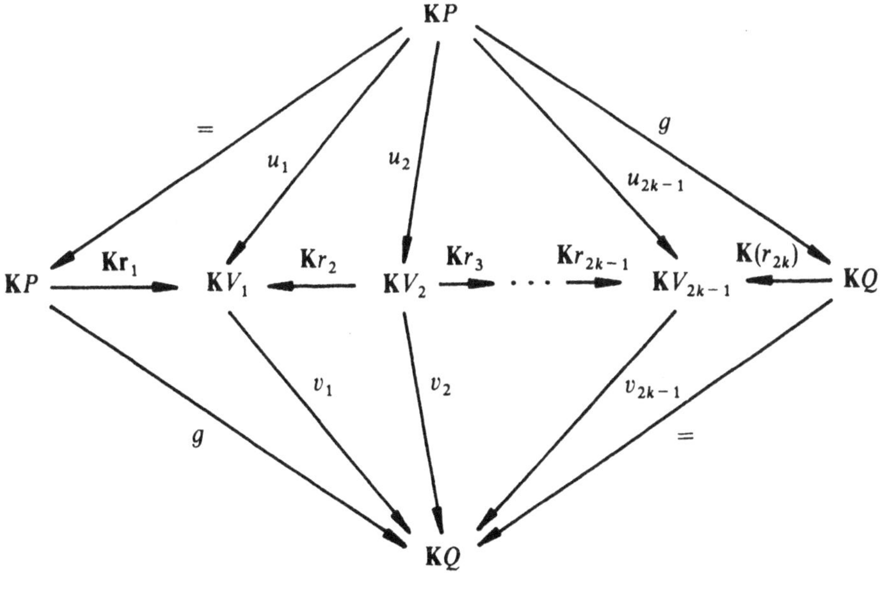

**Proposition 4**
*For a functor* $\mathbf{K}: \mathbf{A} \to \mathbf{B}$, *one has*

(a) $\mathbf{K}$ *full* $\Rightarrow \mathbf{K}$ *very rich*
(b) $\mathbf{K}$ *very rich* $\Rightarrow \mathbf{K}$ *rich*
(c) $\mathbf{K}$ *rich* $\Rightarrow \mathbf{K}$ *opaque*
(d) $\mathbf{K}$ *opaque* $\Rightarrow \mathbf{K}$ *formal*

*Proof*
(a), (b) and (c) are immediate from the definitions; we are thus left to prove (d), for which we use the following lemma.

**Lemma**
*Given any commutative diagram*

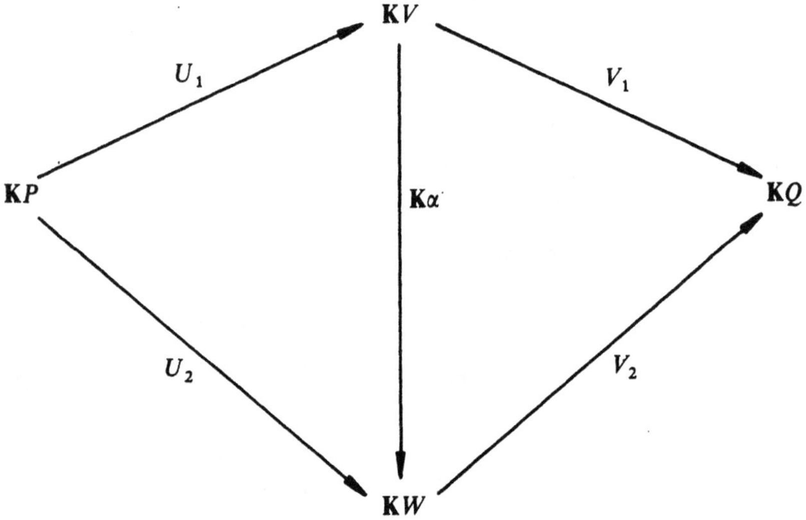


*and any* $s \in \mathrm{Nat}(\mathbf{B}(\mathbf{K}P, \mathbf{K}-), \mathbf{B}(X, \mathbf{K}-))$,

$$V_1 s(U_1) = V_2 s(U_2).$$

*Proof*
Given $\alpha: V \to W$ in $\mathbf{A}$, we note that $s(\mathbf{K}(\alpha)U_1) = \mathbf{K}(\alpha)s(U_1)$, thus

$$\begin{aligned} V_1 s(U_1) &= V_2 \mathbf{K}(\alpha) s(U_1) \\ &= V_2 s(\mathbf{K}(\alpha)U_1) \\ &= V_2 s(U_2) \quad \text{as required.} \end{aligned}$$

Returning to the proof of (d) and using the lantern diagram of the definition of opaque, we find for $g\colon \mathbf{K}P \to \mathbf{K}Q$ by repeated use of the lemma

$$\begin{aligned} gs(\mathrm{Id}_{\mathbf{K}P}) &= V_1 s(U_1) \\ &= V_2 s(U_2) \\ &= \cdots \\ &= s(g), \end{aligned}$$

i.e.

$$s = S(s(\mathrm{Id}_{\mathbf{K}P})).$$

## NOTES ON SOURCES

Bicategories were introduced by Bénabou [7] in 1967. His notes on distributors [8] date from 1973, but it seems that his ideas were fairly well 'distributed' before, as the Kleisli category construction is given in Thiébaud's Thesis [98] in 1971. Also in 1973, Gouzou and Grunig published [49], and Harting gave a talk on distributors at Oberwolfach. The latter was published later [55].

The connection with shape theory and nearly all the material in section 4.3 is taken from Bourn and Cordier's paper [14]. Richness as a notion seems to have first appeared in Deleanu and Hilton's paper [27]. Frei [43] noted that richness implied axiom (S2) for a shape theory (his condition C). 'Opaque functors' were introduced by Guitart in [54] and there he mentioned that an opaque functor always gives a Holsztyński shape functor satisfying (S2). The calculation that opaque implies formal is a simple extension of these results.

# 5

# Functors between Shape Theories

As we have investigated quite fully the properties of the shape category $\mathbf{S_K}$ constructed from a functor $\mathbf{K}: \mathbf{A} \to \mathbf{B}$, it is a natural extension of this process to question to what extent changing the functor $\mathbf{K}$ will induce a functor between the corresponding shape theories.

The fullest answer to this question involves the use of exact squares as introduced by Guitart [54]. These are most naturally described using the language of distributors that we have already introduced in Chapter 4.

Examples of exact squares abound: not only generic examples arise such as those in adjointness situations, but also some particular topological examples first exploited in a shape-like context by Deleanu and Hilton [23]. These examples lead to a simple proof of the compatibility of suspension with shape constructions.

## 5.1 CHANGE OF MODELS

Since in considering the shape theory relative to a functor $\mathbf{K}: \mathbf{A} \to \mathbf{B}$, one is considering approximations to objects in $\mathbf{B}$ by objects from $\mathbf{A}$, it is natural to call the objects of $\mathbf{A}$ the 'models' for the shape theory. This accords well with the idea implicit in the term 'model-induced monad' that we have noted is the term used for the construction dual to that of a codensity monad.

The simplest situation in which a functor is induced between shape theories is when one changes the models in the following way.

We suppose that we are given a functor $\mathbf{L}: \mathbf{C} \to \mathbf{B}$ with associated shape theory $(\mathbf{S_L}, S_\mathbf{L})$. Suppose $\mathbf{F}: \mathbf{A} \to \mathbf{C}$ is another functor and write $\mathbf{K} = \mathbf{LF}$.

**Proposition 1**
$\mathbf{F}$ *induces a functor*

$$\mathbf{F}^*: \mathbf{S_L} \to \mathbf{S_K}$$

*such that* $\mathbf{F}^* S_\mathbf{L} = S_\mathbf{K}$.

*Proof*

Since both $S_{\mathbf{L}}$ and $S_{\mathbf{K}}$ are the identity on objects, it must be the same for $\mathbf{F}^*$; thus we set $\mathbf{F}^*(B) = B$ for each object $B$ of $\mathbf{S_L}$.

If $u: \mathbf{B}(Y, \mathbf{L}-) \to \mathbf{B}(X, \mathbf{L}-)$ is an $\mathbf{S_L}$-morphism, we will specify $[\mathbf{F}^*(u)(A)](f)$ for $f \in \mathbf{B}(Y, \mathbf{K}A)$.

As $\mathbf{B}(Y, \mathbf{K}A) = \mathbf{B}(Y, \mathbf{LF}(A))$, defining

$$[\mathbf{F}^*(u)(A)](f) = [u(A)](f)$$

works and satisfies $\mathbf{F}^* S_{\mathbf{L}} = S_{\mathbf{K}}$ as required.

**Examples**

1. One has for any category $\mathbf{B}$, the shape theory of $\mathbf{Id}: \mathbf{B} \to \mathbf{B}$, the identity functor, which is of course $(\mathbf{B}, \mathbf{Id})$.

Given any functor $\mathbf{K}: \mathbf{A} \to \mathbf{B}$, we thus obtain an induced

$$\mathbf{K}^*: \mathbf{B} \to \mathbf{S_K}$$

and, as $\mathbf{K}^* S_{\mathbf{Id}} = S_{\mathbf{K}}$, $\mathbf{K}^*$ is, in fact, just $S_{\mathbf{K}}$ itself.

2. The usual situation in which one obtains induced functors via change of models is via restriction to a subcategory of models.

If $\mathbf{F}$ is the inclusion of a subcategory $\mathbf{A}$ into $\mathbf{C}$ then $\mathbf{F}^*$ can be loosely interpreted as a sort of '$\mathbf{A}$-localisation' of the $\mathbf{L}$-shape, but note no adjointness conditions on $\mathbf{F}$ are required to get this 'localisation' to exist.

3. Returning to the case when $\mathbf{K}: \mathbf{A} \to \mathbf{B}$ and $\mathbf{L}: \mathbf{C} \to \mathbf{B}$ have left adjoints, suppose $\mathbf{A}$ and $\mathbf{C}$ are actually monadic over $\mathbf{B}$ and that $\mathbf{K}$ and $\mathbf{L}$ are the underlying $\mathbf{B}$-object functors. We write $\mathbf{K}_1$ and $\mathbf{L}_1$ for the left adjoints of $\mathbf{K}$ and $\mathbf{L}$ respectively and $\mathrm{K}$ and $\Lambda$ for the monads they generate. Thus by the results of section 2.2 example (b), we have

$$\mathbf{S_K} \cong \mathbf{B}_{\mathrm{K}} \quad \text{and} \quad \mathbf{S_L} \cong \mathbf{B}_{\Lambda}.$$

Proposition 1 above gives us a functor

$$\mathbf{F}^*: \mathbf{B}_{\Lambda} \to \mathbf{B}_{\mathrm{K}}$$

satisfying $\mathbf{F}^*\mathbf{F}_{\Lambda} \to \mathbf{F}_{\mathrm{K}}$. This functor is easy to describe explicitly.

Suppose $f: X \to \mathbf{LL}_1 Y$ is a map from $X$ to $Y$ in $\mathbf{B}_{\Lambda}$; then one has $F^*(f): X \to \mathbf{KK}_1 Y$ is the composite

$$X \xrightarrow{f} \mathbf{LL}_1 Y \xrightarrow{\mathbf{L}\eta'_Y} \mathbf{LFK}_1 Y = \mathbf{KK}_1 Y$$

where

$$\eta'_Y: \mathbf{L}_1 Y \to \mathbf{FK}_1 Y$$

is the morphism corresponding in the adjunction to

$$\eta_Y: Y \to \mathbf{LFK}_1 Y = \mathbf{KK}_1 Y.$$

It is easy to check that $\mathbf{F}^*\mathbf{F}_{\Lambda} = \mathbf{F}_{\mathrm{K}}$.

## 5.2 ABSOLUTE KAN EXTENSIONS

In Chapter 2, we proved that if we have a situation

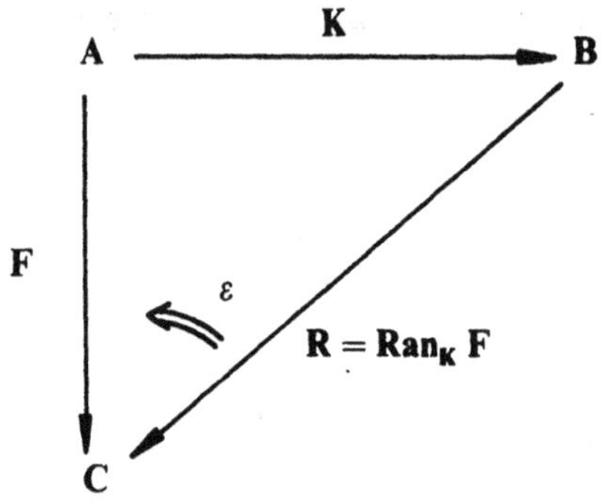


with **C** complete, so that **R** is the right Kan extension of **F** along **K**, then **R** factors as

$$\mathbf{R} = \bar{\mathbf{F}} S_{\mathbf{K}},$$

whence $\bar{\mathbf{F}}$ is a functor from $\mathbf{S_K}$ to **C**, i.e. **R** is a shape-invariant functor.

We can put this another way (using the fact that the shape category, $\mathbf{S_C}$, of the identity functor on **C** is the category **C** itself):

—the square

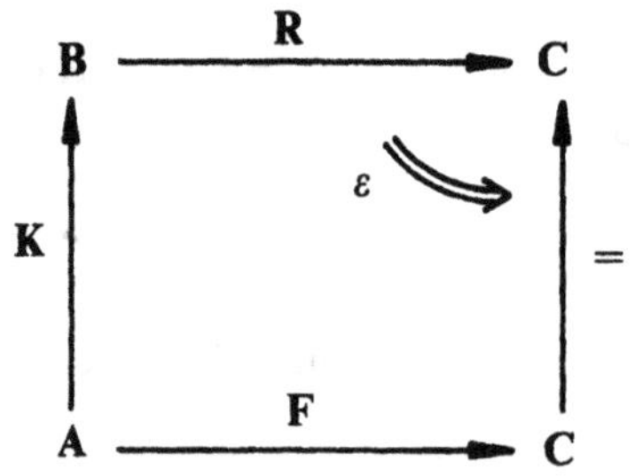


induces a functor from $\mathbf{S_K}$ to $\mathbf{S_C}$ if **C** is complete. This is not covered by our change of models situation. We will find that further investigation of this type of square will pay off, giving a second type of situation where induced functors arise.

To remove the restriction that **C** be complete, we will look in detail at absolute Kan extensions only.

We recall for convenience that $\mathbf{R} = \mathbf{Ran_K\,F}$ means that there is a natural transformation

$$\varepsilon\colon \mathbf{RK} \Rightarrow \mathbf{F}$$

with the universal property that if $\mathbf{S}\colon \mathbf{B} \to \mathbf{C}$ is any functor, there is a natural isomorphism

$$\mathrm{Nat}(\mathbf{S}, \mathbf{R}) \cong \mathrm{Nat}(\mathbf{SK}, \mathbf{F})$$

given by

$$\sigma \to \varepsilon \cdot \sigma\mathbf{K}.$$

We say that $(\mathbf{R}, \varepsilon)$ is an *absolute Kan extension* if for any category $\mathbf{D}$ and functors $\mathbf{H}: \mathbf{C} \to \mathbf{D}$, $\mathbf{S}: \mathbf{B} \to \mathbf{D}$, the induced map

$$\mathrm{Nat}(\mathbf{S}, \mathbf{HR}) \cong \mathrm{Nat}(\mathbf{SK}, \mathbf{HF})$$

is an isomorphism.

Thus briefly $(\mathbf{R}, \varepsilon)$ is absolute if for any $\mathbf{H}$, $\mathbf{HR} \cong \mathbf{Ran_K}\, \mathbf{HF}$, i.e. $\mathbf{R}$ is preserved by all functors.

Unlike on the previous occasion when we studied Kan extensions, we now have distributors at our disposal and they will greatly simplify the calculations. Before we use them, however, we need to prove some lemmas. The first two of these are the analogues for distributors of the result used in the proof of Proposition 2 of section 2.6, namely that if $\mathbf{F} \dashv \mathbf{G}$ then for any $\mathbf{H}$

$$\mathrm{Nat}(\mathbf{S}, \mathbf{HF}) \cong \mathrm{Nat}(\mathbf{SG}, \mathbf{H}).$$

**Lemma 1**

*For any diagram in* **Dist**

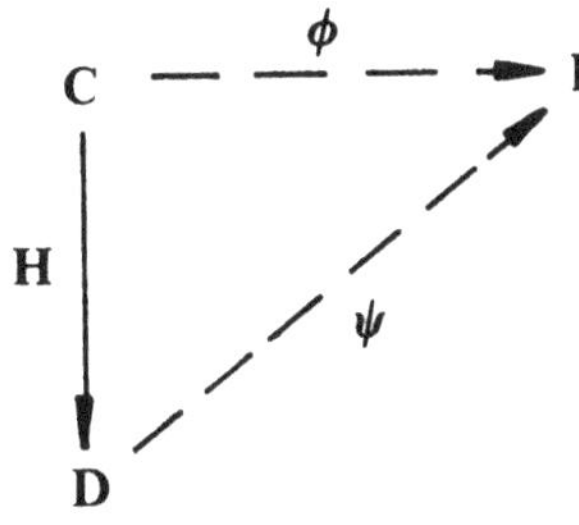


*there is a natural isomorphism*

$$\mathbf{Dist}(\phi \otimes \phi^{\mathbf{H}}, \psi) \cong \mathbf{Dist}(\phi, \psi \otimes \phi_{\mathbf{H}}).$$

*Proof*

Let $\beta: \phi \to \psi \otimes \phi_{\mathbf{H}}$ in **Dist**, then

$$\beta \otimes \phi^{\mathbf{H}}: \phi \otimes \phi^{\mathbf{H}} \to (\psi \otimes \phi_{\mathbf{H}}) \otimes \phi^{\mathbf{H}} \cong \psi \otimes (\phi_{\mathbf{H}} \otimes \phi^{\mathbf{H}}).$$

Now there are natural transformations (see Proposition 1 of section 4.1)

$$\eta: \mathbf{Id} \to \phi^{\mathbf{H}} \otimes \phi_{\mathbf{H}}$$

$$\varepsilon: \phi_{\mathbf{H}} \otimes \phi^{\mathbf{H}} \to \mathbf{Id}$$

so we may form

$$\underset{\sim}{\beta} = (\psi \otimes \varepsilon) \circ (\beta \otimes \phi^{\mathbf{H}}) \in \mathbf{Dist}(\phi \otimes \phi^{\mathbf{H}}, \psi).$$

Next take $\alpha \in \mathbf{Dist}({}^{\phi} \otimes \phi_{\mathbf{H}}, \psi)$ and form

$$\tilde{\alpha} = (\alpha \otimes \phi_{\mathbf{H}}) \circ (\phi \otimes \eta).$$

These two constructions give the desired natural isomorphism. The proof of this is not difficult, but as distributors were only introduced in the last chapter, it is probably advisable to check explicitly at least one of the two directions. We will check that $\alpha$ is isomorphic to $(\tilde{\alpha})_{\sim}$:

$$\begin{aligned}(\tilde{\alpha})_{\sim} &= (\boldsymbol{\psi} \otimes \varepsilon) \circ ((\alpha \otimes \boldsymbol{\phi}_{\mathbf{H}}) \otimes \boldsymbol{\phi}^{\mathbf{H}}) \circ ((\boldsymbol{\phi} \otimes \eta) \otimes \boldsymbol{\phi}^{\mathbf{H}}) \\ &\cong (\boldsymbol{\psi} \otimes \varepsilon) \circ (\alpha \otimes (\boldsymbol{\phi}_{\mathbf{H}} \otimes \boldsymbol{\phi}^{\mathbf{H}})) \circ (\boldsymbol{\phi}^{\mathbf{R}} \otimes (\eta \otimes \boldsymbol{\phi}^{\mathbf{H}})) \\ &\cong (\mathbf{Id} \otimes \alpha) \circ (\boldsymbol{\phi} \otimes \boldsymbol{\phi}^{\mathbf{H}} \otimes \varepsilon) \circ (\boldsymbol{\phi} \otimes \eta \otimes \boldsymbol{\phi}^{\mathbf{H}})\end{aligned}$$

(by the Godement interchange law, quoted in Chapter 4 in the form

$$(t_2 \otimes \phi') \circ (\psi \otimes t_1) = (\psi' \otimes t_1) \circ (t_2 \otimes \phi)).$$

Thus

$$\begin{aligned}(\tilde{\alpha})_{\sim} &\cong \alpha \circ \boldsymbol{\phi} \otimes ((\boldsymbol{\phi}^{\mathbf{H}} \otimes \varepsilon) \circ (\eta \otimes \boldsymbol{\phi}^{\mathbf{H}})) \\ &\cong \alpha \circ (\boldsymbol{\phi} \otimes \mathbf{Id}) \\ &\cong \alpha,\end{aligned}$$

as required.

**Lemma 2**

*For any diagram in* **Dist**

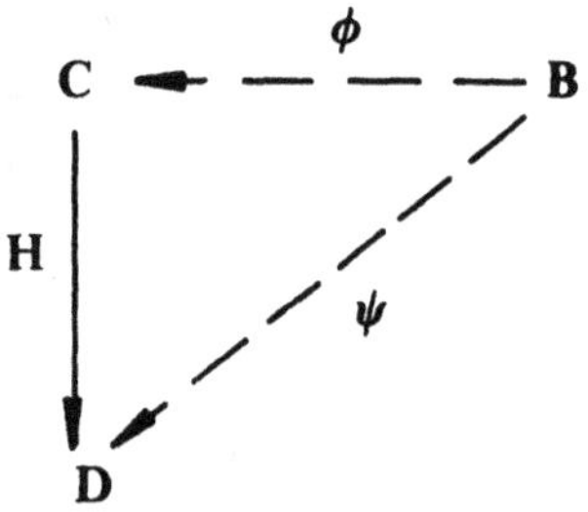


*there is a natural isomorphism*

$$\mathbf{Dist}(\boldsymbol{\phi}_{\mathbf{H}} \otimes \boldsymbol{\phi}, \boldsymbol{\psi}) \cong \mathbf{Dist}(\boldsymbol{\phi}, \boldsymbol{\phi}^{\mathbf{H}} \otimes \boldsymbol{\psi}).$$

The proof is similar to that of lemma 1, so we will leave it out.

We will also need the following factorisation lemma of Bénabou.

**Lemma 3**

*For any distributor* $\boldsymbol{\phi}: \mathbf{A} \dashrightarrow \mathbf{B}$, *there are categories* **C**, **D** *and functors*

$$\mathbf{F}: \mathbf{A} \to \mathbf{C}, \qquad \mathbf{F}': \mathbf{D} \to \mathbf{A}$$

$$\mathbf{G}: \mathbf{B} \to \mathbf{C}, \qquad \mathbf{G}': \mathbf{D} \to \mathbf{B}$$

*such that*

$$\boldsymbol{\phi} \cong \boldsymbol{\phi}^{\mathbf{G}} \otimes \boldsymbol{\phi}_{\mathbf{F}} \cong \boldsymbol{\phi}_{\mathbf{G}'} \otimes \boldsymbol{\phi}^{\mathbf{F}'}.$$

*Proof*

We first construct **C**. We take **C** to have as its class of objects the disjoint union of the classes of objects of **A** and **B**.

—If $A, A'$ are objects of $\mathbf{A}$ considered as objects of $\mathbf{C}$, then $\mathbf{C}(A, A') = \mathbf{A}(A, A')$. Similarly

$$\mathbf{C}(B, B') = \mathbf{B}(B, B').$$

if $B, B'$ are in $\mathbf{B}$.

If $A$ is in $\mathbf{A}$ and $B$ in $\mathbf{B}$ then

$$\mathbf{C}(B, A) = \phi(B, A).$$

The compositions are given by the actions. Thus if $\alpha: A \to A'$ in $\mathbf{A}$ and $x \in \phi(B, A)$, the composite in $\mathbf{C}$

$$B \xrightarrow{x} A \xrightarrow{\alpha} A'$$

is given by

$$\phi(B, \alpha)(x) \in \phi(B, A').$$

Similarly if $\beta: B' \to B$, the composite

$$B' \xrightarrow{\beta} B \xrightarrow{x} A$$

is

$$\phi(\beta, A)(x) \in \phi(B', A).$$

The functors $\mathbf{F}$ and $\mathbf{G}$ are the obvious inclusion functors of $\mathbf{A}$ and $\mathbf{B}$ into $\mathbf{C}$. The relation

$$\phi \cong \phi_{\mathbf{G}} \otimes \phi_{\mathbf{F}}$$

follows from the fact mentioned earlier that in any situation

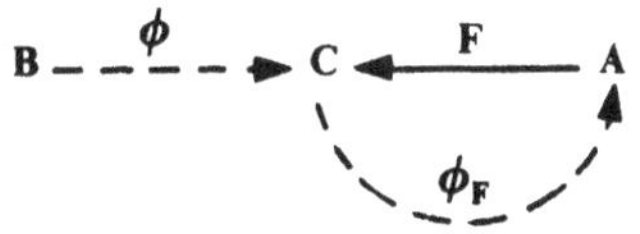


one has

$$(\phi^{\mathbf{F}} \otimes \phi)(A, B) = \phi(\mathbf{F}A, B).$$

This is, of course, trivially checked in our situation here. We next turn to the construction of $\mathbf{D}$. This is essentially a comma-category construction.

An object of $\mathbf{D}$ is a triplet $(B, A, \gamma)$, formed from an object $B$ of $\mathbf{B}$, an object $A$ of $\mathbf{A}$ and an element $\gamma \in \phi(B, A)$. A morphism in $\mathbf{D}$ from $(B_1, A_1, \gamma_1)$ to $(B_2, A_2, \gamma_2)$ is a pair $(\beta, \alpha)$ such that '$\alpha\gamma_1 = \gamma_2\beta$', or more exactly we have $\alpha: A_1 \to A_2$, $\beta: B_1 \to B_2$,

$$\gamma_1 \in \phi(B_1, A_1), \qquad \gamma_2 \in \phi(B_2, A_2),$$

and we require

$$\phi(\beta, A_2)(\gamma_2) = \phi(B_1, \alpha)(\gamma_1) \in \phi(B_1, A_2);$$

the functors $\mathbf{F}'$ and $\mathbf{G}'$ are the projections

—$\mathbf{F}'\colon \mathbf{D} \to \mathbf{A}$ is given by $\mathbf{F}'(B, A, \gamma) = A$

$$\mathbf{F}'(\beta, \alpha) \quad = \alpha$$

—$\mathbf{G}'\colon \mathbf{D} \to \mathbf{B}$ is given by $\mathbf{G}'(B, A, \gamma) = B$

$$\mathbf{G}'(\beta, \alpha) \quad = \beta.$$

We next verify that

$$\phi \cong \phi_{\mathbf{G}'} \otimes \phi^{\mathbf{F}'}.$$

Let $A$ be in $\mathbf{A}$, $B$ be in $\mathbf{B}$:

$$\phi_{\mathbf{G}'} \otimes \phi^{\mathbf{F}'}(B, A) \cong \left( \bigsqcup_{(B', A, \gamma')} \mathbf{B}(B, \mathbf{G}'(\gamma')) \times \mathbf{A}(\mathbf{F}'(\gamma'), A) \right) \Big/ R$$

$$\cong \left( \bigsqcup_{(B', A', \gamma')} \mathbf{B}(B, B') \times \mathbf{A}(A, A) \right) \Big/ R.$$

Now $R$ is the equivalence relation generated by the relation

$$\text{‘}(\phi_{\mathbf{G}'}(B, b)(g), f) \sim (g, \phi^{\mathbf{F}'}(b, A)(f))\text{’}$$

(see the definition of composition in general given in section 4.1) where

$$f \in \phi^{\mathbf{F}'}(\gamma', A), \qquad g \in \phi_{\mathbf{G}'}(B, \gamma'') \quad \text{and} \quad b \in \mathbf{D}(\gamma'', \phi').$$

Of course we may interpret $f$, $g$ and $b$ as

$$f \in \mathbf{A}(A', A), \qquad g \in \mathbf{B}(B, B'') \quad \text{and} \quad b = (\beta, \alpha)\colon (B'', A'', \gamma'') \to (B', A', \gamma')$$

as a pair $\beta\colon B'' \to B'$, $\alpha\colon A'' \to A'$ such that

$$\phi(\beta, A')(\gamma') = \phi(B'', \alpha)(\gamma'').$$

Thus the relation amounts to

$$(\mathbf{B}(B, \beta)g, f) \sim (g, \mathbf{A}(\alpha, A)f)$$

or

$$(\beta g, f) \sim (g, f\alpha).$$

We may thus represent any element of $\phi_{\mathbf{G}'} \otimes \phi^{\mathbf{F}'}(B, A)$ as an equivalence class $[(g, f), \gamma']$ where $g \in \mathbf{B}(B, B')$, $f \in \mathbf{A}(A', A)$, $\phi' \in \phi(B', A')$, but then we can map $((g, f), \gamma')$ to

$$\gamma'' = \phi(g, A') \circ \phi(B, f)(\phi') \in \phi(B, A).$$

This map is compatible with the equivalence relation and so defines a map

$$\phi_{\mathbf{G}'} \otimes \phi_{\mathbf{F}}{}'(B, A) \to \phi(B, A)$$

which is split by

$$\gamma \to [(\mathrm{Id}_B, \mathrm{Id}_A), \gamma].$$

It is, however, easily checked that

$$((g, f), \gamma) \sim ((\mathrm{Id}_B, \mathrm{Id}_A), \gamma'')$$

so the above map is an isomorphism. Its naturality being obvious, we have therefore completed the proof of Lemma 3.

We can now return to the formulation of absolute Kan extensions; there we had the isomorphism

$$\mathrm{Nat}(\mathbf{S}, \mathbf{HR}) \cong \mathrm{Nat}(\mathbf{SK}, \mathbf{HF}).$$

We have already used the morphism

$$\mathbf{Cat} \to \mathbf{Dist}$$

given by

$$\mathbf{F} \to \phi_{\mathbf{F}}.$$

We now need to use the morphism

$$\mathbf{Cat}^{\mathrm{op}} \to \mathbf{Dist}$$

$$\mathbf{F} \to \phi^{\mathbf{F}}.$$

This is an embedding, i.e.

$$\mathrm{Nat}(\mathbf{F}, \mathbf{G}) \cong \mathbf{Dist}(\phi^{\mathbf{G}}, \phi^{\mathbf{F}})$$

(by the Yoneda Lemma essentially), so we can apply it to translate our absolute Kan extension condition to one in **Dist**, namely

$$\mathbf{Dist}(\phi^{\mathbf{R}} \otimes \phi^{\mathbf{H}}, \phi^{\mathbf{S}}) \cong \mathbf{Dist}(\phi^{\mathbf{F}} \otimes \phi^{\mathbf{H}}, \phi^{\mathbf{K}} \otimes \phi^{\mathbf{S}}).$$

Using Lemma 1 we obtain

$$\mathbf{Dist}(\phi^{\mathbf{R}}, \phi^{\mathbf{S}} \otimes \phi_{\mathbf{H}}) \cong \mathbf{Dist}(\phi^{\mathbf{F}}, \phi^{\mathbf{K}} \otimes \phi^{\mathbf{S}} \otimes \phi_{\mathbf{H}})$$

and then applying Lemma 2,

$$\mathbf{Dist}(\phi^{\mathbf{F}}, \phi^{\mathbf{K}} \otimes \phi^{\mathbf{S}} \otimes \phi_{\mathbf{H}}) \cong \mathbf{Dist}(\phi_{\mathbf{K}} \otimes \phi^{\mathbf{F}}, \phi^{\mathbf{S}} \otimes \phi_{\mathbf{H}}).$$

Thus

$$\mathbf{Dist}(\phi^{\mathbf{R}}, \phi^{\mathbf{S}} \otimes \phi_{\mathbf{H}}) \cong \mathbf{Dist}(\phi_{\mathbf{K}} \otimes \phi^{\mathbf{F}}, \phi^{\mathbf{S}} \otimes \phi_{\mathbf{H}}).$$

By Lemma 3, any distributor $\phi$ can be written in the form $\phi^{\mathbf{S}} \otimes \phi_{\mathbf{H}}$, so we have an isomorphism

$$\mathbf{Dist}(\phi^{\mathbf{R}}, \phi) \cong \mathbf{Dist}(\phi_{\mathbf{K}} \otimes \phi^{\mathbf{F}}, \phi)$$

for arbitrary $\phi$. Of course this is equivalent to

$$\phi_{\mathbf{K}} \otimes \phi^{\mathbf{F}} \cong \phi^{\mathbf{R}}.$$

Thus we have proved

**Proposition 1**
**R** *is the absolute Kan extension of* **F** *along* **K** *if and only if* $\varepsilon$ *induces an isomorphism*

$$\phi_{\mathbf{K}} \otimes \phi^{\mathbf{F}} \cong \phi^{\mathbf{R}}.$$

This provides us with the possibility of giving an 'equational' description of absolute Kan extensions by interpreting the 'tensor' as a colimit. (Explicitly we will use the formula

$$(\psi \otimes \phi)(C, A) = \left( \bigsqcup_{B \in |\mathbf{B}|} \psi(C, B) \times \phi(B, A) \right) \Big/ \sim$$

where $\sim$ is generated by

$$(\psi(C, b)(\beta), \alpha) \sim (\beta, \phi(b, A)(\alpha))$$

for

$$\alpha \in \phi(B, A), \qquad \beta \in \psi(C, B') \quad \text{and} \quad b \in \mathbf{B}(B', B).)$$

If we evaluate the tensor $\phi_{\mathbf{K}} \otimes \phi^{\mathbf{F}}$ at $(B, C)$, we obtain

$$(\phi_{\mathbf{K}} \otimes \phi^{\mathbf{F}})(B, A) = \left( \bigsqcup_{A \in |\mathbf{A}|} \mathbf{B}(B, \mathbf{K}A) \times \mathbf{C}(\mathbf{F}A, C) \right) \Big/ \sim .$$

If $\alpha: B \to \mathbf{K}A$, $\beta: \mathbf{F}A \to C$, then it is clear how to define an element of $\phi^{\mathbf{R}}(B, C) = \mathbf{C}(\mathbf{R}B, C)$, namely the composite

$$\mathbf{R}B \xrightarrow{\mathbf{R}\alpha} \mathbf{RK}A \xrightarrow{\varepsilon(A)} \mathbf{F}A \xrightarrow{\beta} C,$$

i.e. we define a natural 2-cell between distributors,

$$\bar{\varepsilon}: \phi_{\mathbf{K}} \otimes \phi^{\mathbf{F}} \to \phi^{\mathbf{R}}$$

by

$$\bar{\varepsilon}(\alpha \otimes \beta) = \beta\varepsilon(A)\mathbf{R}\alpha$$

where $\alpha: B \to \mathbf{K}A$. The naturality of $\varepsilon$ ensures that this is well defined. Of course this is exactly the map given by taking $\mathbf{S} = \mathbf{R}$, $\mathbf{H} = \text{Identity}$, using the isomorphism

$$\mathbf{Dist}(\phi^{\mathbf{R}}, \phi^{\mathbf{R}}) \cong \mathbf{Dist}(\phi_{\mathbf{K}} \otimes \phi^{\mathbf{F}}, \phi^{\mathbf{R}}),$$

as can be easily checked on following the identity on $\phi^{\mathbf{R}}$ through the various isomorphisms that we have been using.

Thus $\mathbf{R}$ is an absolute Kan extension if and only if $\bar{\varepsilon}$ is an isomorphism, but as we have an explicit description of $\bar{\varepsilon}$, it is now easy to give an explicit condition for the absoluteness of the extension.

Thus to obtain conditions for $\bar{\varepsilon}$ to be an isomorphism, we must note when $\bar{\varepsilon}(B, C)$ is one–one and when it is onto.

Clearly '$\bar{\varepsilon}(B, C)$ is onto' is equivalent to saying: Given any $\gamma: \mathbf{R}B \to C$ in $\mathbf{C}$, there are maps

$$\alpha: B \to \mathbf{K}A, \qquad \beta: \mathbf{F}A \to C$$

such that

$$\beta\varepsilon(A)\mathbf{R}(\alpha) = \gamma.$$

Since $\bar{\varepsilon}$ is natural in both $B$ and $C$, it will suffice to consider the case when $\phi$ is the identity on $\mathbf{R}B$ and, of course, $C = \mathbf{R}B$.

To ensure that $\bar{\varepsilon}(B, C)$ is one–one, we must have that any two possible factorisations of a $\gamma$ as above are linked by a chain of elementary relationships $\sim$ (see the definition of $\psi \otimes \phi$ recalled above). Thus we obtain the following:

**Proposition 2**

*$\mathbf{R}$ is an absolute (right) Kan extension of $\mathbf{F}$ along $\mathbf{K}$ if and only if the following two conditions are satisfied:*

(i) *Given $B$ in $\mathbf{B}$, there is an object $\mathbf{L}B$ in $\mathbf{A}$ and maps $\phi_B: B \to \mathbf{KL}B$, $\psi_B: \mathbf{FL}B \to \mathbf{R}B$ such that*

$$\psi_B \varepsilon(\mathbf{L}B)\mathbf{R}(\phi_B) = \text{identity on } \mathbf{R}B$$

(ii) *Given $\gamma: \mathbf{R}B \to C$ in $\mathbf{C}$ and $\alpha: B \to \mathbf{K}A$, $\beta: \mathbf{F}A \to C$ such that*

$$\beta\varepsilon(\mathrm{A})\mathbf{R}(a) = \gamma,$$

*there is a zigzag*

$$A \xrightarrow{r_1} A_1 \xleftarrow{r_2} \cdots \xrightarrow{r_{2k-1}} A_{2k-1} \xleftarrow{r_{2k}} \mathbf{L}B$$

*in $\mathbf{A}$ and 'lanterns'*

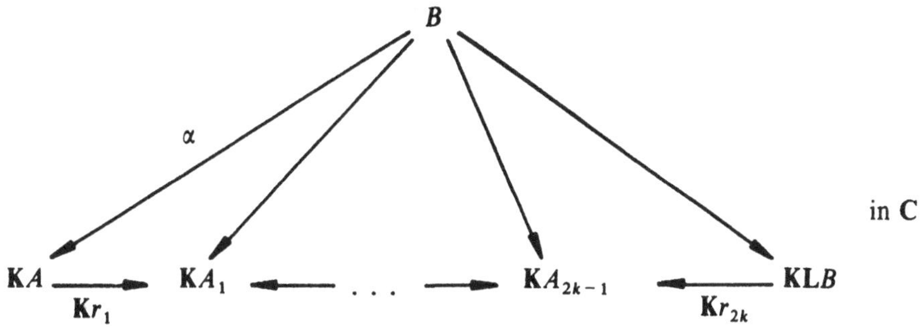


*and*

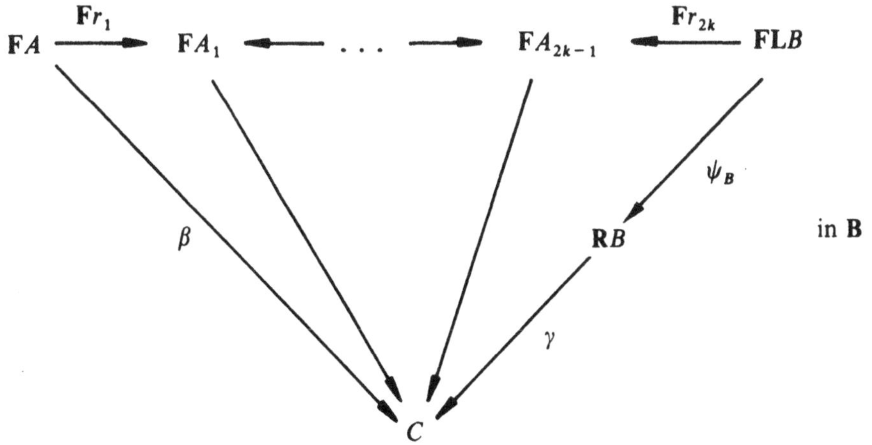


Finally as yet another way of describing absolute Kan extensions, we give a description using comma categories.

Given an object $B$ in $\mathbf{B}$, we can define a functor

$$\mathbf{H}_\varepsilon: B\downarrow\mathbf{K} \to \mathbf{R}B\downarrow\mathbf{C}$$

which sends $(B \xrightarrow{\alpha} \mathbf{K}A, A)$ to $(\mathbf{R}B \xrightarrow{\varepsilon(A)\mathbf{R}(\alpha)} \mathbf{F}A, \mathbf{F}A)$.

**Proposition 3**
*The extension* **R** *is absolute if and only if* $\mathbf{H}_\varepsilon$ *is initial for all B.*

*Proof*
Initiality translates as saying that given any $\gamma = (\mathbf{R}B \xrightarrow{\gamma} C, C)$ in $\mathbf{R}B\downarrow\mathbf{C}$, the category $\mathbf{H}_\varepsilon\downarrow\gamma$ is non-empty and connected. However 'non-empty' is equivalent to condition (1) of Proposition 2, whilst 'connected' is just the zigzag/lantern condition, so Proposition 3 is really just a restatement of Proposition 2.

So far we have restricted attention to absolute right Kan extensions, as these were special cases of the pointwise Kan extensions that we have considered earlier. Of course the comma-category description above suggests that **R** should induce a functor

$$\mathbf{S_K} \to \mathbf{C}$$

even though **C** is not necessarily complete. We shall later on show that this is in fact the case; however, before that, we must mention the corresponding results for absolute left Kan extensions. The proofs are nearly always dual to those given above and so will be omitted.

In a diagram of the form

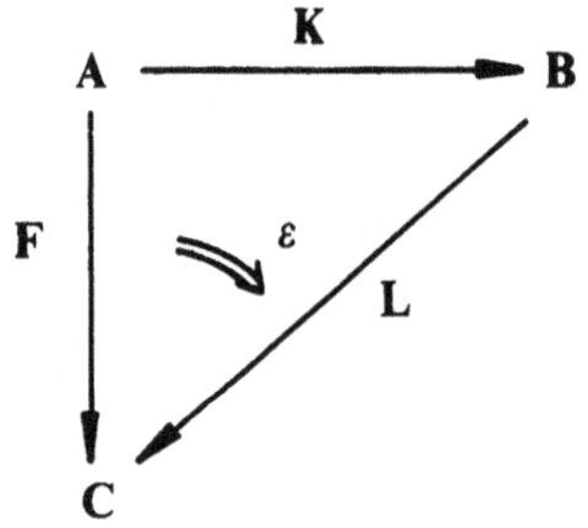


**L** is the absolute left Kan extension of **F** along **K** if for any functor $\mathbf{H}: \mathbf{B} \to \mathbf{D}$, **HL** is the left Kan extension of **HF** along **J** (i.e. $\mathbf{HR} = \mathbf{Lan_K}\, \mathbf{HF}$), i.e. $\varepsilon$ induces a natural isomorphism

$$\mathrm{Nat}(\mathbf{HL}, \mathbf{S}) \cong \mathrm{Nat}(\mathbf{HF}, \mathbf{SK}).$$

**Proposition 1***
**L** *is the absolute left Kan extension of* **F** *along* **K** *if and only if* $\varepsilon$ *induces an isomorphism*

$$\phi_\mathbf{F} \otimes \phi^\mathbf{K} \cong \phi_\mathbf{L}.$$

**Proposition 2***
**L** *is the absolute left Kan extension of* **F** *along* **K** *if and only if the following two conditions are satisfied:*

(i) *Given B in* **B**, *there is an object* $\mathbf{R}B$ *in* **A** *and maps*

$$\phi_B: \mathbf{KR}B \to B, \qquad \psi_B: \mathbf{L}B \to \mathbf{FR}B$$

*such that*

$$\mathbf{L}(\phi_B)\varepsilon(\mathbf{R}B)\psi_B = \text{identity on } \mathbf{L}B.$$

(ii) *Given any* $\gamma: C \to \mathbf{L}B$, *and* $\alpha: \mathbf{K}A \to B$, $\beta: C \to \mathbf{F}A$ *such that*

$$\mathbf{L}(\alpha)\varepsilon(A)\beta = \gamma,$$

*there is a zigzag*

$$A \xrightarrow[r_1]{} A_1 \xleftarrow[r_2]{} \cdots \longrightarrow A_{2k-1} \xleftarrow[r_{2n}]{} \mathbf{R}B$$

*in* **A** *and 'lanterns'*

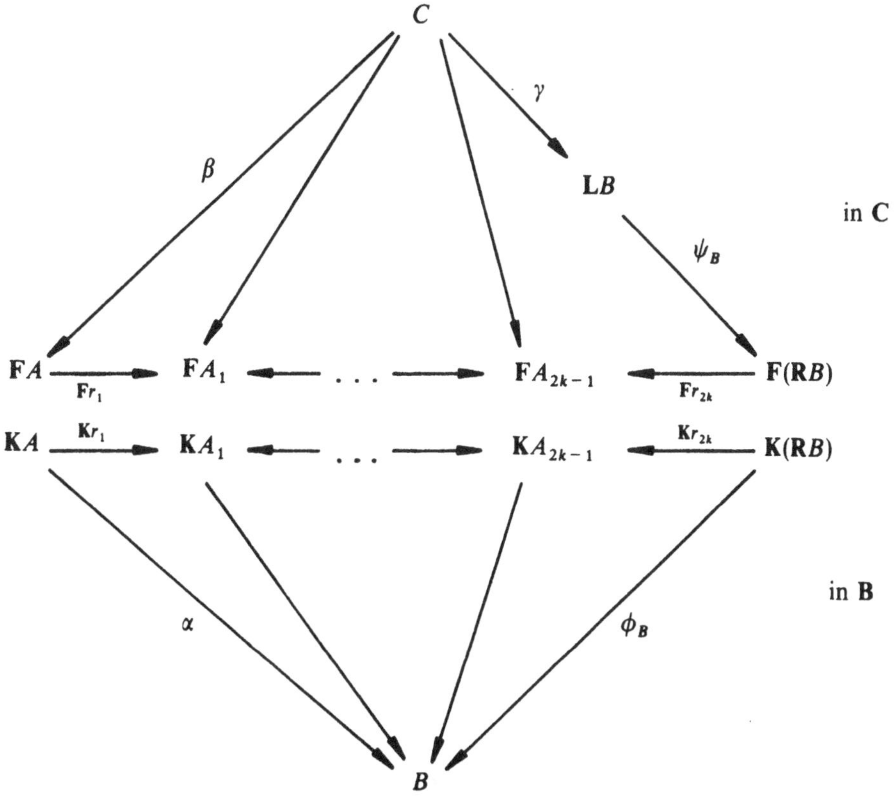


Although there is a dual of Proposition 3 giving a finality condition equivalent to absoluteness of **L**, it is of more use to twist things around to obtain a different functor induced by $\varepsilon$ and an initiality condition.

Given $C$ in **C**, $\varepsilon$ induces a functor

$$\mathbf{H}_\varepsilon: C\downarrow\mathbf{F} \to C\downarrow\mathbf{L}$$

given by

$$\mathbf{H}_\varepsilon(C \xrightarrow{\beta} \mathbf{F}A, A) = (C \xrightarrow{\varepsilon(A)\beta} \mathbf{LK}A, \mathbf{K}A).$$

**Proposition 4**

**L** *is absolute if and only if* $\mathbf{H}_\varepsilon$ *is initial for all* $C$.

Again the proof is simply a reduction to the zigzag/lantern conditions of Proposition 2*.

The use of zigzags and lanterns may have reminded the reader of the definition of a opaque functor given towards the end of Chapter 4. So to make it easier to study the relationship between opaque functors and absoluteness, we recall the definition of opacity below.

The functor $\mathbf{K}: \mathbf{A} \to \mathbf{B}$ is *opaque* if for all $P$, $Q$ in $\mathbf{A}$ and any $g: \mathbf{K}P \to \mathbf{K}Q$ in $\mathbf{B}$, there is a zigzag

$$P \xrightarrow{r_1} A_1 \xleftarrow{r_2} A_2 \longrightarrow \cdots \longrightarrow A_{2k-1} \xleftarrow{r_{2k}} Q$$

*in* $\mathbf{A}$, whose image by $\mathbf{K}$ can be incorporated in a commutative 'lantern':

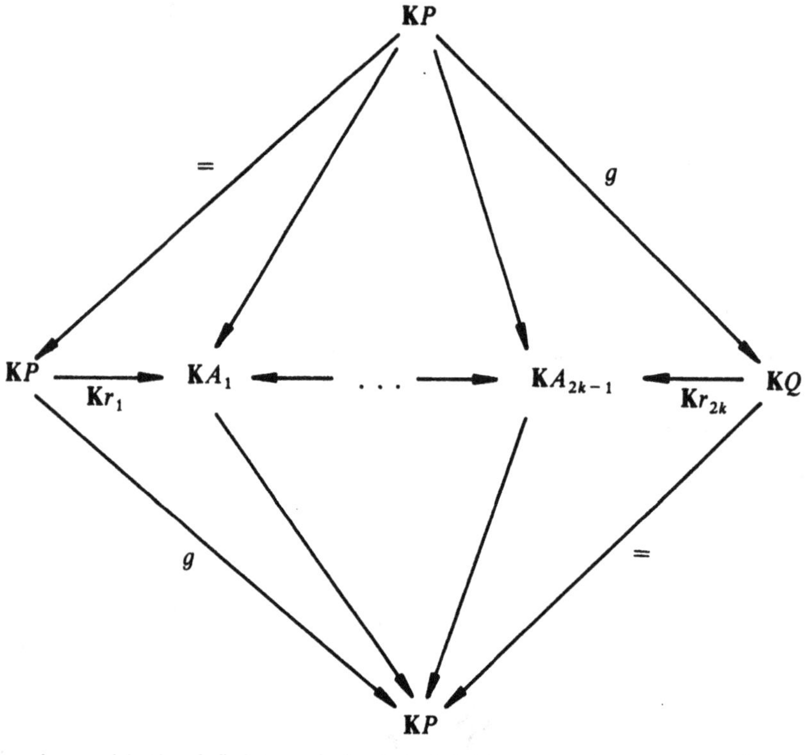


Comparison with the definition of absolute left Kan extension suggests that one is looking at the following factorisation diagram for $\mathbf{K}$:

Let $\mathbf{B_K}$ be the full subcategory of $\mathbf{B}$ with objects the $\mathbf{K}A$ for $A$ in $\mathbf{A}$ (i.e. the full image of $\mathbf{K}$); then there is a diagram

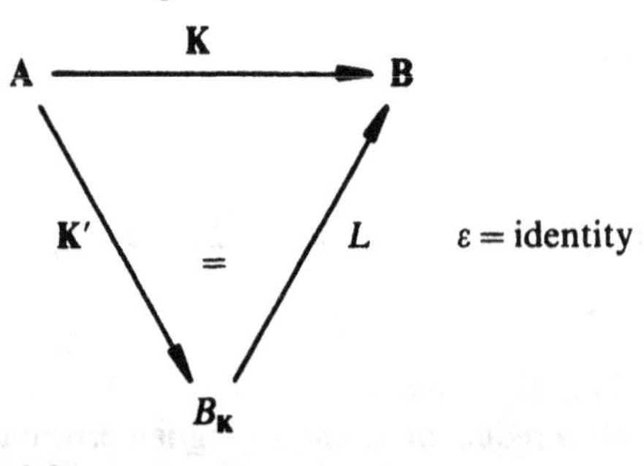


where $\mathbf{K}'$ is the induced functor and $\mathbf{L}$ is the inclusion of $\mathbf{B_K}$ into $\mathbf{B}$.

Suppressing the difference between **K** and **K′L**, we obtain the following:

**Proposition 5**

*$\mathbf{K}$ is opaque if and only if the factorisation diagram (above) makes $\mathbf{L}$ into the absolute left Kan extension of $\mathbf{K}$ along $\mathbf{K}'$.*

*Proof*

Firstly we will assume that **K** is opaque; we therefore have, by Proposition 2*, to verify two conditions. The first is rather trivial—in fact, it states that for any $\mathbf{K}A$, there is a factorisation of the identity on $\mathbf{K}A$ as the composite of two maps $\mathbf{K}A \xrightarrow{\alpha} \mathbf{K}A$, and $\mathbf{K}A \xrightarrow{\beta} \mathbf{K}A$ with $A$ in $\mathbf{A}$, but, of course, taking both $\alpha$ and $\beta$ to be the identity works perfectly well.

The second condition is slightly more subtle. Suppose given $\gamma: B \to \mathbf{K}A$ in $\mathbf{B}$ and maps $\alpha: \mathbf{K}A' \to \mathbf{K}A$ and $\beta: B \to \mathbf{K}A'$ such that

$$\alpha\beta = \gamma,$$

we need to find a zigzag

$$A' \to A_1 \leftarrow \cdots \to A_{2k-1} \leftarrow A$$

such that there is a 'lantern'

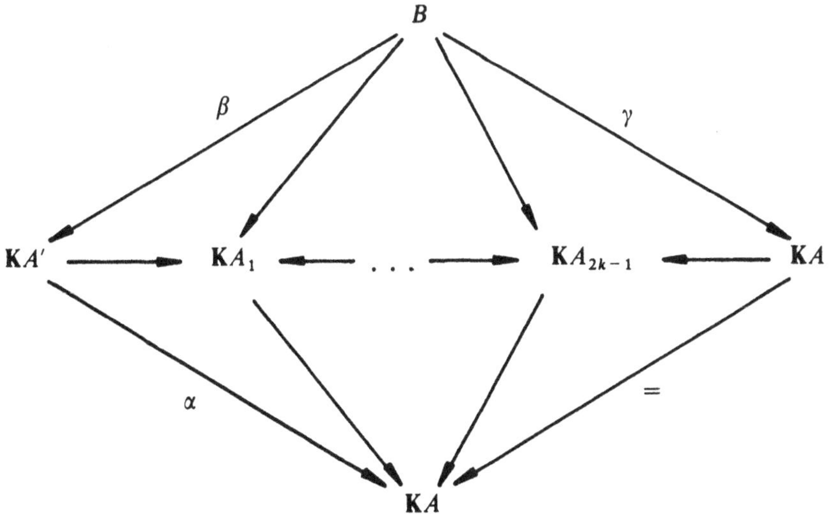


The opaqueness condition gives us a lantern

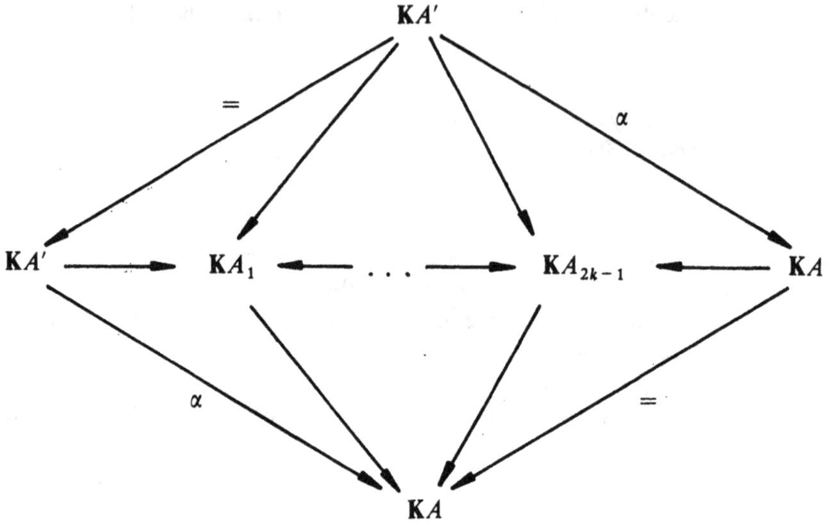


corresponding to just such a zigzag and precomposition with $\beta: B \to KA'$ gives the required lantern as in (*).

Conversely suppose $\mathbf{L}$ is the absolute left Kan extension of $\mathbf{K}$ along $\mathbf{K}'$. Given $P$, $Q$ in $\mathbf{A}$ and $g: \mathbf{K}P \to \mathbf{K}Q$ in $\mathbf{B}$, we can choose $\phi_{\mathbf{K}Q}$ and $\psi_{\mathbf{K}Q}$ both to be the identity on $\mathbf{K}Q$. We then take $\beta$ to be the identity on $\mathbf{K}P$ and $\alpha = g$ to give a zigzag

$$P \to A_1 \leftarrow \cdots \to A_{2k-1} \leftarrow Q$$

and lantern

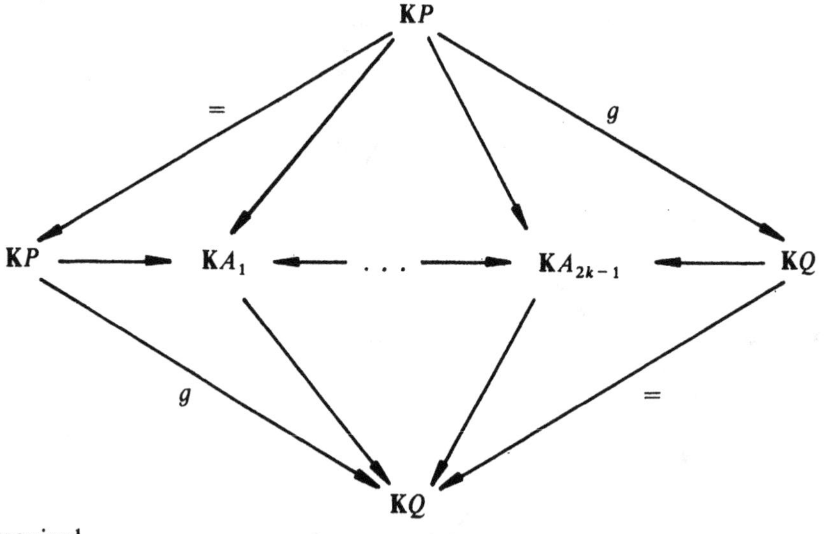


as required.

This result, of course, provides several alternative formulations of the idea of an opaque functor. Since we know that an opaque functor is formal (Proposition 4 of section 4.3), it is of particular note that opacity is equivalent to the condition

$$\phi^{\mathbf{K}} \otimes \phi_{\mathbf{K}'} \cong \phi_{\mathbf{L}}$$

(see the definition of *formal* in section 4.3).

We have already noted that absolute right Kan extensions with complete codomain induce shape functors

$$\mathbf{S_K} \rightarrow \mathbf{C}$$

since this is just a special case of earlier results on the shape invariance of Kan extensions (section 2.5). If one wishes for an induced functor

$$\mathbf{S_K} \rightarrow \mathbf{S_L}$$

between two shape categories, consistent with the shape functors $\mathbf{S_K}$ and $\mathbf{S_L}$, then one might ask for some sort of 'relative' form of Kan extension, i.e. a square

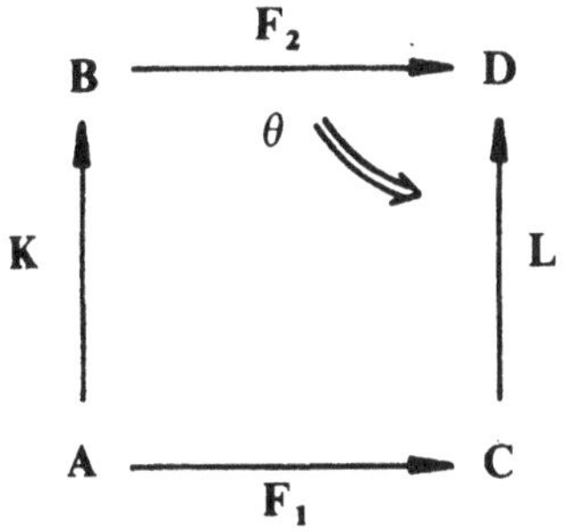


with some universal property. One would hope that absolute right extensions would correspond to the case when $\mathbf{C} = \mathbf{D}$, and $\mathbf{L}$ is the identity functor. Precisely this is true, as will be shown in the next section, but better still, the formulation of exact squares given there also contains the case of absolute left Kan extension and some other interesting constructions.

## 5.3 EXACT SQUARES

We consider a diagram

$(\mathbb{F})$

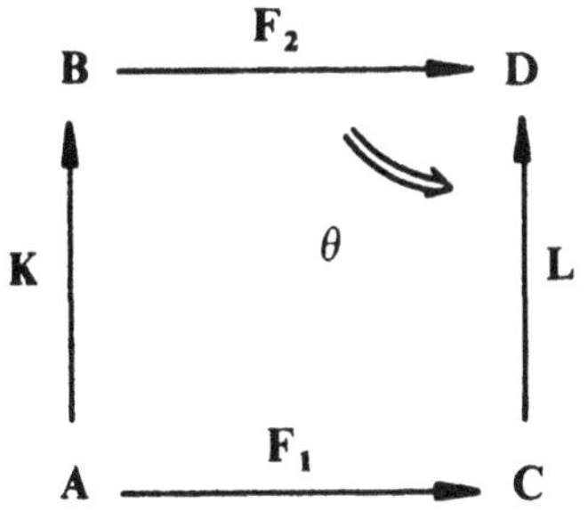


where $\theta: \mathbf{F_2K} \rightarrow \mathbf{LF_1}$ is a natural transformation.

For each $B$ in $\mathbf{B}$, $C$ in $\mathbf{C}$ and $A$ in $\mathbf{A}$, one has a canonical map

$$\theta_{B,C}(A): \mathbf{B}(B, \mathbf{K}A) \times \mathbf{C}(\mathbf{F_1}A, C) \rightarrow \mathbf{D}(\mathbf{F_2}B, \mathbf{L}C)$$

given by

$$\theta_{B,C}(A)(f, g) = \mathbf{L}(g)\theta(A)\mathbf{F_2}(f).$$

Thus $\theta_{B,C}$ induces a map

$$\tilde{\theta}_{B,C}\colon \phi_{\mathbf{K}} \otimes \phi^{\mathbf{F}_1}(B, C) \to \phi^{\mathbf{F}_2} \otimes \phi_{\mathbf{L}}(B, C)$$

since if $\alpha\colon A \to A'$, $f\colon B \to \mathbf{K}A$, $g'\colon \mathbf{F}_1 A' \to C$

$$\begin{aligned}\theta_{B,C}(A)(f, g'\mathbf{F}_1(\alpha)) &= \mathbf{L}(g'\mathbf{F}_1(\alpha))\theta(A)\mathbf{F}_2(f)\\ &= \mathbf{L}(g')\mathbf{L}\mathbf{F}_1(\alpha)\theta(A)\mathbf{F}_2(f)\\ &= \mathbf{L}(g')\theta(A')\mathbf{F}_2\mathbf{K}(\alpha)\mathbf{F}_2(f)\\ &= \mathbf{L}(g')\theta(A')\mathbf{F}_2(\mathbf{K}(\alpha)f)\\ &= \theta_{B,C}(A')(\mathbf{K}(\alpha)f, g'),\end{aligned}$$

so $\theta_{B,C}$ is compatible with the relation $\sim$ used in defining the composition of distributors.

**Definition**
The square ($\mathbb{F}$) is said to be *exact* if, for each $B, B$, $\theta_{B,C}$ is an isomorphism, i.e. $\theta$ induces an isomorphism

$$\tilde{\theta}\colon \phi_{\mathbf{K}} \otimes \phi^{\mathbf{F}_1} \xrightarrow{\cong} \theta^{\mathbf{F}_2} \otimes \phi_{\mathbf{L}}.$$

We write $\mathbb{F} = (\mathbf{F}, \mathbf{L}, \theta, \mathbf{K}, \mathbf{F}_2)$ for the exact square considered above.

Of course our previous section provides us with some examples of exact squares.

The square

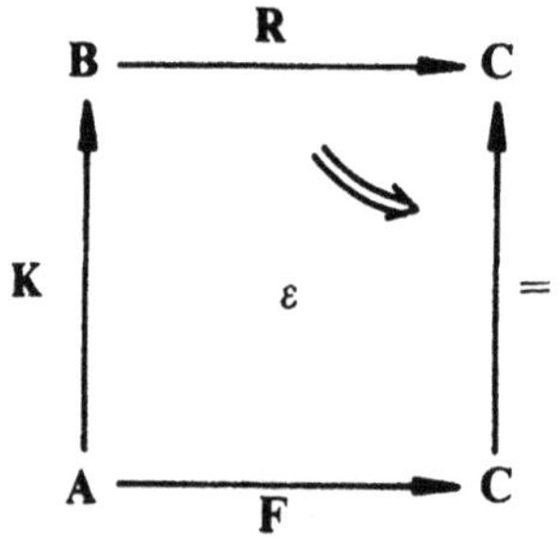


is exact if and only if **R** is the absolute right Kan extension of **F** along **K**, whilst

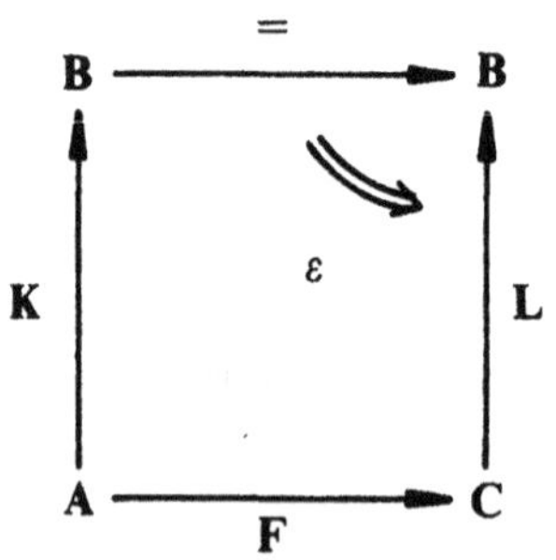


is exact if and only if **L** is the absolute left Kan extension of **K** along **L**.

This suggests that there should be simple extensions of Propositions 2, 2*, 3 and 4 to the case of exact squares. As both the initiality and zigzag/lantern conditions are merely restatements of the existence of an isomorphism of distributors, it is clear that this is so. We state the results without proof, as the special case proved in the last section should enable the diligent reader to prove them himself.

**Proposition 1**
*The square*

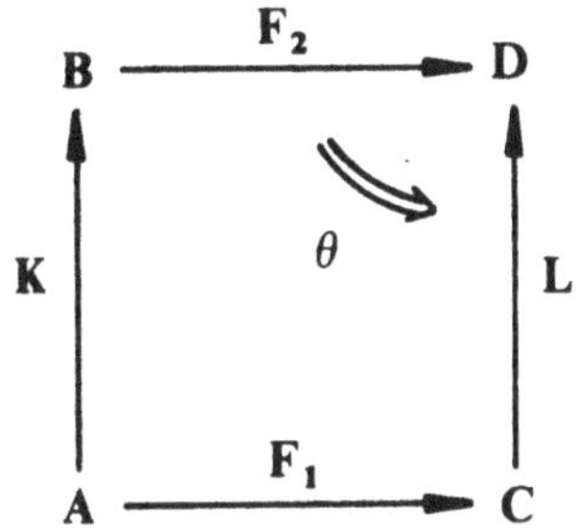


*is exact if and only if the following two conditions are satisfied:*

(i) *Given any objects $B$ in* **B**, *$C$ in* **C** *and a map $\gamma\colon \mathbf{F}_2B \to \mathbf{L}C$ in* **D**, *there is an object $A$ in* **A** *and maps $m\colon B \to \mathbf{K}A$, $n\colon \mathbf{F}_1A \to C$ such that*

$$\left(\mathbf{F}_2B \xrightarrow{\mathbf{F}_2(m)} \mathbf{F}_2\mathbf{K}A \xrightarrow{\theta(A)} \mathbf{L}\mathbf{F}_1A \xrightarrow{\mathbf{L}(n)} \mathbf{L}C\right) = \gamma,$$

*and*

(ii) *Given $m, n, A$ as above and $m', n', A'$ such that $\gamma = \mathbf{L}(n')\theta(A')\mathbf{F}_2(n')$, there is a zigzag*

$$A \to A_1 \leftarrow \cdots \to A_{2k-1} \leftarrow A'$$

*in* **A** *and a lantern*

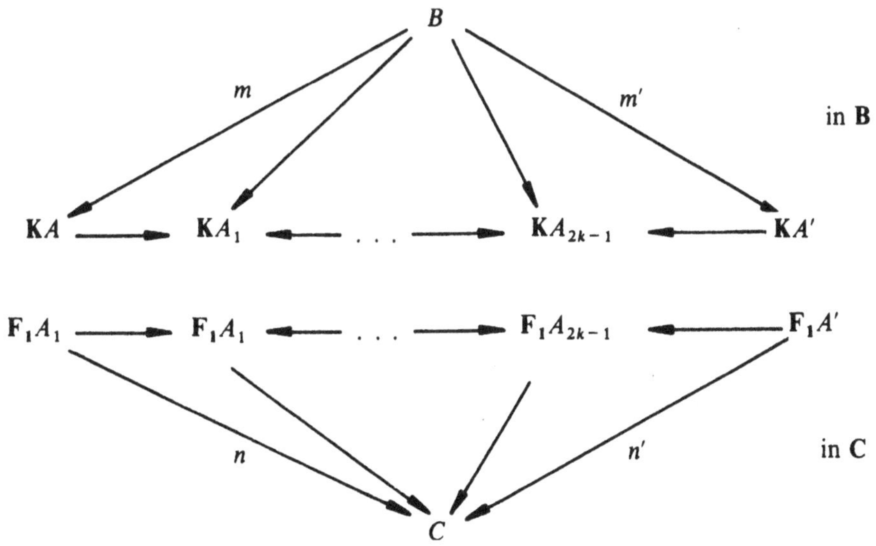

**Proposition 2**

*Let $B$ in an object of* $\mathbf{B}$, *and let*

$$\mathbf{H_B}\colon B\downarrow\mathbf{K} \to \mathbf{F}_2 B\downarrow\mathbf{L}$$

*be the functor which sends $(f, A)$ to $(\theta(A)\mathbf{F}_2(F), \mathbf{F}_1(A))$; then the square*

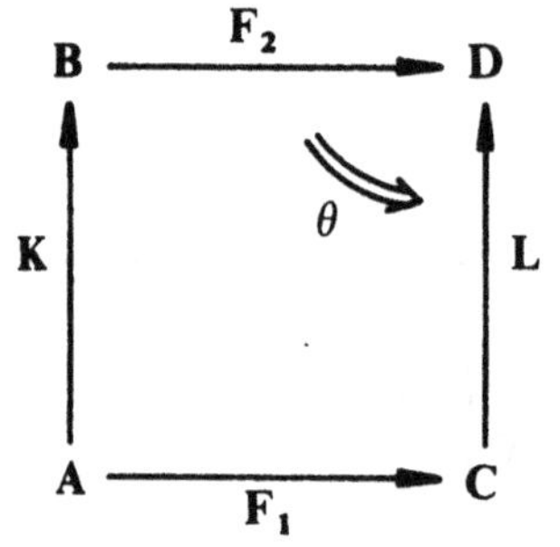


*is exact if and only if for each $B$ in* $\mathbf{B}$, $\mathbf{H_B}$ *is initial.*

Exact squares are exactly what is needed for inducing functors between shape categories. What one hopes for is that given an exact square

$$\mathbb{F} = (\mathbf{F}_1, \mathbf{L}, \theta, \mathbf{K}, \mathbf{F}_2),$$

one will obtain an induced functor

$$\mathbb{F}_*\colon \mathbf{S_K} \to \mathbf{S_L}$$

so that

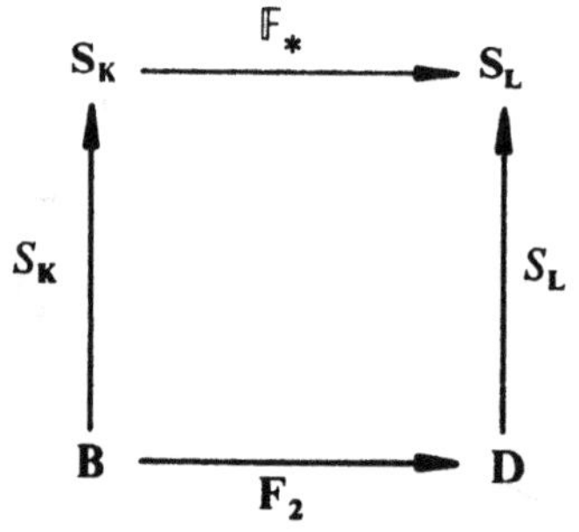


commutes. Of course, this tells us how $\mathbb{F}_*$ must be defined on the objects of $\mathbf{S_K}$ since $S_\mathbf{K}$ is the identity on objects. So we must take $\mathbb{F}^*(B) = \mathbf{F}_2(B)$ for each $B$ in $\mathbf{B}$. Now suppose $u\colon B \to B'$ is a morphism in $\mathbf{S_K}$, i.e. it is, in the Holsztyński model, a natural transformation

$$u\colon \mathbf{B}(B', \mathbf{K}-) \to \mathbf{B}(B, \mathbf{K}-);$$

this induces for each $A$ in $\mathbf{A}$, and $C$ in $\mathbf{C}$, a function

$$\mathbf{B}(B', \mathbf{K}A) \times \mathbf{C}(\mathbf{F}_1 A, C) \to \mathbf{B}(B, \mathbf{K}A) \times \mathbf{C}(\mathbf{F}_1 A, C),$$

hence via the formula for the composition of distributors, a natural morphism

$$\phi_\mathbf{K} \otimes \phi^{\mathbf{F}_1}(B', C) \to \phi_\mathbf{K} \otimes \phi^{\mathbf{F}_2}(B, C).$$

As the square is exact,

$$\phi_\mathbf{K} \otimes \phi^{\mathbf{F}_1} \cong \phi^{\mathbf{F}_2} \otimes \phi_\mathbf{L},$$

the isomorphism being induced by $\theta$, so we obtain a natural transformation

$$\phi^{\mathbf{F}_2} \otimes \phi_{\mathbf{L}}(B', -) \rightarrow \phi^{\mathbf{F}_2} \otimes \phi_{\mathbf{L}}(B, -),$$

i.e.

$$\mathbf{D}(\mathbf{F}_2 B', \mathbf{L}-) \rightarrow \mathbf{D}(\mathbf{F}_2 B, \mathbf{L}-)$$

and hence a shape morphism from $\mathbf{F}_2 B$ to $\mathbf{F}_2 B'$ in $\mathbf{S}_{\mathbf{L}}$. It is easy to check that this gives a functor

$$\mathbb{F}_*: \mathbf{S}_{\mathbf{K}} \rightarrow \mathbf{S}_{\mathbf{L}}$$

as hoped for. We summarise the result in the following proposition for future reference.

**Proposition 3**
*If* $\mathbb{F} = (\mathbf{F}_1, \mathbf{L}, \theta, \mathbf{K}, \mathbf{F}_2)$ *is an exact square,* $\mathbb{F}$ *induces a functor*

$$\mathbb{F}_*: \mathbf{S}_{\mathbf{K}} \rightarrow \mathbf{S}_{\mathbf{L}}$$

*satisfying* $\mathbb{F}^* S_{\mathbf{K}} = S_{\mathbf{L}} \mathbf{F}_2$.

If one prefers the description of $\mathbf{S}_{\mathbf{K}}$ using comma categories, $\mathbb{F}_*$ can easily be defined using the initial functors $\mathbf{H}_B$ and $\mathbf{H}_{B'}$, introduced in Proposition 2. It is also not too difficult to rephrase the above proof purely in terms of Kleisli categories of monad distributors. We leave both these descriptions to the reader.

It is useful to be able to compose exact squares. The functor induced by a composite square is the composite of their corresponding induced functors.

Given $\mathbb{F}_1 = (\mathbf{F}_{1,1}, \mathbf{K}_2, \theta_1, \mathbf{K}_1, \mathbf{F}_{2,1})$ and $\mathbb{F}_2 = (\mathbf{F}_{1,2}, \mathbf{K}_3, \theta_2, \mathbf{K}_2, \mathbf{F}_{2,2})$ are two exact squares, we can compose them in a natural and rather obvious way

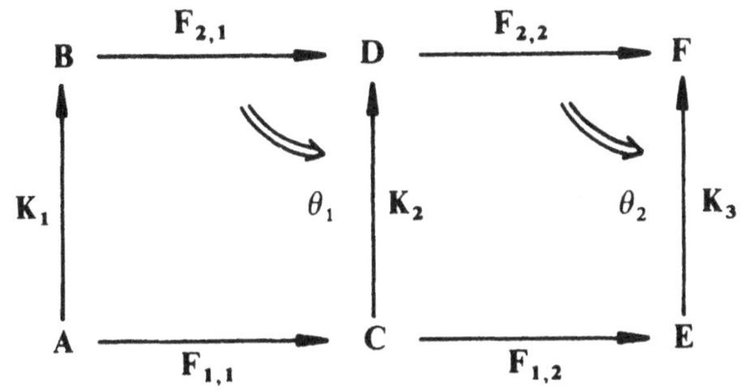


to obtain

$$\mathbb{F}_2 * \mathbb{F}_1 = (\mathbf{F}_{1,2}\mathbf{F}_{1,1}, \mathbf{K}_3, \theta_3, \mathbf{K}_1, \mathbf{F}_{2,2}\mathbf{F}_{2,1})$$

where $\theta_3 = \theta_2 \mathbf{F}_{1,1} \circ \mathbf{F}_{2,2}\theta_1$. The proof that $\mathbb{F}_2 * \mathbb{F}_1$ is exact is fairly simple:

$$\begin{aligned}
\phi_{\mathbf{K}_1} \otimes \phi^{\mathbf{F}_{1,2}\mathbf{F}_{1,1}} &\cong \phi_{\mathbf{K}_1} \otimes (\phi^{\mathbf{F}_{1,1}} \otimes \phi^{\mathbf{F}_{1,2}}) \\
&\cong (\phi_{\mathbf{K}_1} \otimes \phi^{\mathbf{F}_{1,1}}) \otimes \phi^{\mathbf{F}_{1,2}} \\
&\cong (\phi^{\mathbf{F}_{2,1}} \otimes \phi_{\mathbf{K}_2}) \otimes \phi^{\mathbf{F}_{1,2}} \\
&\cong \phi^{\mathbf{F}_{2,1}} \otimes (\phi_{\mathbf{K}_2} \otimes \phi^{\mathbf{F}_{1,2}}) \\
&\cong \phi^{\mathbf{F}_{2,1}} \otimes (\phi^{\mathbf{F}_{2,2}} \otimes \phi_{\mathbf{K}_3}) \\
&\cong (\phi^{\mathbf{F}_{2,1}} \otimes \phi^{\mathbf{F}_{2,2}}) \otimes \phi_{\mathbf{K}_3} \\
&\cong \phi^{\mathbf{F}_{2,2}\mathbf{F}_{2,1}} \otimes \phi_{\mathbf{K}_3}.
\end{aligned}$$

This isomorphism is easily checked to be that given by $\theta_3$, but as the calculation involves many applications of the coherence properties of the associativity isomorphisms, we leave such verification to the ultra diligent reader.

**Proposition 4**

*If $\mathbb{F}_2 * \mathbb{F}_1$ is defined, $(\mathbb{F}_2 * \mathbb{F}_1)_* = \mathbb{F}_{2*} \circ \mathbb{F}_{1*}$ and the identity exact square* $(\mathbf{Id}, \mathbf{K}, \mathrm{Id}, \mathbf{K}, \mathbf{Id})$ *induces the identity functor on $\mathbf{S_K}$.*

This result should be clear, given the construction of the induced functors.

This proposition can be viewed as saying that the shape category construction of the Holsztyński is, in fact, functorial when considered as having domain the category **Ex** whose objects are functors and whose morphisms from **K** to **L** are the exact squares with verticals **K** and **L**. In fact one can extend this functoriality to a 2-category level.

Using less high powered language, we are claiming the following:

**Proposition 5**

*Given exact squares*

$$\mathbb{F} = (\mathbf{F}_1, \mathbf{K}, \theta, \mathbf{L}, \mathbf{F}_2)$$

*and*

$$\mathbb{G} = (\mathbf{G}_1, \mathbf{K}, \phi, \mathbf{L}, \mathbf{G}_2)$$

*and two natural transformations*

$$\nu_1 : \mathbf{F}_1 \to \mathbf{G}_1, \qquad \nu_2 : \mathbf{F}_2 \to \mathbf{G}_2$$

*satisfying*

$$\mathbf{L}(\nu_1)\theta = \phi\nu_2(\mathbf{K})$$

*then there is a natural transformation $\nu_* : \mathbb{F}_* \to \mathbb{G}_*$ such that for any $B$ is $\mathbf{S_K}$, $\nu_*(\mathrm{B}) = S_{\mathbf{L}}(\nu_2(B))$.*

*Proof*

The conditions on $\nu_1$ and $\nu_2$ guarantee that the diagram

$$\begin{array}{ccc}
\phi^{\mathbf{F}_2} \otimes \phi_{\mathbf{L}} & \xrightarrow{\cong} & \phi_{\mathbf{K}} \otimes \phi^{\mathbf{F}_1} \\
\Big\downarrow \nu_2 \otimes \phi_{\mathbf{L}} & & \Big\downarrow \phi_{\mathbf{K}} \otimes \nu_1 \\
\phi^{\mathbf{G}_2} \otimes \phi_{\mathbf{L}} & \xrightarrow{\cong} & \phi_{\mathbf{K}} \otimes \phi^{\mathbf{G}_1}
\end{array}$$

commutes. The proof now follows by evaluating this square on $B$ and $B'$ linked by a **K**-shape morphism $u: B \to B'$. The details are again left to the reader.

## 5.4 EXAMPLES AND APPLICATIONS OF EXACT SQUARES

We have already seen two types of examples of exact square coming from absolute left and right Kan extensions. We start by looking at these in more detail.

**(*i*) *Absolute right Kan extensions***
The following is an alternative version of Proposition 4 of section 2.5. We have removed the restriction that the codomain be complete, but as a result have to require that the Kan extension be absolute.

**Proposition 1**
*Suppose* $\mathbf{R}$ *is the absolute right Kan extension of* $\mathbf{F}: \mathbf{A} \to \mathbf{C}$ *along* $\mathbf{K}: \mathbf{A} \to \mathbf{B}$; *then* $\mathbf{R}$ *is a shape invariant functor.*

*Proof*
The square $\mathbb{F} = (\mathbf{F}, \mathbf{K}, \varepsilon, \mathbf{Id}, \mathbf{R})$ is exact, so there is a functor

$$\mathbb{F}_*: \mathbf{S_K} \to \mathbf{S_{Id}} \cong \mathbf{C}$$

satisfying

$$\mathbb{F}_*\mathbf{S_K} = \mathbf{R}S_{\mathbf{Id}} = \mathbf{R},$$

but this merely states $\mathbf{R}$ is shape invariant in the sense of the definition given in section 2.5.

**(*ii*) *Absolute left Kan extensions***
If

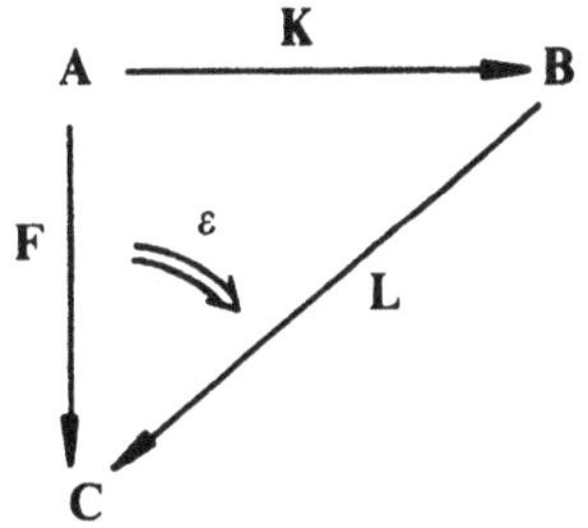


expresses $\mathbf{L}$ as an absolute left Kan extension of $\mathbf{F}$ along $\mathbf{K}$, then the square $\mathbb{F} = (\mathbf{K}, \mathbf{L}, \varepsilon, \mathbf{F}, \mathbf{Id})$, i.e.

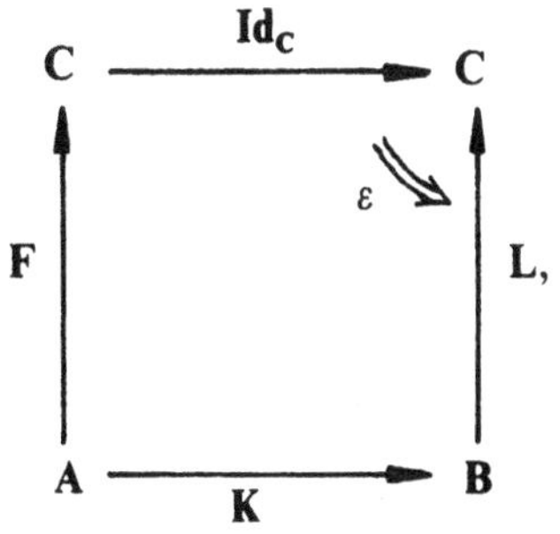

is exact, so one obtains an induced functor

$$\mathbb{F}_*: \mathbf{S_F} \to \mathbf{S_L}$$

so that $\mathbb{F}_* S_{\mathbf{F}} = S_{\mathbf{L}}$.

There are one or two interesting special cases of this. Firstly the case when $\varepsilon$ is the identity natural transformation and so $\mathbf{F} = \mathbf{LK}$. This is a typical change of models situation resulting in an induced functor

$$\mathbf{K}^*: \mathbf{S_L} \to \mathbf{S_F}.$$

It is natural to ask what the composition $\mathbb{F}_*\mathbf{K}^*$ is.

**Proposition 2**
*If* $\mathbb{F} = (\mathbf{K}, \mathbf{L}, \mathrm{Id}, \mathbf{F}, \mathbf{Id})$ *is exact, then* $\mathbb{F}_*: \mathbf{S_F} \to \mathbf{S_L}$ *is an isomorphism of categories with inverse* $\mathbf{K}^*$.

*Proof*
Let $u: \mathbf{C}(Y, \mathbf{F}-) \to \mathbf{C}(X, \mathbf{F}-)$ represent a morphism in $\mathbf{S_F}$; then by definition

$$\mathbf{K}^*(\mathbb{F}_*(u))(A) = \mathbb{F}_*(u)(\mathbf{K}A),$$

but the defining diagram for this latter map is

$$\begin{array}{ccc}
\phi_{\mathbf{F}} \otimes \phi^{\mathbf{K}}(Y, \mathbf{K}A) & \xrightarrow[\bar{\theta}(Y, \mathbf{K}A)]{\cong} & \phi_{\mathbf{L}}(Y, \mathbf{K}A) = \mathbf{C}(Y, \mathbf{F}A) \\
\Big\downarrow \tilde{u}(Y, \mathbf{K}A) & & \Big\downarrow \mathbb{F}_*(u)(\mathbf{K}A) \\
\phi_{\mathbf{F}} \otimes \phi^{\mathbf{K}}(X, \mathbf{K}A) & \xrightarrow[\bar{\theta}(X, \mathbf{K}A)]{\cong} & \phi_{\mathbf{L}}(X, \mathbf{K}A) = \mathbf{C}(X, \mathbf{F}A)
\end{array}$$

where for $x \in \mathbf{C}(Y, \mathbf{F}A_1))$, $y \in \mathbf{B}(\mathbf{K}A_1, \mathbf{K}A)$ (so that $x \otimes y \in \phi_{\mathbf{F}} \otimes \phi^{\mathbf{K}}(Y, \mathbf{K}A)$)

$$\tilde{u}(Y, \mathbf{K}A)(x \otimes y) = (u(A_1)x) \otimes y \in \phi_{\mathbf{F}} \otimes \phi^{\mathbf{K}}(X, \mathbf{K}A).$$

As taking the dotted arrow to be $u(A)$ itself also makes the diagram commutative, we must have

$$\mathbf{K}^*(\mathbb{F}_*(u)) = u.$$

Similarly for $v: \mathbf{C}(Y, \mathbf{L}-) \to \mathbf{C}(X, \mathbf{L}-)$, we calculate $\mathbb{F}_*\mathbf{K}^*(v)$. The diagram this time is

$$\begin{array}{ccc}
\phi_{\mathbf{F}} \otimes \phi^{\mathbf{K}}(Y, B) & \xrightarrow{\cong} & \phi_{\mathbf{L}}(Y, B) = \mathbf{C}(Y, \mathbf{L}B) \\
{\scriptstyle \tilde{v}(Y, C)} \downarrow & & \downarrow {\scriptstyle \mathbb{F}_* \mathbf{K}^*(v)(B)} \\
\phi_{\mathbf{F}} \otimes \phi^{\mathbf{K}}(X, B) & \xrightarrow{\cong} & \phi_{\mathbf{L}}(Y, B) = \mathbf{C}(X, \mathbf{L}B)
\end{array}$$

where for

$$x \in \mathbf{C}(Y, \mathbf{F}A) = \mathbf{C}(Y, \mathbf{LK}A),$$

$$y \in \mathbf{B}(\mathbf{K}A, B),$$

$$\tilde{v}(Y, C)(x \otimes y) = (v(\mathbf{K}A)x) \otimes y \in \phi_{\mathbf{F}} \otimes \phi^{\mathbf{K}}(X, B)$$

but again $v(B)$ makes the diagram commute so

$$\mathbb{F}_* \mathbf{K}^* v = v.$$

As both functors are the identities on objects, this completes the proof.

**Corollary**
*Suppose* $\mathbb{F} = (\mathbf{K}, \mathbf{L}, \mathrm{Id}, \mathbf{F}, \mathbf{Id})$ *is exact, that* $\mathbf{G} \dashv \mathbf{F}(\eta, \varepsilon)$ (i.e. $\mathbf{G}$ *is left adjoint to* $\mathbf{F}$ *with counit* $\eta$ *and unit* $\varepsilon$) *and that*

$$\mathbb{G} = (\mathbf{G}, \mathbf{F}, \mathbf{L}\eta, \mathbf{L}, \mathbf{Id})$$

*is exact; then* $\mathbb{G}_* = \mathbf{F}^*$.

*Proof*
By Proposition 5 of the last section, the transformation of exact squares $(\eta, \mathrm{Id})$: $\mathbf{Id} \to \mathbb{F}\mathbb{G}$ yields a natural transformation

$$\eta_* : \mathbf{Id} \to \mathbb{F}_* \mathbb{G}_*$$

and

$$\eta_*(C) = S_{\mathbf{L}}(\mathrm{Id}_C) = \mathrm{Id}_C$$

so $\eta_*$ is the identity natural transformation and $\mathbb{F}^* = \mathbb{G}_*$.

**Remark**
These last two results should not come as any great surprise, since if $\mathbb{F} = (\mathbf{K}, \mathbf{L}, \mathrm{Id}, \mathbf{F}, \mathbf{Id})$ is exact, then each morphism from $B$ to $\mathbf{L}C$ factors through one from $B$ to some $\mathbf{LK}A = \mathbf{F}A$, i.e. objects from $\mathbf{A}$ give better approximations than those from $\mathbf{C}$. Thus all $\mathbf{L}$-shape theoretic information on $\mathbf{B}$ is contained in its $\mathbf{F}$-shape.

Of special interest in the above is the case of an opaque functor. If $\mathbf{K}: \mathbf{A} \to \mathbf{B}$ is opaque, then clearly

$$\begin{array}{ccc} \mathbf{B} & \xrightarrow{=} & \mathbf{B} \\ {\scriptstyle \mathbf{K}}\uparrow & \diagdown\!\!\diagdown & \uparrow{\scriptstyle \mathbf{L}} \\ \mathbf{A} & \xrightarrow[]{\mathbf{K}'} & \mathbf{B_K} \end{array}$$

is exact and Proposition 2 gives that $\mathbf{S_K}$ and $\mathbf{S_L}$ are isomorphic. As the axioms for a shape theory are fairly trivially satisfied for $\mathbf{L}$ a full inclusion (see the Corollary to Proposition 1 section 2.1), this provides a separate proof that for opaque functors, the Holsztyński construction does yield a shape theory.

**(iii) *'Classical' examples of exact squares***

There are many examples of exact squares coming from fairly elementary category theory. The most obvious ones are the identity squares

$$\begin{array}{ccc} \mathbf{B} & \xrightarrow{=} & \mathbf{B} \\ {\scriptstyle \mathbf{K}}\uparrow & & \uparrow{\scriptstyle \mathbf{K}} \\ \mathbf{A} & \xrightarrow[=]{} & \mathbf{A} \end{array} \quad \text{and} \quad \begin{array}{ccc} \mathbf{A} & \xrightarrow[\mathbf{F}]{} & \mathbf{C} \\ {\scriptstyle =}\uparrow & & \uparrow{\scriptstyle =} \\ \mathbf{A} & \xrightarrow[\mathbf{F}]{} & \mathbf{C} \end{array}$$

These induce the identity functor on $\mathbf{S_K}$ and $\mathbf{F}$ itself respectively.

Of more interest are the exact squares coming from partial adjoints and adjoint pairs.

Given functors

$$\mathbf{A} \xrightarrow{\mathbf{G}} \mathbf{D}, \quad \mathbf{A} \xrightarrow{\mathbf{F}} \mathbf{C}$$

and

$$\mathbf{L}: \mathbf{C} \to \mathbf{D}$$

and a natural transformation $\phi: \mathbf{G} \to \mathbf{LF}$, one says that $\mathbf{F}$ is *partially adjoint* to $\mathbf{L}$ *along* $\mathbf{G}$ (we will write

$$\mathbf{F} \dashv_{\mathbf{G}} \mathbf{L}(\phi))$$

if given any $f: \mathbf{G}A \to \mathbf{L}C$, the category of factorisations of $f$ as $\mathbf{L}(c)\phi(A)$ for $c: \mathbf{F}A \to C$, is non-empty and connected.

If $\mathbf{G}$ is the identity that reduces to the ordinary situation with $\phi: \mathbf{Id} \to \mathbf{LF}$, the unit of the adjunction. (One can also formulate the condition in terms of $\phi$ being a relative counit.)

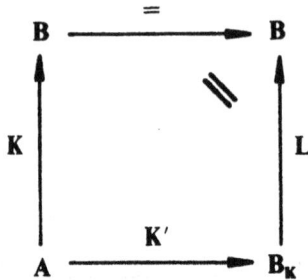

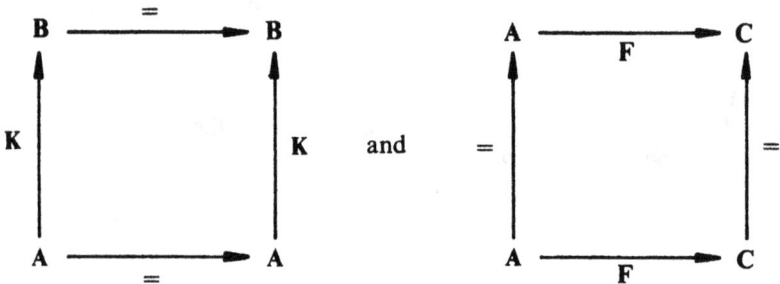

In terms of exact squares

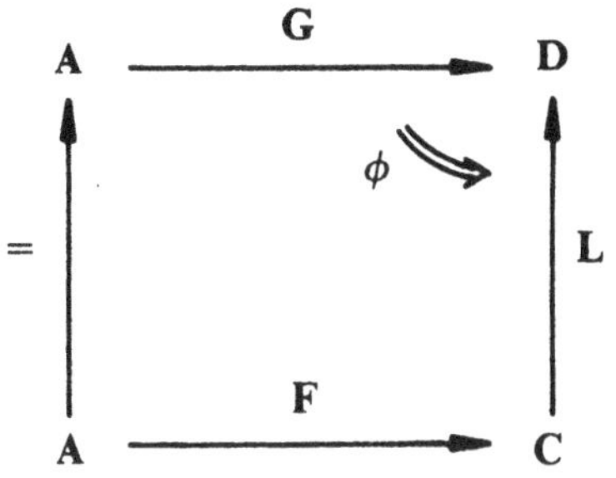

$$
\begin{array}{ccc}
A & \xrightarrow{\ \ G\ \ } & D \\
\big\uparrow{\scriptstyle =} & \overset{\phi}{\Rightarrow} & \big\uparrow{\scriptstyle L} \\
A & \xrightarrow{\ \ F\ \ } & C
\end{array}
$$

is exact if and only if $F \dashv_{G} L(\phi)$, whereas the square

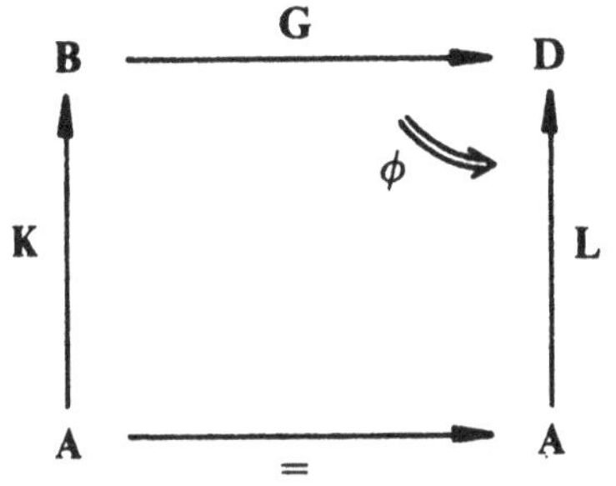

$$
\begin{array}{ccc}
B & \xrightarrow{\ \ G\ \ } & D \\
\big\uparrow{\scriptstyle K} & \overset{\phi}{\Rightarrow} & \big\uparrow{\scriptstyle L} \\
A & \xrightarrow{\ \ =\ \ } & A
\end{array}
$$

is exact if and only if $G \dashv_{L} K$ but with $\phi$ being counit, $\phi: GK \to L$. Thus the ordinary cases are

(i)

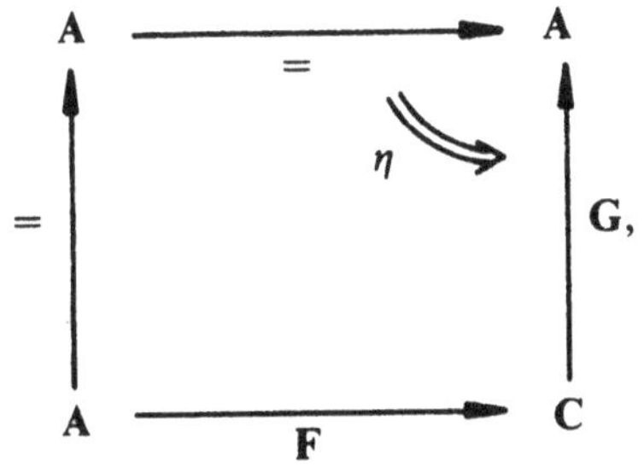

$$
\begin{array}{ccc}
A & \xrightarrow{\ \ =\ \ } & A \\
\big\uparrow{\scriptstyle =} & \overset{\eta}{\Rightarrow} & \big\uparrow{\scriptstyle G,} \\
A & \xrightarrow{\ \ F\ \ } & C
\end{array}
\qquad\text{i.e. } \mathbf{F}\dashv\mathbf{G}(\eta,\varepsilon)
$$

and (ii)

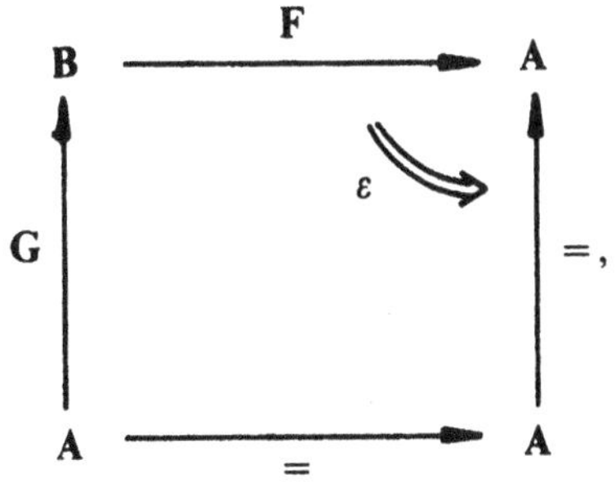

$$
\begin{array}{ccc}
B & \xrightarrow{\ \ F\ \ } & A \\
\big\uparrow{\scriptstyle G} & \overset{\varepsilon}{\Rightarrow} & \big\uparrow{\scriptstyle =,} \\
A & \xrightarrow{\ \ =\ \ } & A
\end{array}
\qquad\text{i.e. } \mathbf{F}\dashv\mathbf{G}(\eta,\varepsilon).
$$

We will denote these exact squares by $\mathbb{F}(\mathrm{i})$ and $\mathbb{F}(\mathrm{ii})$ for the rest of this section.

Note that in each case the shape category is a Kleisli category. In (i), $\mathbf{S_G} \cong \mathbf{A_T}$ for $\mathbf{T} = \mathbf{GF}$ and in (ii), $\mathbf{S_G} = \mathbf{B_T}$ again with $\mathbf{T} = \mathbf{GF}$.

The induced functors are

$$\mathbb{F}(\mathrm{i})_*: \mathbf{A} \to \mathbf{A_T}$$

$$\mathbb{F}(\mathrm{ii})_*: \mathbf{B_T} \to \mathbf{A}.$$

$\mathbb{F}(\mathrm{i})_*$ is the 'free' object functor denoted $\mathbf{F_T}$ in section 2.2(b). To identify $\mathbb{F}(\mathrm{ii})_*$, we compose squares as follows:

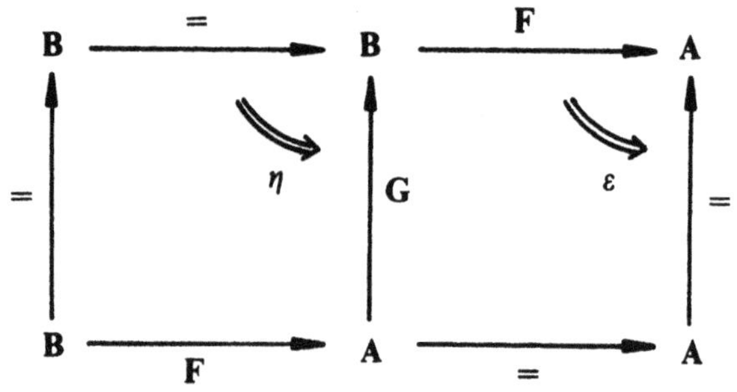


to obtain

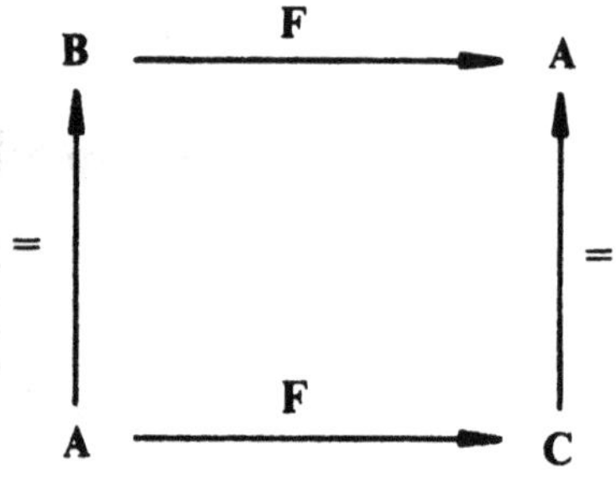


so that

$$\mathbb{F}(\mathrm{ii})_* \mathbf{F_T} = \mathbf{F}.$$

This leads one to suspect that $\mathbb{F}(\mathrm{ii})_*$ is the comparison functor for this situation. Let us pause to recall what this means.

In Chapter 2, we discussed briefly the construction of the Kleisli category of a monad. This was interpretable as the category of free algebras of some type, where 'algebra' itself is to be interpreted in a very general way (see MacLane [71], p. 136). Now given any functor $\mathbf{G}: \mathbf{C} \to \mathbf{D}$ with a left adjoint $\mathbf{F}: \mathbf{D} \to \mathbf{C}$, one has a monad $\mathbf{T}$ generated by the adjunction and there is a diagram

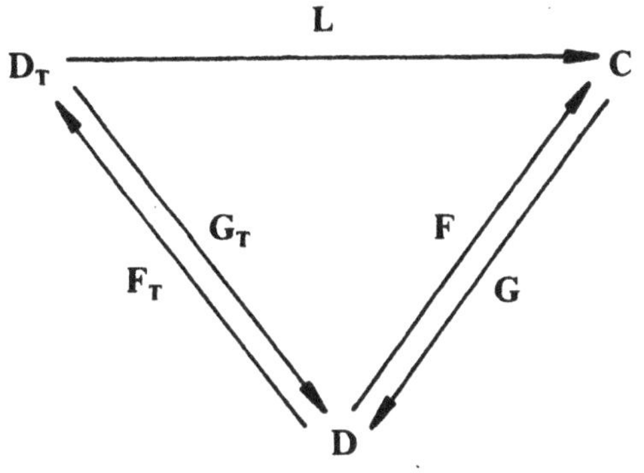


whence $\mathbf{D_T}$ can be considered as the category of free objects (relative to the adjunction) in **C** (see MacLane [71], p. 144 for a more precise statement of this). This functor **L** is called the *comparison functor* for the monad, **T**, and it is the unique functor satisfying $\mathbf{GL} = \mathbf{G_T}$ and $\mathbf{LF_T} = \mathbf{F}$.

To check that $\mathbb{F}(\mathrm{ii})_*$ is the comparison functor, it will thus suffice to prove that $\mathbf{G}\mathbb{F}(\mathrm{ii})_* = \mathbf{G_T}$.

We will consider the composite

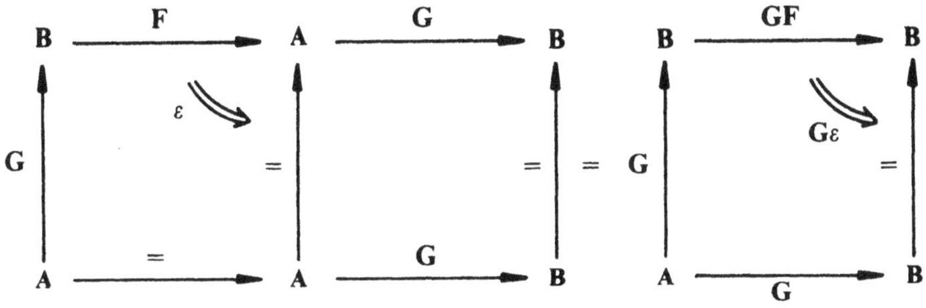


We denote the composite square by $\mathbb{F}(\mathrm{iii})$. We have

$$\mathbf{G}\mathbb{F}(\mathrm{ii}) = \mathbb{F}(\mathrm{iii})$$

so it remains to identify $\mathbb{F}(\mathrm{iii})_*\colon \mathbf{B_T} \to \mathbf{B}$. Recall that $\mathbf{B_T}$ has as objects the objects of **B** and a map $f\colon X \to Y$ in $\mathbf{B_T}$ is a map $f\colon X \to \mathbf{T}Y = \mathbf{GF}Y$ within **B**. $\mathbf{G_T}$ on objects is then given by $\mathbf{G_T}(X) = \mathbf{T}X$ and on morphisms $\mathbf{G_T}(f) = \mu_Y \mathbf{T}(f)\colon \mathbf{T}X \to \mathbf{T}^2 Y \to \mathbf{T}Y$.

Comparing we find $\mathbb{F}(\mathrm{iii})_*$ is certainly $\mathbf{GF} = \mathbf{T}$ on objects and if $f\colon X \to \mathbf{T}Y$ in $\mathbf{B_T}$, we consider $f$ as an object in $X \downarrow \mathbf{G}$, namely $(f, \mathbf{F}Y)$, and apply the initial functor

$$\mathbf{H}_X\colon X \downarrow \mathbf{G} \to \mathbf{GF}X \downarrow \mathbf{B}$$

corresponding to the exact square (Proposition 2 of section 5.3) and given by

$$\mathbf{H}_X(X \xrightarrow{g} \mathbf{G}A, A) = (\mathbf{G}\varepsilon(A)\mathbf{GF}(g), \mathbf{G}(A)).$$

Applied to $(f, \mathbf{F}Y)$, this gives

$$\mathbf{GF}X \xrightarrow{\mathbf{GF}(f)} \mathbf{GFGF}(Y) \xrightarrow{\mathbf{G}\varepsilon\mathbf{F}(Y)} \mathbf{GF}Y$$

or

$$\mathbf{T}X \xrightarrow{\mathbf{T}(f)} \mathbf{T}^2(Y) \xrightarrow{\mu_Y} \mathbf{T}Y$$

since by construction $\mu = \mathbf{G}\varepsilon\mathbf{F}$. Thus $\mathbf{H}_X(f, \mathbf{F}Y)$ is $\mathbf{G}_\mathbf{T}(f)$. It is now clear that $\mathbb{F}(\text{iii})_*(f)$ and $\mathbf{G}_\mathbf{T}(f)$ are the same.

Thus both the free and the comparison functors associated with the Kleisli category can be obtained as functors induced from exact squares.

## 5.5 SOME TOPOLOGICAL EXAMPLES OF EXACT SQUARES WITH APPLICATIONS TO SUSPENSIONS OF (POINTED) SHAPES

Suppose given a functor $\mathbf{K}: \mathbf{W} \to \mathbf{H}$, as we originally considered, where **W** and **H** are homotopy categories of pointed spaces and **K** is an inclusion functor. One natural question to ask if whether familiar constructions such as the suspension of a space are **K**-shape invariant.

Recall that given a pointed space, $(X, *)$, the reduced suspension, $\Sigma X$, of $X$ is the space obtained from $X \times [0, 1]$ by identifying $X \times \{0\}$, $X \times \{1\}$ and $* \times [0, 1]$ to a single point. Put another way, the question asked above is then the following: suppose $X$ and $Y$ are **K**-shape equivalent (i.e. they are isomorphic in $\mathbf{S_K}$), is it true that $\Sigma X$ and $\Sigma Y$ are **K**-shape equivalent?

The very minimum for this to happen must be that $\Sigma$ gives a functor $\Sigma: \mathbf{H} \to \mathbf{H}$ restricting to $\Sigma: \mathbf{W} \to \mathbf{W}$ on **W**, i.e. that both **H** and **W** are closed under suspension. This, in turn, can be rephrased as saying that a commutative diagram

$$\begin{array}{ccc} \mathbf{H} & \xrightarrow{\Sigma} & \mathbf{H} \\ {\scriptstyle \mathbf{K}}\uparrow & \diagdown\!\!\diagdown & \uparrow{\scriptstyle \mathbf{K}} \\ \mathbf{W} & \xrightarrow{\Sigma} & \mathbf{W} \end{array}$$

exists.

This should suggest an obvious condition for suspension to respect **K**-shape, namely that the above square be exact since this will mean that it induces a functor

$$\Sigma: \mathbf{S_K} \to \mathbf{S_K}$$

just as we require.

Thus we are reduced to finding conditions on **H** and **W** which will imply exactness of this square. To do this it helps to 'domesticate' the exactness condition by finding conditions stronger than it that are more easily checked in this topological situation or at least involve concepts which are nearer to well known concepts of topology.

To do this we will use the zigzag/lantern version of exactness (Proposition 1 of section 5.3).

Given the square

(𝔽)

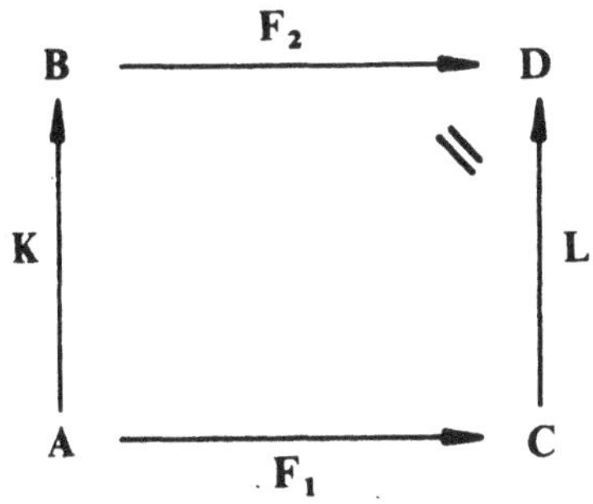


(we assume $\theta$ is the identity, as it is only this case we need), 𝔽 is exact if and only if

(i) for any object $B$ in **B**, $C$ in **C** and map $\gamma: \mathbf{F}_2 B \to \mathbf{L}C$ in **D**, there is an object $A$ in **A** and maps $m: B \to \mathbf{K}A$, $n: \mathbf{F}_1 A \to C$ such that $\gamma = \mathbf{L}(n)\mathbf{F}_2(m)$ and
(ii) the zigzag/lantern condition.

Now the zigzag/lantern condition is very general, but two particular cases immediately strike one as being more intuitive and easier to check:

Given $m$, $n$, $A$ as in (i) above and $m'$, $n'$, $A'$ such that $\gamma = \mathbf{L}(n')\mathbf{F}_2(n')$ there is a zigzag of type

$$\cdot \to \cdot \leftarrow \cdot$$

(resp. of type

$$\cdot \leftarrow \cdot \to \cdot)$$

and a lantern

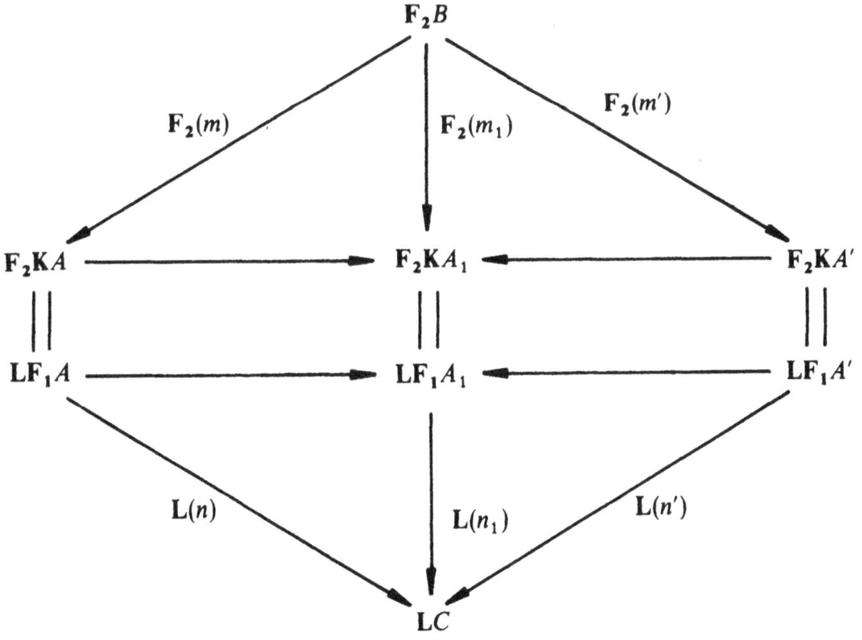


(resp. with the central arrows reversed.)

In our case we will restrict attention to both $\mathbf{F}_1$ and $\mathbf{F}_2$ being suspension $\Sigma$, with $\mathbf{K} = \mathbf{L}$ being the inclusion of a full subcategory, $\mathbf{K}\colon \mathbf{W} \to \mathbf{H}$, of the homotopy category. Thus the conditions are

(i) given any map

$$\gamma\colon \Sigma X \to \mathbf{K}P$$

with $X$ in $\mathbf{H}$ and $P$ in $\mathbf{W}$, $\gamma$ factors are

$$\Sigma X \xrightarrow{\Sigma m} \Sigma\mathbf{K}Q \xrightarrow{\mathbf{K}n} \mathbf{K}P$$

where $m\colon X \to \mathbf{K}Q$ is in $\mathbf{H}$. (Here we use $\Sigma\mathbf{K} = \mathbf{K}\Sigma$.)

(ii) given two such factorisations $(m, n)$ and $(m', n')$, one has one of the two lanterns

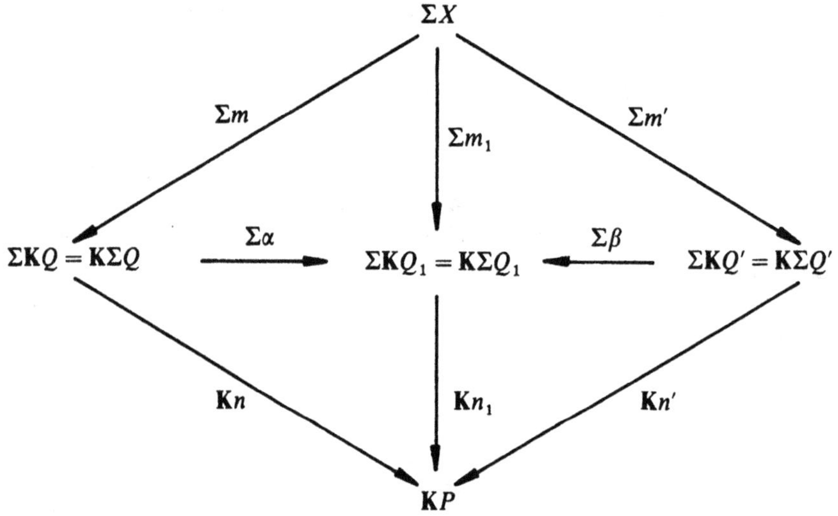


or

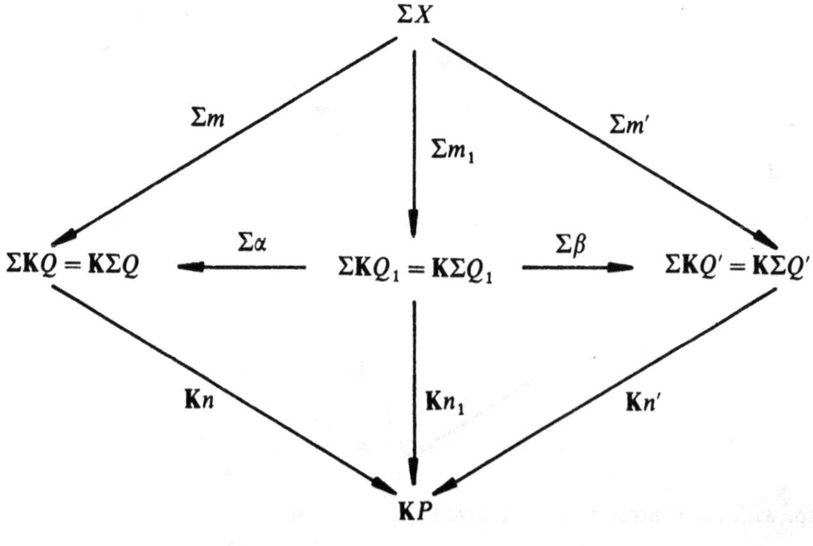

The two lanterns correspond to the two possible attacks on the problem of verifying this condition. The upper one can be viewed as an existence of a pushout-type condition on the corner

$$\Sigma Q \longleftarrow \Sigma X \longrightarrow \Sigma Q'$$

The lower diagram asks for a pullback-type condition on

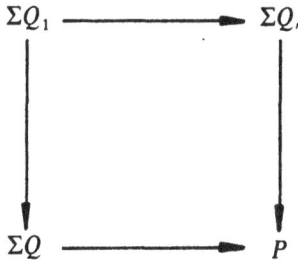

$$\begin{array}{ccc} \Sigma Q_1 & \longrightarrow & \Sigma Q, \\ \downarrow & & \downarrow \\ \Sigma Q & \longrightarrow & P \end{array}$$

Note, however, that in practice we are working in homotopy categories **H** and **W**, and so pullbacks and pushouts do not exist. The saving virtue of the above diagrams is that they do not demand uniqueness of some induced map, only existence, that is to say, they correspond to *weak pushout and pullback* conditions.

The factorisation condition suggests the use of loop spaces. We recall that on many categories of pointed topological spaces, $\Sigma$ has a right adjoint $\Omega$, the loopspace functor. Explicitly $\Omega X$ is the space of loops in $X$, that is, the set of all maps from the unit interval, $I$, to $X$ mapping both endpoints of $I$ to the basepoint of $X$ together with a suitable topology (e.g. compact open).

The classical adjunction

$$[\Sigma X, P] \cong [X, \Omega P]$$

ensures that $\gamma\colon \Sigma X \to \mathbf{K}P$ can be factored as

$$\Sigma X \xrightarrow[\Sigma f']{} \Sigma\Omega P \xrightarrow[\varepsilon_P]{} P$$

where $f'$ is adjoint to $f$ and $\varepsilon_P$ is the counit of the adjunction. For this factorisation to be suitable for us, we need that $\Omega P$ is in **W** if $P$ is there. This together with a weak pullback-type condition seems to give a reasonable set of conditions which should guarantee that $\Sigma$ induces a functor on shape. However, we can reduce these conditions still further and by weakening them make them applicable to a wider class of pairs (**H**, **W**). This we will do next.

Our first reduction uses the following terminology (due to Deleanu and Hilton in this context):

We say that a full subcategory $\mathbf{T}_1$, of the category $\mathbf{Top}_*$ of based topological spaces, is an *admissible subcategory* if it contains the one-point space, $*$, is closed under the formation of mapping cones and satisfies the condition that if $X$ is an object in $\mathbf{T}_1$ and $Y$ has the same based homotopy type as $X$, then $Y$ is also in $\mathbf{T}_1$ (we might briefly say that $\mathbf{T}_1$ contains entire (based) homotopy types).

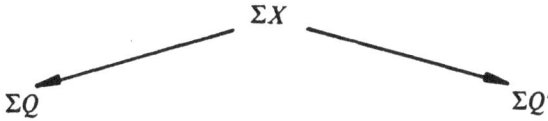

It probably pays to recall the definition of a mapping cone on a map $f: X \to Y$. The *reduced cone* on $X$ is the space $CX = X \times I/(X \times \{1\} \cup \{x_0\} \times I)$. $X$ is embeddable in $CX$ via the map sending $x$ to the equivalence class of $(x, 0)$. The mapping cone $C_f$ is given by the pushout

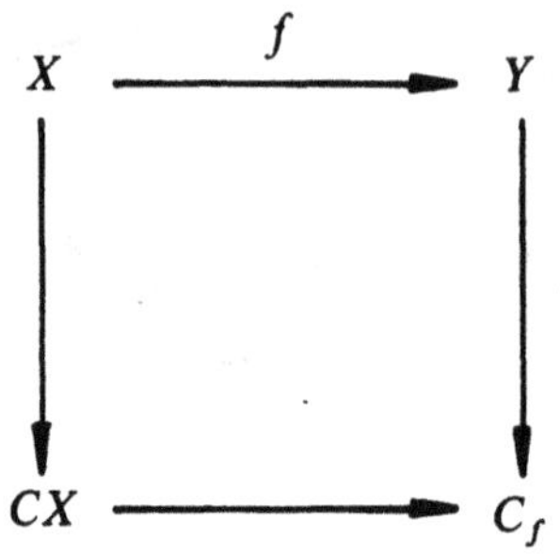

$$
\begin{array}{ccc}
X & \xrightarrow{\ f\ } & Y \\
\downarrow & & \downarrow{\scriptstyle j} \\
CX & \longrightarrow & C_f
\end{array}
$$

We assume $(T_0, T_1)$ is a pair of admissible subcategories of **Top$_*$** with $T_0 \subset T_1$ and we are looking at the shape theory of the functor

$$\mathbf{K} \colon \mathbf{HoT_0} \to \mathbf{HoT_1}$$

induced by inclusion on the corresponding homotopy categories.

We shall say that $T_0$ has property I with respect to $T_1$ if given a diagram

$$
\begin{array}{ccc}
 & & X \\
 & & \downarrow{\scriptstyle f} \\
Z & \xrightarrow{\ g\ } & Y
\end{array}
$$

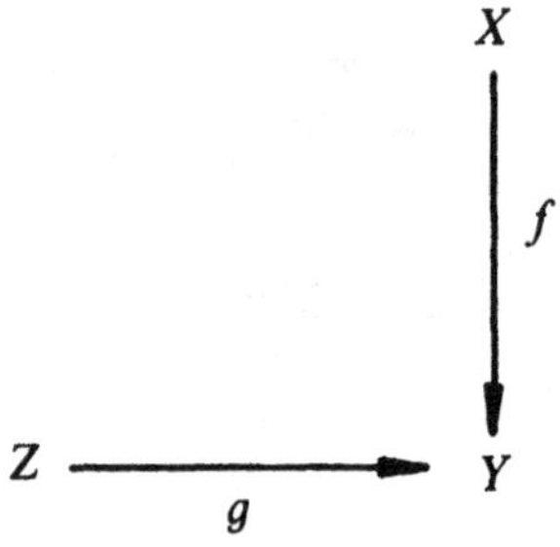

with $f, g$ both fibrations, then the pullback in $T_1$ exists and is in $T_0$.

**Lemma**
*If $T_0$ is admissible and has property I with respect to $T_1$, then $T_0$ is closed with respect to finite products and to construction of loop spaces.*

*Proof*
Of course, one can obtain $X \times Y$ by pulling back

$$
\begin{array}{ccc}
 & & X \\
 & & \downarrow \\
Y & \longrightarrow & *
\end{array}
$$

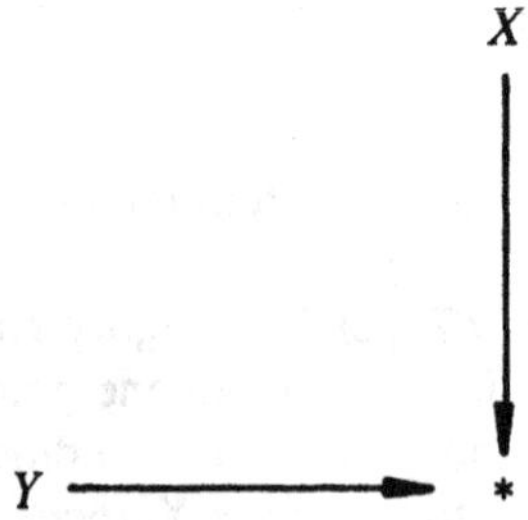

so the first part follows.

Since $*$ is in $\mathbf{T}_0$ and $\mathbf{T}_0$ contains entire homotopy types, it contains the contractible space $\Gamma X$ of based paths in $X$. Pulling back in the diagram

$$\begin{array}{ccc} & & \Gamma X \\ & & \big\downarrow{\scriptstyle \mathrm{ev}_1} \\ \Gamma X & \xrightarrow[\mathrm{ev}_1]{} & X \end{array}$$

(where if $\lambda\colon I \to X$, with $\lambda(0) = *$, is an element of $\Gamma X$, $\mathrm{ev}_1(\lambda) = \lambda(1)$), by property I, the pullback, which has the homotopy types of $\Omega X$, is also in $\mathbf{T}_0$.

**Proposition 1**
*If* $\mathbf{T}_0$ *has property I with respect to* $\mathbf{Top}_*$, *then* $\mathbf{HoT}_0$ *has weak local pullbacks relative to* $\mathbf{HoTop}_*$.

*Proof*
We start by making explicit what we have to prove. Suppose we are given a corner

$$\begin{array}{ccc} & & A \\ & & \big\downarrow{\scriptstyle \phi} \\ B & \xrightarrow[\psi]{} & X \end{array}$$

in $\mathbf{HoT}_0$, we have to show that there is a commutative square in $\mathbf{HoT}_0$

$$\begin{array}{ccc} T & \xrightarrow{\alpha} & A \\ {\scriptstyle \beta}\big\downarrow & & \big\downarrow{\scriptstyle \phi} \\ B & \xrightarrow[\psi]{} & X \end{array}$$

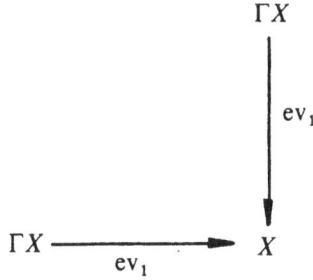

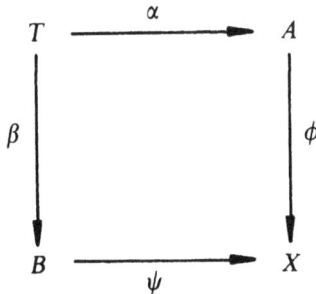

such that given any commutative square in **HoTop**$_*$

$$\begin{array}{ccc} T' & \xrightarrow{\gamma} & A \\ {\scriptstyle\delta}\downarrow & & \downarrow{\scriptstyle\phi} \\ B & \xrightarrow[\psi]{} & X \end{array}$$

there is a $\rho: T' \to T$ satisfying $\gamma = \alpha\rho$, $\delta = \beta\rho$.

We start by picking representative maps

$$f: A \to X$$

$$g: B \to X$$

in the classes $\phi$, $\psi$ respectively (so $\phi = [f]$, $\psi = [g]$).

It is well known that a map such as $f: A \to X$ can be factored as

$$A \xrightarrow{r} M^f \xrightarrow{f'} X$$

with $r$ a homotopy equivalence and $f'$ a fibration. We have to check that this is possible within $\mathbf{T_0}$ (this will also serve as an introduction to the general result for those unfamiliar with it, although we will omit verification of some of the details).

We take a path space, $X^I$, with base point the constant path at $*$. ($X^I$ is in $\mathbf{T_0}$ since it has the same homotopy type as $X$.) Using the evaluation map

$$\mathrm{ev}_0: X^I \to X,$$

which is a fibration, we form a pullback (within **Top**$_*$)

$$\begin{array}{ccc} M^f & \xrightarrow[\pi^f]{} & X^I \\ {\scriptstyle j^f}\downarrow & & \downarrow{\scriptstyle \mathrm{ev}_0} \\ A & \xrightarrow[f]{} & X \end{array}$$

Using the constant path map $X \xrightarrow{c} X^I$, which is left inverse to $\mathrm{ev}_0$, one can construct (within **Top**$_*$) using pullbacks, a left inverse $p^f$ for $j^f$ and a homotopy: $p^f j^f \simeq$ Identity on $M^f$. Thus $j^f$ and $p^f$ are homotopy inverse to each other and so $M^f$ is in $\mathbf{T_0}$. $p^f$ will be the $r$ in our factorisation. (One can give an explicit description of $M^f$

$$M^f = \{(a, \lambda) | f(a) = \lambda(0)\} \subseteq A \times X^I,$$

and $p^f(a) = (a$, constant path at $f(a))$.) Now taking $f' = \mathrm{ev}_1 \pi^f$ gives

$$\begin{aligned} f'p^f(a) &= f'(a, \sigma(f(a))) \\ &= \mathrm{ev}_1(\sigma(f(a))) \\ &= f(a), \end{aligned}$$

hence $f'r = f$ as required. We omit the proof that $f'$ is a fibration as this is easy to find in standard texts on algebraic topology.

Thus we have

$$A \xrightarrow{r} M^f \xrightarrow{f'} X = A \xrightarrow{f} X,$$

similarly for $g$,

$$B \xrightarrow{s} M^g \xrightarrow{g'} X = B \xrightarrow{g} X.$$

We next apply property I to the corner of fibrations

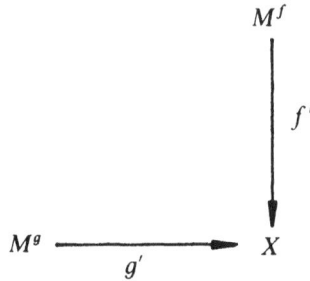

$$\begin{array}{ccc} & & M^f \\ & & \downarrow{\scriptstyle f'} \\ M^g & \xrightarrow[g']{} & X \end{array}$$

to obtain

$$\begin{array}{ccc} T & \xrightarrow{a'} & M^f \\ {\scriptstyle b'}\downarrow & & \downarrow{\scriptstyle f'} \\ M^g & \xrightarrow[g']{} & X \end{array}$$

We set $\alpha = [j^f a']$, $\beta = [j^g b']$. Of course one has $\phi\alpha = [fj^f a'] = [\mathrm{ev}_0 \pi^f a'] = [\mathrm{ev}_1 \pi^f a'] = [f'a']$. Similarly $\psi\beta = [g'b']$, so the square is commutative. It remains to check that it is a weak pullback.

If as earlier $\gamma\phi = \psi\delta$, then picking $k\colon T' \to A$, $l\colon T' \to B$ with $\gamma = [k]$, $\delta = [l]$, we have

$$fk \simeq gl$$

or

$$f'rk \simeq gl.$$

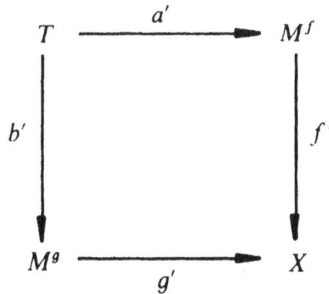

Now $f'$ is a fibration, so there is some $t$ with $rk \simeq t$ and

$$f't = gl = g'sl.$$

Since the diagram defining $T$ is a pullback, there is a map $m: T' \to T$ with $a'm = t \simeq rk$, $b'm = sl$. We set $\rho = [m]$,

$$\begin{aligned}\alpha\rho &= [j^f a'm] = [j^f p^f k] \sim \text{since we took } r = p^f \\ &= [k] = \gamma\end{aligned}$$

and $\beta\rho = \delta$ similarly.

**Corollary**

*If* $(\mathbf{T}_0, \mathbf{T}_1)$ *is an admissible pair of categories and* $\mathbf{T}_0$ *has property* I, *then if* $X$, $Y$ *have the same pointed* **K**-*shape for* $\mathbf{K}: \mathbf{HoT}_0 \to \mathbf{HoT}_1$, $\Sigma X$ *and* $\Sigma Y$ *have the same (pointed)* **K**-*shape*

*Proof*

Using Proposition 1, it remains only to add in the details of the idea of a verification of conditions (i) and (ii) that we gave earlier. Firstly since $\mathbf{T}_0$ is closed under loop spaces, we do have a factorisation of any

$$\gamma: \Sigma X \to \mathbf{K}P$$

as

$$\Sigma X \xrightarrow{\Sigma m} \Sigma\mathbf{K}Q (= \mathbf{K}\Sigma Q) \xrightarrow{\mathbf{K}n} \mathbf{K}P$$

by taking $Q = \Omega P$ and $n = \varepsilon_P$, the adjunction map.

Secondly suppose we have a diagram,

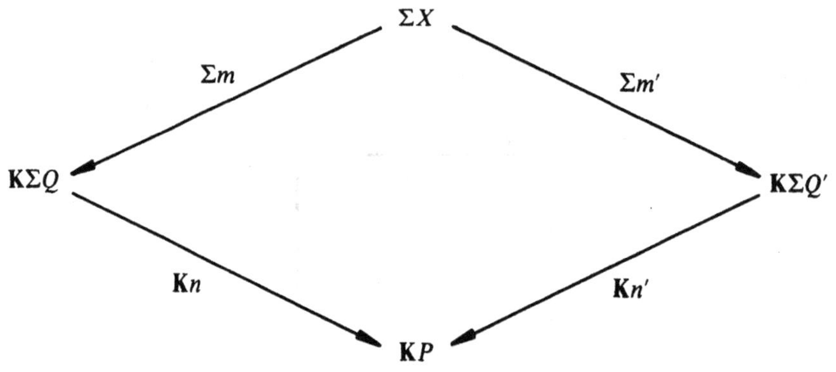


in $\mathbf{HoT}_1$. We can form the weak pullback

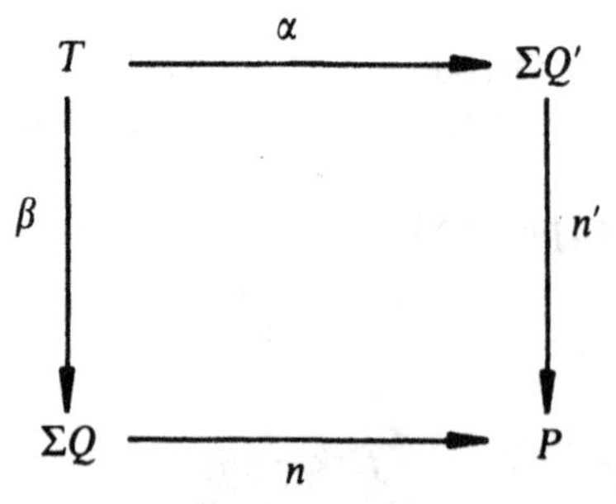

and take $Q_1 = \Omega T$, giving a diagram

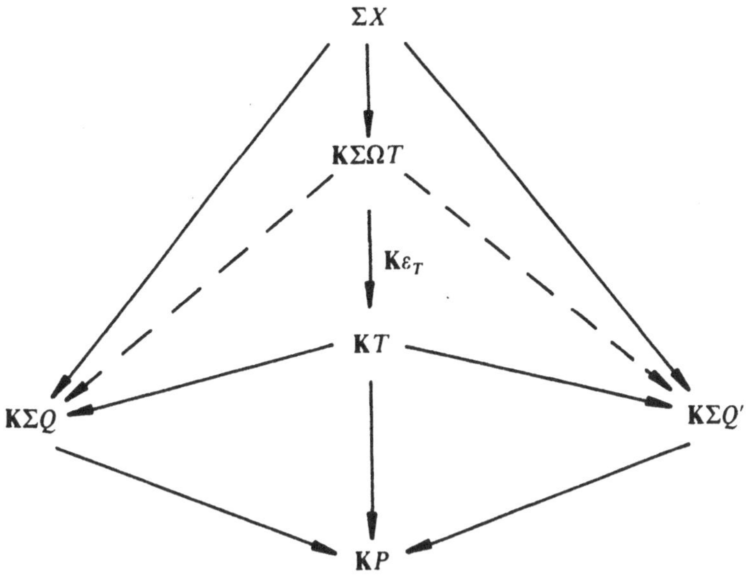


and hence the required lantern. This implies that the square $(\mathbf{\Sigma}, \mathbf{K}, \mathrm{Id}, \mathbf{K}, \mathbf{\Sigma})$ is exact and so $\mathbf{\Sigma}$ induces a functor

$$\Sigma\colon \mathbf{S_K} \to \mathbf{S_K}.$$

Hence if $S_{\mathbf{K}}(X) \cong S_{\mathbf{K}}(Y)$, $S_{\mathbf{K}}(\Sigma X) \cong S_{\mathbf{K}}(\Sigma Y)$.

**Remark**

In the corollary we have limited ourselves to stating the most obvious consequences of the existence of

$$\Sigma\colon \mathbf{S_K} \to \mathbf{S_K};$$

later we will look at some slightly more subtle consequences.

**Examples**

(i) Let $\mathbf{T_1}$ be $\mathbf{Top_*}$ and $\mathbf{T_0}$ be the category of pointed spaces of the homotopy type of pointed CW-complexes; then $(\mathbf{T_1}, \mathbf{T_0})$ is an admissible pair and $\mathbf{T_0}$ has property I; hence for this form of shape theory, suspension preserves shape equivalence.

(ii) Let $\mathbf{T_1}$ be $\mathbf{Top_*}$ and $\mathbf{T_0}$ be the category of pointed spaces having the homotopy type of pointed countable CW-complexes.

The example one would most like would be with $\mathbf{T_1}$ = pointed compact spaces, $\mathbf{T_0}$ the full subcategory of pointed spaces having the homotopy type of finite pointed CW-complexes; however, although the above proposition is almost applicable here, one has to adjust the details slightly at one or two points. We rework the proof from the start to avoid confusion.

Suppose $X$ is compact, $Y$ a finite CW-complex and $f\colon \Sigma X \to Y$ a pointed map. Let $f'\colon X \to \Omega Y$ be the adjoint map to $f$. $\Omega Y$ need not be, in fact will not usually be, of the homotopy type of a finite CW-complex, but one knows by Milnor's theorem that it does have the homotopy type of a CW-complex, $Z'$ say. We shall write $f': X \to Z'$ for simplicity. $X$ is compact so there is a finite subcomplex $Z$ of $Z'$ containing $f'(X)$. Let $g: X \to Z$ be the corestriction of $f'$, i.e. obtained by restricting its codomain. $\varepsilon_y: \Sigma\Omega Y \to Y$ restricts to $\varepsilon: \Sigma Z \to Y$ and $f = u\Sigma g$ giving the required factorisation.

We now need to check that the lantern condition holds. Let $g_1: X \to Z_1, g_2: X \to Z_2$, $u_1: \Sigma Z_1 \to Y$ and $u_2: \Sigma Z_2 \to Y$ be such that $u_1\Sigma g_1 = u_2\Sigma g_2$. We denote by $Z'(\simeq \Omega Y)$ a CW-complex homotopically equivalent to $\Omega Y$ and by $u_i': Z_i \to Z'$, $i = 1\ 2$, the maps constructed from the adjoints and the homotopy equivalence; again $\varepsilon: \Omega Z' \to Y$ denotes the map coming from $\varepsilon_Y$.

One obtains the following commutative diagrams in $\mathbf{HoT}_1$

(a)

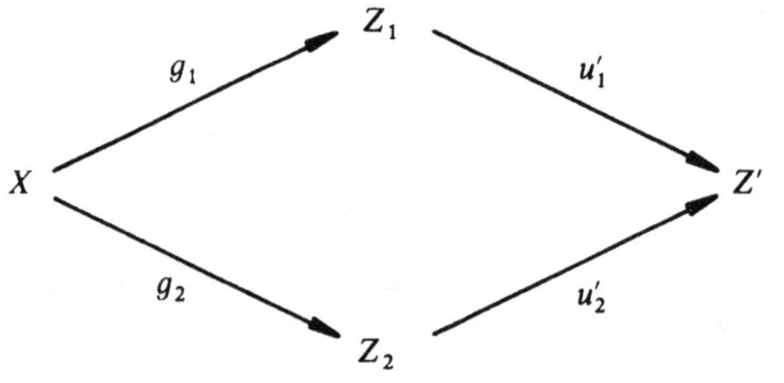


(b)

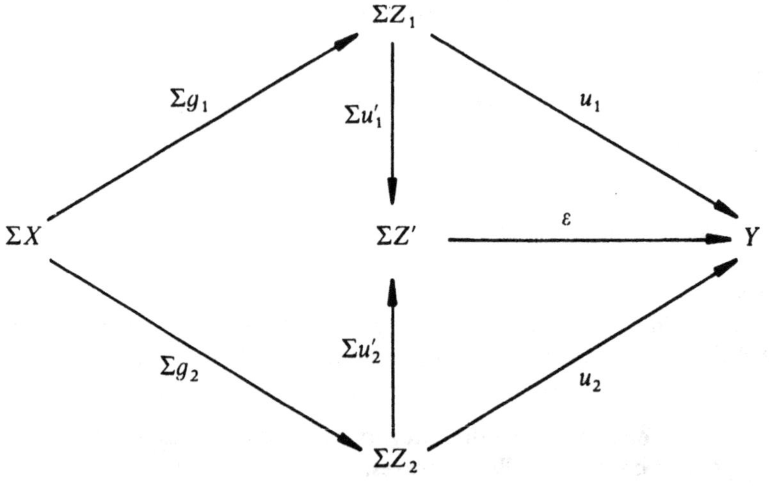


All seems well except that $Z'$ is not a finite complex. However diagram (a) commutes up to homotopy, so there is a homotopy

$$H: X \times [0, 1] \to Z'$$

linking $u_1'g_1$ with $u_2'g_2$. $H(X \times [0, 1])$ is compact so one can find a finite subcomplex $Z$ of $Z'$ containing it. Restricting all maps accordingly, one finds

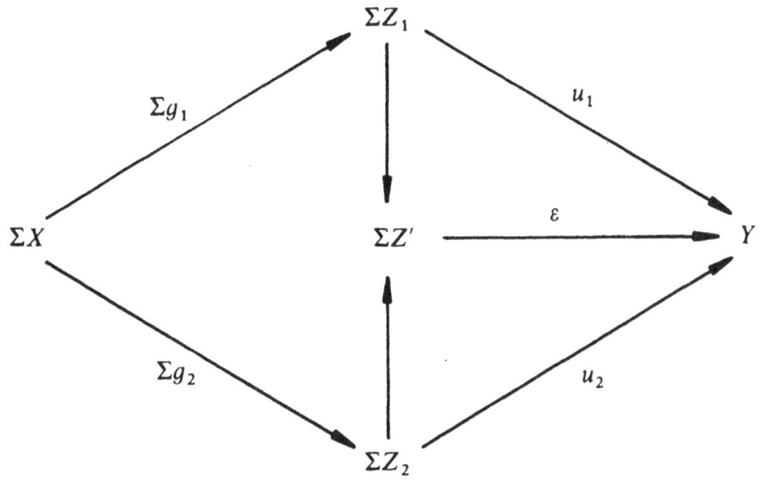


as required.

**Corollary**

(i) *The Borsuk shape of the suspension of a (pointed) compact metric space, $X$, depends only on the Borsuk shape of $X$.*

(ii) *The Mardešić–Segal shape of the suspension $\Sigma X$ of a pointed compact Hausdorff space, $X$, depends only on the Mardešić–Segal shape of $X$.*

In order to extend the applicability of the result given in the Corollary, one can analyse the proof carefully and one finds that to obtain the result, only the following were used:

(a) $\mathbf{T}_0$ is closed under products (of two objects).

(b) If $\tilde{\Omega}$ is right adjoint to $\Sigma$ in $\mathbf{T}_1$, then $\mathbf{T}_0$ is closed under $\tilde{\Omega}$ (i.e. $\tilde{\Omega}$ does not need to be a loop space as such).

(c) $\mathbf{T}_0$ has weak pullbacks relative to $\mathbf{T}_1$.

We will say that $(\mathbf{T}_0, \mathbf{T}_1)$ has property II if (a), (b) and (c) are satisfied. Thus the proof of the Corollary yields the following stronger result:

**Proposition 2**

*If $(\mathbf{T}_0, \mathbf{T}_1)$ is an admissible pair of categories and it satisfies property II then if $X$ and $Y$ have the same (pointed)* **K**-*shape for* $\mathbf{K}: \mathbf{HoT}_0 \to \mathbf{HoT}_1$, $\Sigma X$ *and* $\Sigma Y$ *have the same (pointed)* **K**-*shape.*

The wider applicability of this result can be illustrated by a particular case.

Consider a category, $\mathbf{T}_s$, of simply connected, connected pointed spaces satisfying the following two conditions:

(A) if $X$ is in $\mathbf{T}_s$, the universal covering space, $\tilde{\Omega}X$, of the loop space of $X$ exists and is in $\mathbf{T}_s$.

(B) given any corner of fibrations

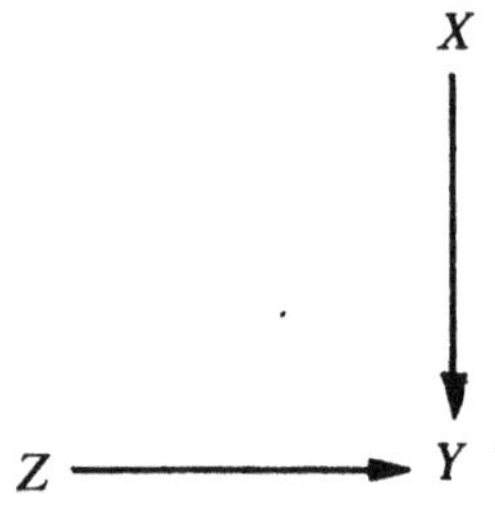


in $\mathbf{T_s}$, the universal covering space, $\tilde{W}$, of the pullback (in $\mathbf{Top_*}$) of the corner exists and is in $\mathbf{T_s}$. (We call $\tilde{W}$ the 1-connected pullback of the corner.)

We note that if $X$ and $Y$ are in $\mathbf{T_s}$ then there is a natural isomorphism

$$\mathbf{T_s}(\Sigma X, Y) \cong \mathbf{T_s}(X, \widetilde{\Omega Y}).$$

Taking $\mathbf{T_1} = \mathbf{T_s}$, let $\mathbf{T_0}$ be a subcategory of $\mathbf{T_s}$. We need to alter property I slightly to fit this situation.

We say the $\mathbf{T_0}$ has property $\mathrm{I}_1$ with respect to $\mathbf{T_1}$ if given a corner of fibrations within $\mathbf{T_0}$, its 1-connected pullback is again in $\mathbf{T_0}$. The analogue of the lemma before Proposition 1 works, giving that if $\mathbf{T_0}$ is admissible and has property $\mathrm{I}_1$, then $\mathbf{T_0}$ is closed with respect to finite products and for each $X$ in $\mathbf{T_0}, \widetilde{\Omega Y}$ is also in $\mathbf{T_0}$.

**Proposition 3**

*If $\mathbf{T_0}$ is an admissible subcategory of $\mathbf{T_s}$, and $\mathbf{T_0}$ has property $I_1$ with respect to $\mathbf{T_s}$, then if $X$ and $Y$ have the same (pointed) $\mathbf{K}$-shape for $\mathbf{K}: \mathbf{HoT_0} \to \mathbf{HoT_s}$, $\Sigma X$ and $\Sigma Y$ also have the same $\mathbf{K}$-shape.*

The proof proceeds by using property $\mathrm{I}_1$ to check property II, and then uses property II. The details are so similar to those of earlier results that we will omit them.

A final and quite interesting set of examples uses the notion of a Serre class of groups.

A non-empty class of Abelian groups, $\mathscr{C}$, is called a *Serre class* if for any exact sequence

$$A \to B \to C$$

of Abelian groups, if $A$ and $C$ are in $\mathscr{C}$, then s is $B$.

The following are typical examples of Serre classes:

(i) all Abelian groups,
(ii) the class containing just the trivial group,
(iii) all finitely generated Abelian groups,
(iv) all finite Abelian groups,
(v) all torsion Abelian groups,
(vi) all Abelian $p$-groups for a given prime $p$,
(vii) all Abelian groups having no element with order a positive power of a given prime $p$.

One says that a pointed connected space $X$ *belongs to* $\mathscr{C}$ if all its homotopy groups $\pi_i(X)$, $i \geq 1$, are in $\mathscr{C}$. We will denote by $\mathbf{T_0}(\mathscr{C})$ the full subcategory of $\mathbf{T_s}$ generated by those spaces $X$ which belong to some fixed given $\mathscr{C}$.

$\mathbf{T}_0(\mathscr{C})$ satisfies property $\mathrm{I}_1$, but one can easily show directly that for each $X$ in $\mathbf{T}_0(\mathscr{C})$, $\tilde{\Omega}X$ is also $\mathbf{T}_0(\mathscr{C})$; in fact, $\pi_i(\Omega X) \cong \pi_{i+1}(X)$ for each $i \geqslant 1$ and so for $\tilde{\Omega}X$, $\pi_i(\tilde{\Omega}X) \in \mathscr{C}$ for $i \geqslant 2$, i.e. $\tilde{\Omega}X$ is in $\mathbf{T}_0(\mathscr{C})$ as we already know $\tilde{\Omega}X$ is in $\mathbf{T}_s$.

To verify that $\mathbf{T}_0(\mathscr{C})$ satisfies property $\mathrm{I}_1$, one uses only the homotopy exact sequences for fibrations to show that $\pi_i(W)$ is in $\mathscr{C}$ for $i \geqslant 1$ and thus that $\pi_i(\tilde{W})$ is in $\mathscr{C}$ for $i \geqslant 2$. As $\tilde{W}$ is in $\mathbf{T}_s$, we have property $\mathrm{I}_1$.

Thus Proposition 3 applies, showing that the Serre class shape theory has suspension functors. This is closely related to various results on localisations of $\mathbf{T}_s$ with respect to a Serre class.

## NOTES ON SOURCES

The notion of exact square in the sense used here is due to Guitart [54]. His idea was a development of earlier work by Hilton and others on generalisations of exactness for sequences in an Abelian category. (The above reference contains some comment about the history of the concept.)

The material on absolute Kan extensions can be found in Harting [55].

The topological examples were implicit in the early work of Deleanu and Hilton [22–25]. The main purpose of their development was to study the Kan and Čech extensions of cohomology theories. An easily accessible reference for this work is Hilton's monograph [56], whereas a development very near to the use of exact squares given here can be found in MacDonald's paper [70].

The explicit use of these 'local right adjunctions' of Deleau and Hilton to prove invariance of suspensions in shape was given by the second author (see Porter [86]), and a section on functors induced between shape theory occurs in the paper by Bourn and the first author [14], in which several other applications of exact squares to categorical shape theory are discussed. As a whole, the chapter is an expanded version of the authors' paper [21].

# 6

# Stability and Movability

## 6.1 THE INTERPRETATION OF CATEGORICAL NOTIONS IN A SHAPE CATEGORY

Clearly the induced functors between shape theories studied in the previous chapter will preserve any notions which are categorically definable within a given shape category. In this section we will consider certain such structures, approaching them from the direction of category theory.

As always, $\mathbf{K}: \mathbf{A} \to \mathbf{B}$ is a functor.

**Proposition 1**

*Let $s, s'$ be two* **K**-*shape morphisms from $B$ to $B'$ in $\mathbf{S_K}$ and $t$ a natural transformation from $s$ to $s'$ (considered as functors from $B'\downarrow\mathbf{K}$ to $B\downarrow\mathbf{K}$). Then there is a* **K**-*shape morphism $s'': B' \to B'$ such that $s''s = s'$.*

*Proof*

Let $(b: B' \to \mathbf{K}A, A)$ be in $B'\downarrow\mathbf{K}$. The morphism $t_{(b,A)}$ is a morphism in $B\downarrow\mathbf{K}$ from $s(b, A)$ to $s'(b, A)$ so $t_{(b,A)}$ is essentially a morphism

$$t_{(b,A)}: A \to A$$

in **A** satisfying $K(t_{(b,A)})s[b, A] = s'[b, A]$ and for $b' = \mathbf{K}(a)b: B' \to \mathbf{K}A'$, $at_{(b,A)} = t_{(b',A')}a$.

We set $s''[b, A] = \mathbf{K}(t_{(b,A)})b$ for $b: B' \to \mathbf{K}A$. Then $s'': B'\downarrow\mathbf{K} \to B'\downarrow\mathbf{K}$ will be defined by $s''(b, A) = (s''[b, A], A)$. Thus defined $s''$ is a functor from $B'\downarrow\mathbf{K}$ to itself since if $a: A \to A'$, we have

$$\begin{aligned}\mathbf{K}(a)s''[b, A] &= \mathbf{K}(a)\mathbf{K}(t_{(b,A)})b \\ &= \mathbf{K}(at_{(b,A)})b \\ &= \mathbf{K}(t_{(b',A')}a)b \\ &= \mathbf{K}(t_{(b',A')})\mathbf{K}(a)b \\ &= s''[\mathbf{K}(a)b, A']\end{aligned}$$

where $\mathbf{K}(a)b = b'$. Clearly therefore $s'' \in \mathbf{S_K}(B', B')$ and it remains to prove that $s''s = s'$, but

$$\begin{aligned}(s''s)[b, A] &= s[s''[b, A], A] \\ &= s[\mathbf{K}(t_{(b,A)})b, a] \\ &= \mathbf{K}(t_{(b,A)})s[b, A] \\ &= s'[b, A]\end{aligned}$$

as required.

**Corollary 1**

*Let $s$ and $s'$ be two* **K***-shape morphisms from $B$ to $B'$ and $t$ a natural equivalence from $s$ to $s'$; then there is a* **K***-shape automorphism, $s''$, of $B'$ such that $s''s = s'$.*

For the next corollary of Proposition 1, we introduce the notion of **K**-shape domination.

**Definition**

Let $B, B'$ be objects of **B**. We will say that $B$ **K**-*dominates* $B'$, and will write $B \underset{\mathbf{K}}{\geqslant} B'$ if there are **K**-shape morphisms $s: B \to B'$, $s': B' \to B$ such that $ss' = \mathrm{Id}_{B'}$.

If $B \underset{\mathbf{K}}{\geqslant} B'$ and $B' \underset{\mathbf{K}}{\geqslant} B$ then we say $B$ and $B'$ are **K**-*equal* and will write $B \underset{\mathbf{K}}{\equiv} B'$.

It is clear that if $B$ dominates $B'$ in **B**, then $B$ **K**-dominates $B'$ and that if $B$ and $B'$ have the same **K**-shape, then they are **K**-equal. Of course, the converse of this latter statement need not be true.

**Corollary 2** (*to* Proposition 1)

*Let $s: B \to B'$, $s': B' \to B$ be two* **K***-shape morphisms and $t: s's \to \mathrm{Id}_B$ a natural transformation; then $B \underset{\mathbf{K}}{\geqslant} B'$.*

*Proof*

There is a **K**-shape morphism $s'': B \to B$ such that $(s''s')s = \mathrm{Id}_B$.

**Corollary 3** (*to* Proposition 1)

*Let $s: B \to B'$, $s': B' \to B$ be two* **K***-shape morphisms and $t: s's \to \mathrm{Id}_B$, $r: ss' \to \mathrm{Id}_B$, two natural transformations; then $B \underset{\mathbf{K}}{\equiv} B'$.*

**Remark**

The last result shows that, if $B{\downarrow}\mathbf{K}$ and $B'{\downarrow}\mathbf{K}$ are equivalent over **A**, then $B$ and $B'$ are **K**-equal. If $B{\downarrow}\mathbf{K}$ and $B'{\downarrow}\mathbf{K}$ are isomorphic over **A**, $B$ and $B'$ clearly have the same **K**-shape. Thus the difference between '$B \underset{\mathbf{K}}{\equiv} B'$' and '$B \cong B'$ in $\mathbf{S_K}$' is somewhat the same as that between equivalence and isomorphism of categories.

We record here an obvious result, applying induced functors to the transference of **K**-domination; in future we will not state results such as these, as such results are easy to manufacture using the building blocks provided.

**Proposition 2**
*Let*

(𝔽) 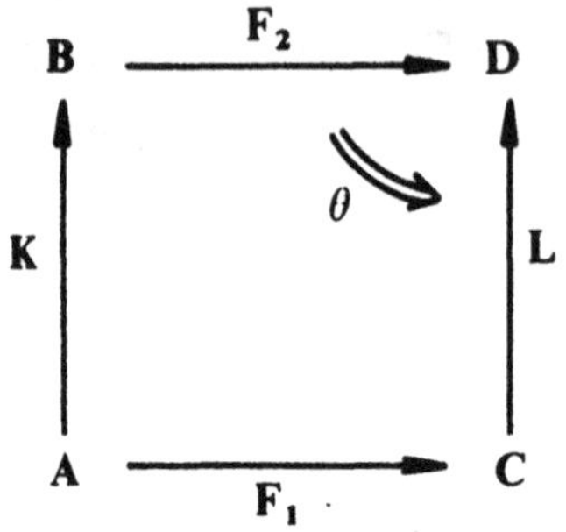

*be an exact square and $B$, $B'$ objects of* **B**. *Then*

(i) if $B \underset{\mathbf{K}}{\geqslant} B'$, $F_2B \underset{\mathbf{L}}{\geqslant} F_2B'$

(ii) if $B \underset{\mathbf{K}}{\equiv} B'$, $F_2B \underset{\mathbf{L}}{\equiv} F_2B'$.

This includes, of course, statement about the preservation of domination, etc. by suspension in those situations handled by section 5.5.

**Proposition 3**
*Let $s: B \to B'$, $s': B' \to B$ be two* **K**-*shape morphisms such that $s$ and $s'$ are a pair of adjoint functors (with corresponding natural transformations, $\varepsilon: \mathrm{Id}_B \to s's$ and $r: ss' \to \mathrm{Id}_{B'}$); then $B$ and $B'$ have the same* **K**-*shape.*

*Proof*
Let $s'': B'\downarrow\mathbf{K} \to B\downarrow\mathbf{K}$ be defined as follows:

$$\text{if } b': B' \to \mathbf{K}A,\ s''[b', A] = \mathbf{K}(r_{(b',A)})s[b', A].$$

This assignment makes $s''$ into a **K**-shape morphism if we set $s''(b, A) = (s''[b, A], A)$; moreover $s''$ is such that

$$\begin{aligned}(s''s')[b', A] &= s'[s''[b', A], A]\\ &= s'[\mathbf{K}(r_{(b',A)})s[b', A], A]\\ &= \mathbf{K}(r_{(b',A)})s'[s[b', A], A]\\ &= b',\end{aligned}$$

since $r$ is a natural transformation from $ss'$ to $\mathrm{Id}_{B'}$.

Similarly if $b: B \to \mathbf{K}A$ in **B**,

$$\begin{aligned}(s''s')[b, A] &= s''[s'[b, A], A]\\ &= [\mathbf{K}r_{(s'[b,A],A)})s[s'[b, A], A]\\ &= \mathbf{K}(r_{(s'[b,A],A)})\mathbf{K}(\varepsilon_{(b,a)})b\\ &= \mathbf{K}(r_{(s'[b,A],A)}\varepsilon_{(b,a)})b\\ &= b,\end{aligned}$$

since $\varepsilon$ is a natural transformation from $\mathrm{Id}_B$ to $s's$, and $\varepsilon$ and $r$ satisfy the adjunction equations.

Consequently $s''s = \mathrm{Id}_B$ and $s's'' = \mathrm{Id}_{B'}$, so that $B$ and $B'$ have the same **K**-shape.

## 6.2 STABILITY

We continue our study of the interpretation of categorical notions within a shape category as follows:

**Theorem 1** (Stability Theorem)
*Let $B$ be an object of* **B**. *The comma category, $B\downarrow\mathbf{K}$, has an initial object, $\eta_B: B \to \mathbf{K}A$, say, if and only if $B$ and $\mathbf{K}A$ have the same* **K**-*shape for some $A$ in* **A**. *In fact $S_{\mathbf{K}}(\eta_B)$ is an isomorphism in $\mathbf{S_K}$.*

*Proof*
Assume that $\eta_B: B \to \mathbf{K}A$ is an initial object in $B\downarrow\mathbf{K}$ then $S_{\mathbf{K}}(\eta_B)$ is a **K**-shape morphism from $B$ to $\mathbf{K}A$ defined by:

$$\text{if } b: \mathbf{K}A \to \mathbf{K}A',\ S_{\mathbf{K}}(\eta_B)(b, A') = (b\eta_B, A').$$

If $b$ is any morphism from $B$ to $\mathbf{K}A'$ then as $\eta_B$ is initial, there is a unique $a: A \to A'$ such that $\mathbf{K}(a)\eta_B = b$. Thus let

$$s': B\downarrow\mathbf{K} \to \mathbf{K}A\downarrow\mathbf{K}$$

be defined by

$$s'[b, A] = \mathbf{K}(a)$$

where $a$ is the unique morphism satisfying $\mathbf{K}(a)\eta_B = b$. It is clear that $s'$ is a shape morphism from $\mathbf{K}A$ to $B$ inverse to $S_{\mathbf{K}}(\eta_B)$.

Conversely as $\mathbf{K}A\downarrow\mathbf{K}$ has an initial object $(\mathrm{Id}_{\mathbf{K}A}, A)$ it follows that if $B$ has the same shape, $B\downarrow\mathbf{K}$, which is isomorphic to $\mathbf{K}A\downarrow\mathbf{K}$, also has one.

**Definition**
An object $B$ in **B** is said to be **K**-*stable* if $B\downarrow\mathbf{K}$ has an initial object. If $\eta_B: B \to \mathbf{K}A$ is an initial object of $B\downarrow\mathbf{K}$ then we say that $B$ **K**-*stabilises* to $\mathbf{K}A$.

**Remark**
Consider the situation studied in section 2.4 in which **K** is the insertion of a full subcategory **A** of a category **B** such that **K** has a proadjoint. Thus to each object $B$ of **B**, there is assigned a **K**-associated pro-object $\mathbf{F}_B: \mathbf{I} \to \mathbf{A}$.

It is clear from Proposition 1 of section 2.4, that given an object of the form $\mathbf{K}A$, one may choose $\mathbf{F}_{\mathbf{K}A}$ to be the trivial 'constant' pro-object $A$ indexed by the category **1** having exactly one morphism.

It now should be obvious from our previous results linking shape with procategories, that $B$ is **K**-stable if and only if there is an object $A$ in **A** such that $(\mathbf{I}, \mathbf{F}_B)$ is isomorphic in **pro**(**A**) to $A$.

In general, if **A** is a category, we will say that a pro-object $(\mathbf{I}, \mathbf{F})$, in **pro**(**A**) is *stable* or *essentially constant* if there is some $A$ in **A** itself such that $A$ (or more exactly $\mathbf{c}A$) and $(\mathbf{I}, \mathbf{F})$ are isomorphic in **pro**(**A**). (This was briefly introduced earlier in section 2.3.)

**Proposition 1**
Let $\mathbf{K}: \mathbf{A} \to \mathbf{B}$, $\mathbf{L}: \mathbf{C} \to \mathbf{D}$ be two functors and let $\mathbf{F}_1, \mathbf{F}_2$ be functors such that

($\mathbb{F}$)

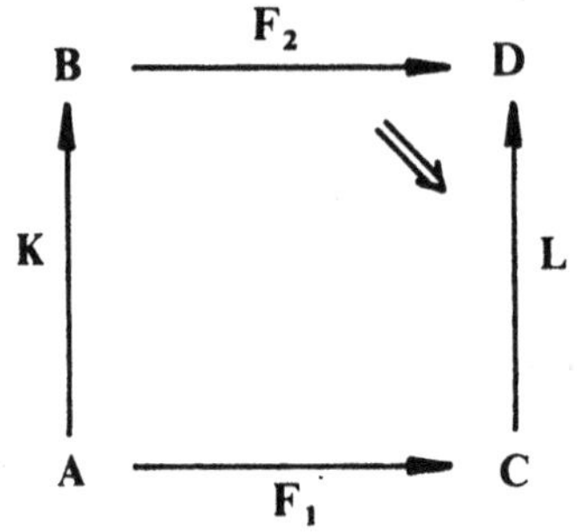

$$
\begin{array}{ccc}
\mathbf{B} & \xrightarrow{\ \ \mathbf{F_2}\ \ } & \mathbf{D} \\
\Big\uparrow{\scriptstyle\mathbf{K}} & \searrow & \Big\uparrow{\scriptstyle\mathbf{L}} \\
\mathbf{A} & \xrightarrow[\ \mathbf{F_1}\ ]{} & \mathbf{C}
\end{array}
$$

is exact.

If $B$ is a $\mathbf{K}$-stable object of $\mathbf{B}$, which $\mathbf{K}$-stabilises to $\mathbf{K}A$, then $\mathbf{F}_2 B$ is an $\mathbf{L}$-stable object of $\mathbf{D}$, $\mathbf{L}$-stabilising to $\mathbf{LF}_1 A$.

*Proof*
The proof should be obvious. ($\mathbb{F}$) induces a functor

$$\mathbb{F}_*: \mathbf{S_K} \to \mathbf{S_L};$$

$B$ and $\mathbf{K}A$ are assumed to be isomorphic in $\mathbf{S_K}$; hence $\mathbf{F}_2 B$ and $\mathbf{F}_2 \mathbf{K}A = \mathbf{LF}_1 A$ are isomorphic in $\mathbf{S_L}$.

**Corollary**
Let $\mathbf{A}$ be the homotopy category of pointed finite $CW$-complexes, $\mathbf{B}$ the homotopy category of pointed compact spaces and $\mathbf{K}$ the inclusion of $\mathbf{A}$ into $\mathbf{B}$. If $B$ in $\mathbf{B}$ is $\mathbf{K}$-stable, then its reduced suspension $\Sigma B$ is also $\mathbf{K}$-stable.
The proof follows from the proposition given in our calculations of section 5.5
The above shows that $\mathbf{K}$-stability can be transformed into $\mathbf{L}$-stability under suitable exactness conditions. We also have reflection of stability under change of models.

**Proposition 2**
Let $\mathbf{K}: \mathbf{A} \to \mathbf{B}$, $\mathbf{L}: \mathbf{C} \to \mathbf{B}$ and $\mathbf{F}: \mathbf{A} \to \mathbf{C}$ be functors so that $\mathbf{K} = \mathbf{LF}$. Then if $B$ in $\mathbf{B}$ $\mathbf{L}$-stabilises to $\mathbf{LF}A$ for some $A$ in $\mathbf{A}$, $B$ $\mathbf{K}$-stabilises to $\mathbf{K}A$.

*Proof*
As $B$ and $\mathbf{LF}A$ are isomorphic in $\mathbf{S_L}$, by applying $\mathbf{F}^*$ (the change of models functor—see section 5.1), one obtains that $B$ and $\mathbf{K}A = \mathbf{LF}A$ have the same $\mathbf{K}$-shape, as required.

**Proposition 3**
Let $\mathbf{K}: \mathbf{A} \to \mathbf{B}$, $\mathbf{L}: \mathbf{C} \to \mathbf{B}$, $\mathbf{F}: \mathbf{A} \to \mathbf{C}$ be functors such that

(i) $\mathbf{LF} = \mathbf{K}$
(ii) the square

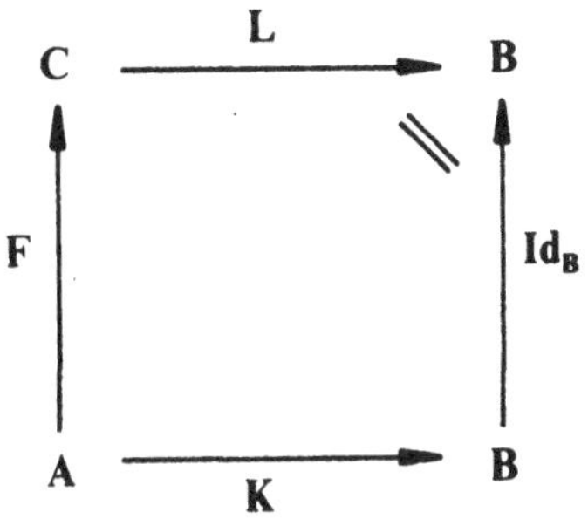


*is exact.*

*Then if $B$ and* $\mathbf{L}C$ *are isomorphic in* $\mathbf{S_L}$, *and $C$ and* $\mathbf{F}A$ *are isomorphic in* $\mathbf{S_L}$, *one has that $B$ and* $\mathbf{K}A$ *are isomorphic in* $\mathbf{S_K}$.

*Proof*
As $B$ and $\mathbf{L}C$ are isomorphic in $\mathbf{S_L}$, applying $\mathbf{F}^*$ one obtains that $B$ and $\mathbf{L}C$ are isomorphic in $\mathbf{S_K}$.

As $C$ and $\mathbf{F}A$ are isomorphic in $\mathbf{S_F}$, and

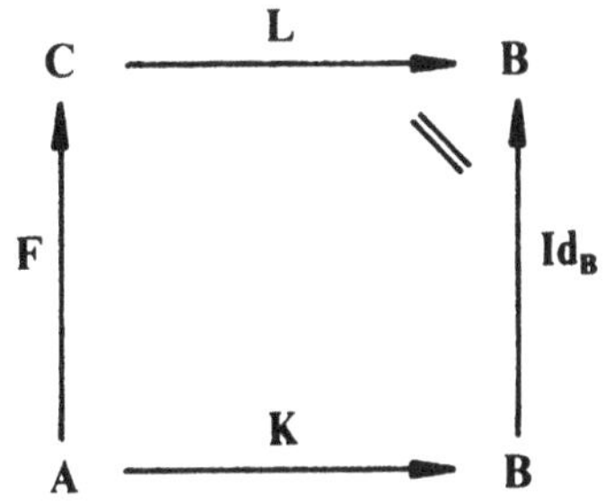


is exact, $\mathbf{L}C$ and $\mathbf{LF}A$ are isomorphic in $\mathbf{S}_{\mathbf{Id_B}}$, as they are the images of $C$ and $\mathbf{F}A$ under the induced functor from $\mathbf{S_F}$ to $\mathbf{S}_{\mathbf{Id_B}}$. However, this implies that $\mathbf{L}C \cong \mathbf{LF}A$ in $\mathbf{B}$ since $\mathbf{B}$ is isomorphic to $\mathbf{S}_{\mathbf{Id_B}}$. Hence $\mathbf{L}C \cong \mathbf{LF}A$ in $\mathbf{S_K}$ and consequently

$$B \cong \mathbf{L}C \cong \mathbf{LF}A = \mathbf{K}A$$

as required.

**Remark**
This proposition is significant inasmuch as it gives information on how stability reacts under change of models. Thus if one is using a full subcategory of **K**-stable objects in **B**, then the proposition states that if $\mathbf{L}\colon \mathbf{C} \to \mathbf{B}$ is the inclusion, the **L**-stable objects in **B** are already **K**-stable, i.e. one does not obtain better approximations in such a case.

On the other hand, if there is a full subcategory **A** of a full subcategory **C** of models within **B** such that the square is exact, and if each $C$ in **C** is **F**-stable for $\mathbf{F}\colon \mathbf{A} \to \mathbf{C}$, then one can reduce the size of the collection of models to **A** without disturbing the class of stable objects in **B**. As the stable objects are those for which one can obtain the most information, for, in some sense, the 'approximations' by models to a stable object are all subsumed within a single best 'approximation', these consequences of changes of models are significant.

To illustrate the 'nice' behaviour of stable objects we will finish this section on stability by showing the usefulness of stable objects in the Morita shape theory of Chapter 3.

So, as before, we let **H** be the homotopy category of topological spaces, **W** the full subcategory of **H** having for its objects those spaces having the homotopy type of a CW-complex, and **K** the insertion of **W** into **H**. Similarly one takes $\mathbf{H}_0$ (resp. $\mathbf{H}^2$, resp. $\mathbf{H}_0^2$) the homotopy category of pointed spaces (resp. of pairs of spaces, resp. of pairs of pointed spaces), $\mathbf{W}_0$ (resp. $\mathbf{W}^2$, resp. $\mathbf{W}_0^2$) the corresponding full subcategories with $\mathbf{K}_0$ (resp. $\mathbf{K}^2$, resp. $\mathbf{K}_0^2$) the insertion. We will denote the corresponding shape theories by $(\mathbf{S}_0, S_0)$, $(\mathbf{S}^2, S^2)$ and $(\mathbf{S}_0^2, S_0^2)$ respectively.

One of the most useful but still relatively simple techniques of algebraic topology is the long homology (or homotopy) exact sequence. It is well known and 'classical' that the homology defined by Čech for a pair of spaces (most usually for a pair of metric compacta) does not give a long exact sequence. The reason is simple: the limit functor used in its construction is left exact but not right exact. This is a grave complication when it comes to calculating the Čech groups (even though the continuity of Čech homology does compensate for this). Thus it is of interest to know conditions on a pair $(X, A)$ or on a group, $G$, of coefficients which will imply that the long Čech homology sequence of $(X, A)$ with coefficients in $G$ is exact. Conditions on $G$ are well known; for example, taking $G$ to be a compact group or a vector space suffices if $(X, A)$ is a compact pair. Without conditions on $G$, the results are less well known, but it should be clear from our previous work that stability should help here, as it would imply that the whole of the shape theoretic information (including the Čech homology) can be read off from one of the 'approximations'. Let us see how this works out in detail.

Let $\mathbf{H}_n(-, G)\colon \mathbf{W}^2 \to \mathbf{Ab}$ be the $n$th homology functor, $G$ being, of course, the coefficients. Using the Čech extension process described in section 2.5, or alternatively, what is here equivalent to it (by Proposition 5 of section 2.5)—the Kan extension—one defines the $n$th Čech homology group of $(X, A)$ with coefficients in $G$ by the formula

$$\check{\mathbf{H}}_n(X, A; G) = \operatorname{Lim} \mathbf{H}_n(-; G)\boldsymbol{\delta}_{(X,A)}$$

where $\mathbf{H}_n(\ ; G)\delta_{(X,A)}\colon (X, A)\downarrow \mathbf{K}^2 \to \mathbf{Ab}$. If one uses the Čech extension, one obtains the more classical description

$$\check{\mathbf{H}}_n(X, A; G) = \operatorname*{Lim}_{\mathscr{U} \in \operatorname{cov}(X)} \mathbf{H}_n(\mathrm{N}(X, \mathscr{U}), \mathrm{N}(A, \mathscr{U}|A); G).$$

As the projective systems used are pro-objects in $\mathbf{W}^2$, one has an extension of $\mathbf{H}_n(-; G)$ to a functor

$$\mathbf{proH}_n(\ ; G)\colon \mathbf{pro}(\mathbf{W}^2) \to \mathbf{pro}(\mathbf{Ab}).$$

If $\mathbf{C}^2$ is the functor from $\mathbf{H}^2$ to $\mathbf{pro}(\mathbf{W}^2)$ given by the 'pair of nerves' functor, one has

$$\check{\mathbf{H}}_n(X, A; G) = \operatorname{Lim} \mathbf{proH}_n(\mathbf{C}^2(X, A); G)$$

where Lim: $\mathbf{pro}(\mathbf{Ab}) \to \mathbf{Ab}$ is the inverse limit functor which can, by well known results on $\mathbf{pro}(\mathbf{Ab})$, be written without an index.

**Proposition 4**
*Let $(X, A)$ be a $\mathbf{K}^2$-stable pair of topological spaces. Then the long Čech homology sequence of $(X, A)$ is exact for any coefficients.*

*Proof*
We have seen in Chapter 2 that the Čech–Kan extension is a shape invariant; consequently, if $(X, A)$ is $\mathbf{K}^2$-stable, there is a pair $(P, Q)$ in $\mathbf{W}^2$ such that $(P, Q)$ and $(X, A)$ have the same $\mathbf{K}^2$-shape, and the Čech homology groups of $(X, A)$ and $(P, Q)$ are isomorphic, but also

$$\check{\mathbf{H}}_n(P, Q; G) \cong \mathbf{H}_n(P, Q; G).$$

One has two functors $\mathbf{O}_1$ and $\mathbf{O}_2$: $\mathbf{H}^2 \to \mathbf{H}$ defined by $\mathbf{O}_1(X, A) = X, \mathbf{O}_2(X, A) = A$. $\mathbf{O}_1$ and $\mathbf{O}_2$ induce comparison functors $\mathbf{O}_{1*}$ and $\mathbf{O}_{2*}$ from $\mathbf{S}^2$ to $\mathbf{S}$ and as $(X, A)$ is $\mathbf{K}^2$-stable, $X$ and $A$ are $\mathbf{K}$-stable and the isomorphisms

$$\check{\mathbf{H}}_n(P, Q; G) \xrightarrow{\cong} \mathbf{H}_n(P, Q; G)$$

$$\check{\mathbf{H}}_n(X, G) \xrightarrow{\cong} \mathbf{H}_n(P; G)$$

$$\check{\mathbf{H}}_n(A; G) \xrightarrow{\cong} \mathbf{H}_n(Q; G)$$

are compatible with the homomorphism of the long homology sequences of $(X, A)$ and $(P, Q)$. In other words, the long Čech homology sequence is functorial on $\mathbf{S}^2$ and the isomorphism between $(X, A)$ and $(P, Q)$ in $\mathbf{S}^2$ induces an isomorphism of long homology sequences. As the long homology sequence for $(P, Q)$ is exact, the same must be true of that of $(X, A)$, which is what was required.

In a similar way to that by which the Čech homology group were defined, one can define Čech homotopy groups of a pointed space, or a pair of pointed spaces, by

—if $(X, A)$ is in $\mathbf{H}_0^2$, $\check{\pi}_n(X, A) = \text{Lim}\ \pi_n(-, -)\delta_{(X,A)}$ where $\pi_n(-, -)\delta_{(X,A)}$ is a functor from $(X, A)\downarrow\mathbf{K}_0^2$ to **Ab** if $n \geqslant 3$ and from $(X, A)\downarrow\mathbf{K}_0^2$ to **Groups** if $n = 2$;
—if $X$ is a pointed space and $n \geqslant 1$, one also has $\check{\pi}_n(X)$ defined in essentially the same way but with $\check{\pi}_1(X)$ not necessarily being Abelian.

The corresponding proposition on the exactness of the long Čech homotopy sequence of a $\mathbf{K}_0^2$-stable pointed pair $(X, A)$ can be proved in a more or less identical way to that of the above Proposition 4; we leave out the details.

**Proposition 5**
*Let $(X, A)$ be a $\mathbf{K}_0^2$-stable pointed pair of spaces. The long Čech homotopy sequence of $(X, A)$ is exact.*

**Examples**
1. The Warsaw circle (see Example 1 in section 3.3) is stable for all of the usual types of shape. By the results of section 1.2, it has the same shape as $S^1$.

2. If $X$ is stable and $i \geqslant 0$, the $i$th iterated suspension $\Sigma^i X$ of $X$ is stable also (by Proposition 1 and the results on suspension exact squares in Chapter 5).

3. Let $P$ be a sequence of prime numbers. The $P$-adic solenoids $S_P$ (again see section 3.3) are not stable for the usual forms of topological shape theory. The proof of this depends on certain other results; we will thus wait until later, to prove it, when we have these results at our disposal; in fact we will show that $S_P$ is not even movable, a condition which is weaker than stability but is often very useful when handling compact metric spaces.

4. We have proved that if $X$ is stable then $\Sigma X$ is stable. The converse of this is, however, false. Let $X = \mathrm{Lim}\,\underline{X}$ where $\underline{X} = (X_n, p_n^{n+1}, \mathbb{N})$, with $X_n = S^1 \vee S^1$ the pointed coproduct of two copies of a circle (thus essentially a figure 8) and $p_n^{n+1}$ a pointed continuous map

$$p\colon (S^1 \vee S^1, *) \to (S^1 \vee S^1, *)$$

which induces the homomorphism

$$\begin{aligned} p_{\#}(a) &= aba^{-1}b^{-1} \\ &= a^2b^2a^{-2}b^{-2} \end{aligned}$$

on $\pi_1(S^1 \vee S^1)$ (which is isomorphic, by the Van Kampen Theorem, to the free group on two symbols, $a$ and $b$, say). We will show later on that $X$ is not stable but $\Sigma X$ has the shape of a point, and thus $\Sigma X$ is stable.

## 6.3 REINDEXING LEMMAS

Before launching into a detailed discussion of stability in various situations, in this section we will pause and look at a collection of 'reindexing' results. These have proved invaluable in the study of how structure on a category **C** can be extended to **pro**(**C**). This is essential if the mathematical techniques that have been developed for the study of the models are to be of use in the shape category, considered as a subcategory of **pro**(**C**). These reindexing lemmas are thus of general interest. They give the possibility of replacing diagrams of certain types in **pro**(**C**) by pro-objects in the category of diagrams in **C**. The simplest and most widely used is the reindexing lemma for promaps.

**Proposition 1** (Reindexing Lemma)
*Let* $(\mathbf{I}, \mathbf{F})$ *and* $(\mathbf{J}, \mathbf{G})$ *represent objects in* **pro**(**C**) *and* $u\colon \mathbf{F} \to \mathbf{G}$ *be a promap between them. Then there is a (cofiltering) index category* $\mathbf{H} = \mathbf{H}_u$ *with initial functors* $\phi\colon \mathbf{H} \to \mathbf{I}$, $\psi\colon \mathbf{H} \to \mathbf{J}$ *and a natural transformation* $u'\colon \mathbf{F}\phi \to \mathbf{G}\psi$ *of* $\mathbf{H}$*-indexed functors such that the diagram*

$$\begin{array}{ccc} \mathbf{F} & \xrightarrow{\quad u \quad} & \mathbf{G} \\ {\scriptstyle\cong}\big\downarrow & & \big\downarrow{\scriptstyle\cong} \\ \mathbf{F}\phi & \xrightarrow{\quad u' \quad} & \mathbf{G}\psi \end{array}$$

*commutes (where the vertical arrows are the natural isomorphisms induced by the functors* $\phi$ *and* $\psi$ *respectively).*

(In other words, any morphism in **pro**(**C**) may be replaced, up to isomorphism, by a morphism in some functor category, $\mathbf{C}^{\mathbf{H}}$.)

*Proof*

Recall (it is a special case of Proposition 10 in section 2.3) that any pro-object $(\mathbf{I}, \mathbf{F})$ can be considered as the limit of itself if we consider $\mathbf{F}$ to be a functor from $\mathbf{I}$ to $\mathbf{pro}(\mathbf{C})$ via the embedding of $\mathbf{C}$ into $\mathbf{pro}(\mathbf{C})$. This gives that for each object $i$ in $\mathbf{I}$, there is a projection morphism

$$p_i: \mathbf{F} \to \mathbf{F}(i).$$

If we have as here a promorphism $u: \mathbf{F} \to \mathbf{G}$, then we will say that a morphism $f: \mathbf{F}(i) \to \mathbf{G}(j)$ in $\mathbf{C}$ *represents* $u$ provided the diagram

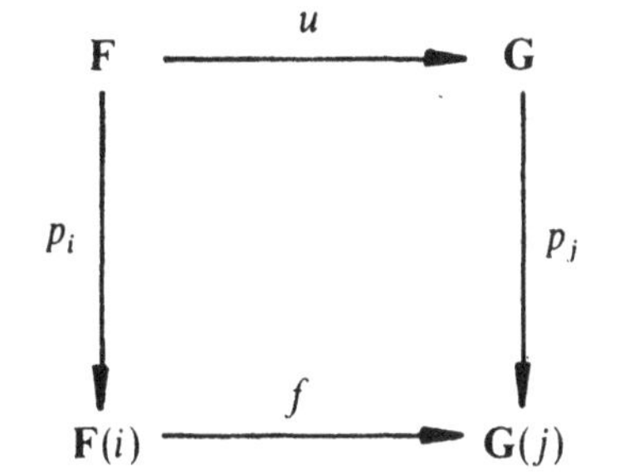


commutes. This is easily checked to be equivalent to stating that $f \in \mathbf{C}(\mathbf{F}(i), \mathbf{G}(j))$ is part of a representative system for $u$ in the description of $\mathbf{pro}(\mathbf{C})(\mathbf{F}, \mathbf{G})$ as Lim Colim $\mathbf{C}(\mathbf{F}(i), \mathbf{G}(j))$.

A morphism $f \to f'$ between morphisms representing $u$ consists of a pair $(\alpha, \beta)$, where $\alpha: i \to i'$ in $\mathbf{I}$ and $\beta: j \to j'$ in $\mathbf{J}$ are such that

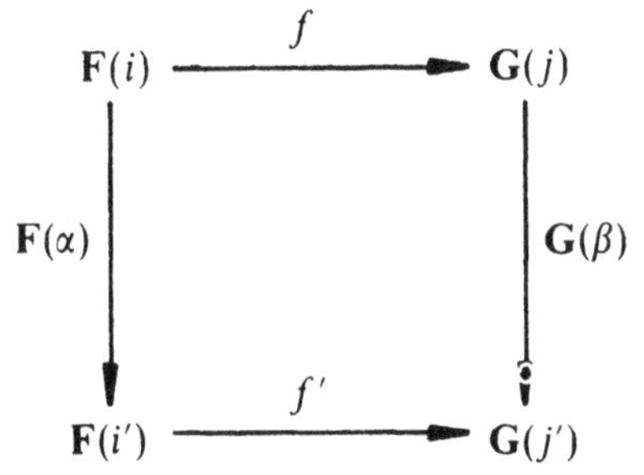


commutes.

Let $\mathbf{H}_u$ be the category of morphisms representing $u$ and morphisms between them. To prevent a certain ambiguity we denote the object of $\mathbf{H}_u$ corresponding to $f: \mathbf{F}(i) \to \mathbf{G}(j)$ by $(i, f, j)$.

There are two functors $\phi: \mathbf{H}_u \to \mathbf{I}$, $\phi(i, f, j) = i$, and $\psi: \mathbf{H}_u \to \mathbf{J}$, $\psi(i, f, j) = j$. The proof will consist of checking $\mathbf{H}_u$ is cofiltering, and $\phi$ and $\psi$ are initial.

We start by a simple lemma.

**Lemma**

*Let* $(\mathbf{I}, \mathbf{F})$ *be an object in* $\mathbf{pro}(\mathbf{C})$, *and* $T$ *an object in* $\mathbf{C}$. *Suppose* $u, v: \mathbf{F}(i) \to T$ *are morphisms for which* $up_i = vp_i$, *then there exists an object* $i'$ *of* $\mathbf{I}$, *and a morphism* $\alpha: i' \to i$ *such that* $u\mathbf{F}(\alpha) = v\mathbf{F}(\alpha)$.

*Proof of Lemma*

Since $up_1 = vp_i$, $u$ and $v$ determine the same element in Colim $\mathbf{C}(\mathbf{F}(i), T)$ but by the usual construction of colimits in sets, this implies the conclusion (since $\mathbf{I}$ is cofiltering).

*Returning to the proof of the Proposition*

Suppose $(i_1, f_1, j_1)$ and $(i_2, f_2, j_2)$ are two objects in $\mathbf{H}_u$. Since $\mathbf{J}$ is cofiltering, there is a $j$ with morphisms $\beta_1: j \to j_1$, $\beta: j \to j_2$ in $\mathbf{J}$. By the alternative description of promorphisms given in Chapter 2 there is an $i$ and an $f: \mathbf{F}(i) \to \mathbf{G}(j)$ such that $(i, f, j)$ is an object of $\mathbf{H}_u$, i.e. $fp_i = p_j u$. Pick an $i'$ with morphisms $\alpha: i' \to i$, $\alpha_1: i' \to i_1$; then $p_i = \mathbf{F}(\alpha)p_{i'}$, and $p_{1_1} = \mathbf{F}(\alpha_i)p_{i'}$, and we obtain $f\mathbf{F}(\alpha)p_{i'} = f_1\mathbf{F}(\alpha_1)p_{i'}$, so by the above lemma, there is an index $k$ in $\mathbf{I}$ and a map $\gamma: k \to i'$ such that $f\mathbf{F}(\alpha\gamma) = f_1\mathbf{F}(\alpha_1\gamma)$.

Repeating with $j$ and $j_2$, we obtain an index $k'$ and morphisms $\phi': k' \to i$, $\gamma_2': k' \to i_2$ so that $f\mathbf{F}(\gamma') = f_2\mathbf{F}(\gamma_2')$. Finally we pick $l$ with maps $\lambda: l \to k$, $\gamma': l \to k'$ to get an object $(l, f\mathbf{F}(\gamma'\lambda), j)$ in $\mathbf{H}_u$ with maps up to the two initially given objects $(i_1, f_1, j_1)$ and $(i_2, f_2, j_2)$.

We next suppose that $(i, f, j)$ and $(i', f', j')$ are two objects of $\mathbf{H}_u$ and that $(\alpha, \beta)$ and $(\alpha', \beta')$ are two morphisms between them. Thus $f'\mathbf{F}(\alpha) = \mathbf{G}(\beta)f$ and similarly for $(\alpha', \beta')$. Since $\mathbf{J}$ is cofiltering we can find a $j_1$ with a map $\gamma: j_1 \to j$ such that $\beta\gamma = \beta'\gamma$.

Using a slight modification of our previous argument, we can find an object $(i_1, f_1, j_1)$ with a map $\delta: i_1 \to i$ such that $\mathbf{G}(\gamma)f_1 = f\mathbf{F}(\delta)$. Thus we obtain, by an easy calculation,

$$f'\mathbf{F}(\alpha\delta) = f'\mathbf{F}(\alpha'\delta).$$

Now we have two maps $\alpha\delta, \alpha'\delta: i_1 \to i$, and as $\mathbf{I}$ is cofiltering there is a $k$ and a morphism $\lambda: k \to i_1$ such that $\alpha\delta\lambda = \alpha'\delta\lambda$. The object $(k, f_1\mathbf{F}(\lambda), j_1)$, together with the map $(\delta\lambda, \gamma)$, is what is needed to finish the proof that $\mathbf{H}_u$ is cofiltering.

The projection $\psi: \mathbf{H}_u \to \mathbf{J}$ is surjective, as we have seen, and so is initial. The projection $\phi$ onto $\mathbf{I}$ is also initial, since if we have $(i, f, j)$ in $\mathbf{H}_u$ and $i'$ in $\mathbf{I}$, we can find $i''$ with maps $\alpha: i'' \to i$, $\alpha': i'' \to i'$ since $\mathbf{I}$ is cofiltering but then $(\alpha, \mathrm{Id}): (i'', f\mathbf{F}(\alpha), j) \to (i, f, j)$ is in $\mathbf{H}_u$ and $\alpha': \phi(i'', f\mathbf{F}(\alpha), j) \to i'$ as required and if $\alpha_1, \alpha_2: i_1 \to i_2$ in $\mathbf{I}$ with $i_1 = \phi(i_1, f_1, j_1)$, then a similar argument gives an object $i = \phi(i, f, j)$ together with a map from $(i, f, j)$ to $(i_1, f_1, j_1)$ whose image in $\mathbf{I}$ equalises $\alpha_1, \alpha_2$. By Proposition 3 of section 2.3, this completes the proof.

**Remark**

It is often useful to have a strengthened form of this Reindexing Lemma. We will not need the result in all generality and so will limit ourselves to proving a very special case. A sketch proof can be found in Artin and Mazur [4] whilst a complete one has been published by Meyer [79]. The proof is by induction.

**Proposition 3** (Uniform Approximation)

*Let* $\mathbf{D}$ *be a finite category with no loops (i.e. each non-identity morphism of* $\mathbf{D}$ *has distinct domain and codomain). Then for any category* $\mathbf{C}$*, the natural functor*

$$\mathbf{F_D}: \mathbf{pro}(\mathbf{F^D}) \to (\mathbf{pro}(\mathbf{F}))^{\mathbf{D}}$$

*is an equivalence of categories.*

**Remark**

In many of the applications of the reindexing lemma and the uniform approximation result, the category **C** has finite limits. In this case, the restriction that **D** has no loops may be lifted. A proof has been given by Meyer [79].

The special case of the above that we will need is where **D** is the category consisting of two composable morphisms and their composite.

Suppose $u: E \to F$ and $v: F \to G$ are promorphisms in **C** then we have seen how to construct categories $H_u$, $H_v$ and $H_{vu}$ with initial functors to the indexing categories of $E$, $F$ and $G$ and natural transformations $\bar{u}: \bar{E} \to \bar{F}$ in **Func**$(H_u, C)$, $\bar{v}: \bar{F} \to \bar{G}$ in **Func**$(H_v, C)$ and $(vu)': E' \to G'$ in **Func**$(H_{vu}, C)$ reindexing $u$, $v$ and $vu$ respectively. (We use a simplified notation.) Thus we have *two* initial functors $H_u \to J$, $H_v \to J$ where $F: J \to C$. Intuitively we want to take their 'intersection' so we take their pullback $H_{(u,v)}$. Thus we have a diagram

$$
\begin{array}{ccc}
H_{(u,v)} & \xrightarrow{\;\;a\;\;} & H_u \\
\Big\downarrow{b} & & \Big\downarrow \\
H_v & \longrightarrow & J
\end{array}
$$

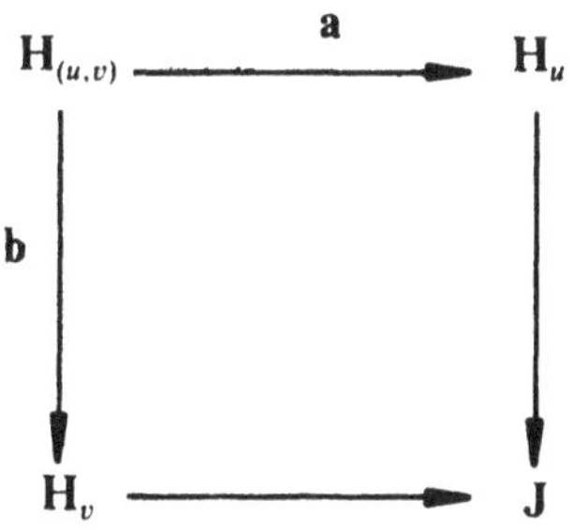

in the category, **Cat**, of small categories. Usually with such a pullback one would not be certain that $H_{(u,v)}$ might not be empty (e.g. pulling back the subcategories of odd and even integers gives their intersection, i.e. the empty category), but as we have explicit constructs of $H_u$, $H_v$, etc., we can find an explicit description of $H_{(u,v)}$.

The objects consist of quintuples $\langle i, f, j, g, k \rangle$ corresponding to representing diagrams

$$
\begin{array}{ccccc}
E & \xrightarrow{\;\;u\;\;} & F & \xrightarrow{\;\;v\;\;} & G \\
\Big\downarrow & & \Big\downarrow & & \Big\downarrow \\
E(i) & \xrightarrow{\;\;f\;\;} & F(j) & \xrightarrow{\;\;g\;\;} & G(k)
\end{array}
$$

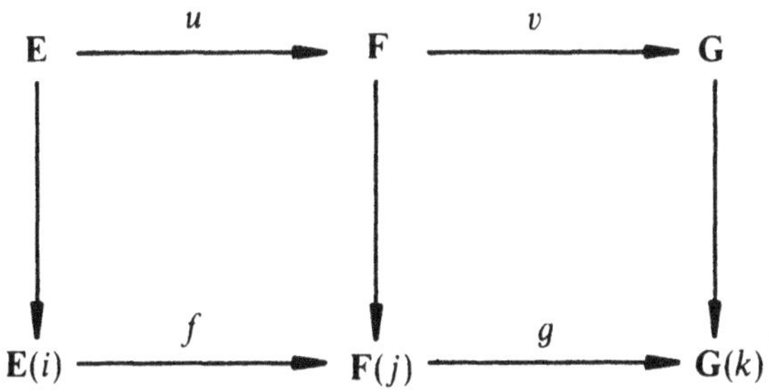

and thus it is clear from our previous arguments (when proving the Reindexing Lemma) that the functor sending $\langle i, f, j, g, k \rangle$ to $k$ is an initial functor from $H_{(u,v)}$ to $k$, and hence $H_{(u,v)}$ is non-trivial. The above representation of the objects of $H_{(u,v)}$ has the advantage that it makes it intuitively clear that the projection functors from $H_{(u,v)}$ to $H_u$, $H_v$ and $H_{vu}$ are initial. (The technical details use the 'alternative description of promorphisms' rather as in the proof of the Reindexing Lemma. As these details add little to one's understanding of the result, they will be omitted.)

The category $\mathbf{H}_{(u,v)}$ is a subcategory of the cofiltering category, $\mathbf{H}_u \times \mathbf{H}_v$, and the inclusion is easily seen to be initial, hence $\mathbf{H}_{(u,v)}$ is cofiltering. Collecting up the various pieces from the above discussion we obtain:

**Lemma**

*For each composable pair $(u, v)$ of arrows in* **pro(C)** *there is a cofiltering index category* $\mathbf{H}_{(u,v)}$ *together with initial functors* $\mathbf{a}\colon \mathbf{H}_{(u,v)} \to \mathbf{H}_u$, $\mathbf{b}\colon \mathbf{H}_{(u,v)} \to \mathbf{H}_v$ *and* $\mathbf{c}\colon \mathbf{H}_{(u,v)} \to \mathbf{H}_{vu}$ *and a natural transformation, $w$, in* $\mathbf{Func}(\mathbf{H}_{(u,v)}, \mathbf{C})$ *such that within* **pro(C)** *the diagram*

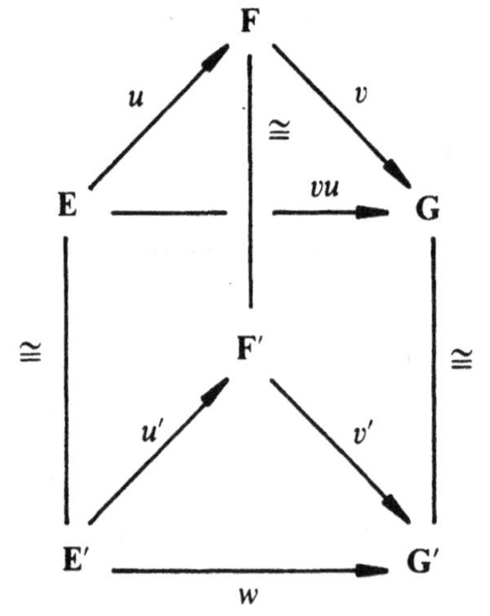


*commutes where $u'$ and $v'$ are natural transformations within* $\mathbf{Func}(\mathbf{H}_{(u,v)}, \mathbf{C})$ *between the reindexed functors* $\mathbf{E}'$, $\mathbf{F}'$, $\mathbf{G}'$.

## 6.4 STABILITY IN PRO-ABELIAN CATEGORIES

In section 6.2, we saw how stability of a pair of spaces leads to good behaviour of certain shape invariants. This raises two points: firstly that it was not, strictly speaking, the stability of $(X, A)$ that was needed to ensure that the long sequence in Čech homology was exact, rather this only depended on the stability in **pro(Ab)** of the objects $\mathbf{proH}_n(\mathbf{C}^2(X, A); G)$, etc.; secondly, we see that there is a stability problem that exists at several different levels; its most general form is that of deciding in a shape theory, $(\mathbf{S_K}, S_\mathbf{K})$, which objects are **K**-stable. We cannot handle this in such generality, but if **K** has a pro-adjoint, the problem becomes more amenable, as it reduces to that of characterising, by internal structure, those pro-objects in a category, **C**, that are isomorphic to constant objects, i.e. essentially constant.

Putting the two points together, we get a stability test. If $\mathbf{H}\colon \mathbf{C} \to \mathbf{D}$ is a functor and $(\mathbf{I}, \mathbf{F})$ is a pro-object in **C**, we say $(\mathbf{I}, \mathbf{F})$ is stable relative to **H** if $(\mathbf{I}, \mathbf{HF})$ is stable in **pro(D)**. Clearly, although such a functor **H** could easily create stability (i.e. $(\mathbf{I}, \mathbf{HF})$ may be stable in **pro(D)** even though $(\mathbf{I}, \mathbf{F})$ is not stable in **pro(C)**), it cannot destroy stability. The usefulness of this technique depends on two aspects of the situation, (i) the efficiency of **H** in revealing structure within **C**, and (ii) the ease with which one can identify stable objects in **pro(D)**.

In many of the algebraic and topological examples of shape theory, there are a wealth of invariants/test functors taking values in Abelian categories, usually but not always categories of modules. Because of this it is useful, as a starting point, to study the problem of stability in categories of the form **pro**(**A**), where **A** is an Abelian category. We will see how the extremely rich and simple basic properties of such categories allow one to describe stability intrinsically, i.e. without reference to some constant pro-object separate from the one being studied. The insights gained here will be extremely useful later on when attempting to repeat the study of stability for a wider class of base categories.

For completeness we recall the definition of an Abelian category. A category **A** is *Abelian* if

(0) it is enriched over **Ab**, i.e. each hom-set $\mathbf{A}(B, C)$ has the structure of an Abelian group in such a way that composition is a bilinear map,
(1) **A** has a zero object, 0,
(2) **A** has binary direct sums,
(3) every morphism in **A** has a kernel and a cokernel,
(4) every monomorphism is a kernel and every epimorphism is a cokernel.

Thus **A** is an additive category having all finite limits and finite colimits. (A good source for the theory and applications of Abelian categories is Popescu's book [84].)

**Proposition 1**
*If* **A** *is an Abelian category, then* **pro**(**A**) *is also Abelian. The canonical embedding* $\mathbf{c}: \mathbf{A} \to \mathbf{pro}(\mathbf{A})$ *is exact, fully faithful and commutes with inductive limits.*

*Proof*
Since if **F**, **G** are objects in **pro**(**A**), we have $\mathbf{pro}(\mathbf{A})(\mathbf{F}, \mathbf{G}) = \text{Lim Colim } \mathbf{A}(\mathbf{F}(i), \mathbf{G}(j))$, the Abelian group structure on each hom-set is easily checked, as is bilinearity of composition.

Any zero object in **A** will yield a zero, also denoted 0, in **pro**(**A**). For any two objects $(\mathbf{I}, \mathbf{F})$ and $(\mathbf{J}, \mathbf{G})$ in **pro**(**A**), define $\mathbf{F} \oplus \mathbf{G}: \mathbf{I} \times \mathbf{J} \to \mathbf{A}$ by $(\mathbf{F} \oplus \mathbf{G})(i, j) = \mathbf{F}(i) \oplus \mathbf{G}(j)$ to obtain the direct sum $(\mathbf{I} \times \mathbf{J}, \mathbf{F} \oplus \mathbf{G})$ of $(\mathbf{I}, \mathbf{F})$ and $(\mathbf{J}, \mathbf{G})$.

To be able to check statements about kernels and cokernels we construct them explicitly: if $(\mathbf{I}, \mathbf{F})$ and $(\mathbf{J}, \mathbf{G})$ are pro-objects in **A** and $u: \mathbf{F} \to \mathbf{G}$ is a promorphism, we have the indexing category $\mathbf{H}_u$ constructed in the last section together with the initial functors $\phi: \mathbf{H}_u \to \mathbf{I}$, $\psi: \mathbf{H}_u \to \mathbf{J}$ and a natural transformation $\bar{u}: \mathbf{F}\phi \to \mathbf{G}\psi$. We will write $\bar{u}: \bar{\mathbf{F}} \to \bar{\mathbf{G}}$ for simplicity. We have that

$$\begin{array}{ccc} \mathbf{F} & \xrightarrow{\;u\;} & \mathbf{G} \\ {\scriptstyle \phi_\#}\downarrow & & \downarrow{\scriptstyle \psi_\#} \\ \bar{\mathbf{F}} & \xrightarrow{\;\bar{u}\;} & \bar{\mathbf{G}} \end{array}$$

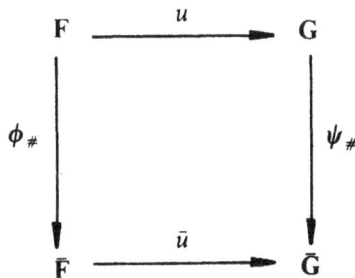

commutes where $\phi_{\#}$, $\psi_{\#}$ are the isomorphisms induced by composition with $\phi$ and $\psi$ respectively. We therefore obtain a pro-object $\mathbf{Ker}\,\bar{u}\colon \mathbf{H}_u \to \mathbf{A}$ given by $(\mathbf{Ker}\,\bar{u})(h) = \operatorname{Ker}\bar{u}(h)$ and with the bonding maps induced by the universal properties of kernels. The kernel of $u$ is then the object $\mathbf{Ker}\,\bar{u}$ together with the composite monomorphism $\mathbf{Ker}\,\bar{u} \to \bar{\mathbf{F}} \xrightarrow{\phi_{\#}^{-1}} \mathbf{F}$.

To see this, suppose $(\mathbf{K}, \mathbf{E})$ and $v\colon \mathbf{E} \to \mathbf{F}$ is such that $vu = 0$. As in the previous section we let $\mathbf{H}_{(u,v)}$ denote the category of representing pairs of maps $\langle k, f, i, g, j\rangle$. Within $\mathbf{H}_{(u,v)}$ is a initial subcategory, which we will denote $\mathbf{H}_{k(u,v)}$ consisting of those $\langle k, f, i, g, j\rangle$ such that $gf = 0$ (that $\mathbf{H}_{k(u,v)}$ is initial follows from the description of equivalence of representing maps and the fact that $vu = 0$). There is an initial functor from $\mathbf{H}_{k(u,v)}$ to $\mathbf{H}_u$ and reindexing

$$\mathbf{Ker}\,\bar{u} \longrightarrow \bar{\mathbf{F}} \xrightarrow{\ \bar{u}\ } \bar{\mathbf{G}}$$

along it, we obtain a unique factorisation of $v$ via the kernel of $\bar{u}$ as required.

The construction of cokernels is entirely similar and is left to the reader.

In order to verify the last axiom: 'all monics are kernels and all epics cokernels', we start by analysing monomorphisms and epimorphisms. We first note that a pro-object $(\mathbf{I}, \mathbf{F})$ is isomorphic to 0 in $\mathbf{pro}(\mathbf{A})$ if and only if there is a 'reindexation' of $\mathbf{F}$, $(\bar{\mathbf{I}}, \bar{\mathbf{F}})$ say, with an initial functor $\phi\colon \bar{\mathbf{I}} \to \mathbf{I}$ such that for any $\alpha\colon i \to j$ in $\bar{\mathbf{I}}$, $\bar{\mathbf{F}}(\alpha)$ is the zero morphism. (The proof, which is a simple application of the alternative definition of promorphism, is left to the reader as an exercise.)

Now suppose $u\colon \mathbf{E} \to \mathbf{F}$ is a monomorphism; by reindexing if necessary we can assume $u$ is realised as a morphism in some Func($\mathbf{I}$, $\mathbf{A}$). As $u$ is monic, $\mathbf{Ker}\,u \cong 0$ and hence for any $i$ in $\mathbf{I}$, there is some $j$ with $\alpha\colon j \to i$ in $\mathbf{I}$ such that $\operatorname{Ker} u(\alpha) = 0$.

We can interpret this diagrammatically as follows:

$$\begin{array}{ccccc}
\operatorname{Ker} u(i) & \longrightarrow & \mathbf{E}(i) & \xrightarrow{\ u(i)\ } & \mathbf{F}(i) \\
\uparrow{\scriptstyle 0} & & \uparrow{\scriptstyle \mathbf{E}(\alpha)} & & \uparrow{\scriptstyle \mathbf{F}(\alpha)} \\
\operatorname{Ker} u(j) & \longrightarrow & \mathbf{E}(j) & \xrightarrow[\ u(j)\ ]{} & \mathbf{F}(j)
\end{array}$$

The morphism from $\operatorname{Ker} u(j)$ to $\operatorname{Ker} u(i)$ is induced by the universal properties of kernels.

Factoring out by the kernels, we obtain

$$\begin{array}{ccccccccc}
0 & \longrightarrow & \operatorname{Ker} u(i) & \longrightarrow & \mathbf{E}(i) & \xrightarrow{\ q(i)\ } & \operatorname{Im} u(i) & \longrightarrow & 0 \\
 & & \uparrow{\scriptstyle 0} & & \uparrow{\scriptstyle \mathbf{E}(u)} & & \uparrow{\scriptstyle \operatorname{Im}(\alpha)} & & \\
0 & \longrightarrow & \operatorname{Ker} u(j) & \longrightarrow & \mathbf{E}(j) & \xrightarrow{\ q(j)\ } & \operatorname{Im} u(j) & \longrightarrow & 0
\end{array}$$

and hence there is a unique morphism $r(i,j)$: $\mathrm{Im}\, u(j) \to \mathbf{E}(i)$ satisfying $\mathrm{Im}(\alpha) = q(i)r(i,j)$ and $\mathbf{E}(\alpha) = r(i,j)u(j)$.

We take $\theta(i) = j$ as above, $r(i) = r(i,j)$ to obtain a representative of a morphism $r$: $\mathbf{Im}\, u \to \mathbf{E}$. It is immediate from the two equations that $r$ is inverse to $q$: $\mathbf{E} \to \mathbf{Im}\, u$, i.e. that $\mathbf{E}$ is isomorphic to $\mathbf{Im}\, u$ and we have a diagram

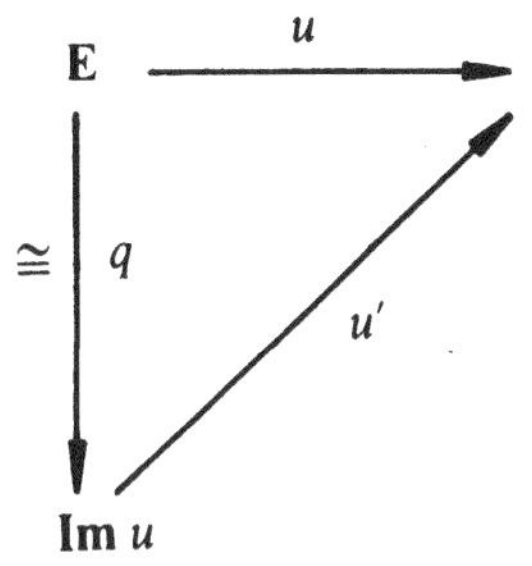


where $u'$: $\mathbf{Im}\, u \to \mathbf{F}$ is given for each $i$ by the 'inclusion' of $\mathbf{Im}\, u(i)$ into $\mathbf{F}(i)$; $u'$ is thus a 'level monomorphism'.

To verify that the monomorphism $u$ is a 'kernel' of a morphism it suffices to note that $u'$ is the kernel of the canonical morphism from $\mathbf{F}$ to $\mathbf{Coker}\, u$ and that $u'$ and $u$ are 'isomorphic' subobjects of $\mathbf{E}$. (As kernels are defined only up to isomorphism, this is all one can expect!)

To complete the verification we can dualise the above argument to show that an epimorphism $v$: $\mathbf{F} \to \mathbf{G}$ is isomorphic to the morphism $v'$: $\mathbf{F} \to \mathbf{Colim}\, v$ where, as usual, $\mathbf{Colim}\, v = \mathbf{F}/\mathbf{Im}\, v$ (formed at each level) and $v'$ is a level epimorphism, hence is $\mathbf{Coker}(\mathbf{Ker}\, v)$.

The properties of the embedding $\mathbf{c}$: $\mathbf{A} \to \mathbf{pro}(\mathbf{A})$ are now easy to check and so are left to the reader.

**Remarks**

There are no doubt easier sets of axioms to check for Abelianness and probably easier ways of doing the verification; however, the explicit checking used above has a great advantage, as it yields with very little effort a series of corollaries, more to the proof than to the Proposition itself.

**Corollary**

(i) *If* $u$: $\mathbf{E} \to \mathbf{F}$ *is a monomorphism in* $\mathbf{pro}(\mathbf{A})$ *with* $\mathbf{A}$ *an Abelian category, then there is a diagram*

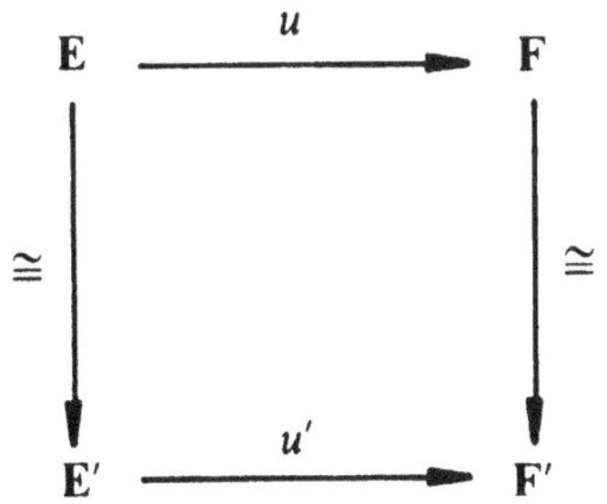

*with $u'$ a level monomorphism, that is a monomorphism in* **Func(I, A)** *for some cofiltering* **I**.

(ii) *If $v: \mathbf{F} \to \mathbf{G}$ is an epimorphism in* **pro(A)**, *then there is a diagram*

$$\begin{array}{ccc} \mathbf{F} & \xrightarrow{\quad v \quad} & \mathbf{G} \\ \cong \big\downarrow & & \big\downarrow \cong \\ \mathbf{F}' & \xrightarrow{\quad v' \quad} & \mathbf{G}' \end{array}$$

*with $v'$ a level epimorphism.*

Putting these together leads, with a bit more checking, to:

**Proposition 2**

*If $0 \to \mathbf{E} \xrightarrow{u} \mathbf{F} \xrightarrow{v} \mathbf{G} \to 0$ is a short exact sequence in* **pro(A)** *with* **A** *Abelian, then there is a cofiltering category* **I** *and a diagram*

$$\begin{array}{ccccccccc} 0 & \longrightarrow & \mathbf{E} & \xrightarrow{\quad u \quad} & \mathbf{F} & \xrightarrow{\quad v \quad} & \mathbf{G} & \longrightarrow & 0 \\ & & \cong \big\downarrow & & \cong \big\downarrow & & \cong \big\downarrow & & \\ 0 & \longrightarrow & \mathbf{E}' & \xrightarrow{\quad u' \quad} & \mathbf{F}' & \xrightarrow{\quad v' \quad} & \mathbf{G}' & \longrightarrow & 0 \end{array}$$

*where the bottom line is an exact sequence in* **Func(I, A)**, *i.e. for each index $i$ in* **I**,

$$0 \longrightarrow \mathbf{E}'(i) \xrightarrow{u(i)} \mathbf{F}'(i) \xrightarrow{v'(i)} \mathbf{G}'(i) \longrightarrow 0$$

*is a short exact sequence in* **A**.

Now that we have the basic properties of **pro(A)** at our disposal together with some useful results on exact sequences, we can start on our analysis of stability in **pro(A)**. The first result characterises subobjects in **pro(A)** of objects of **A**. As usual in this section, **A** will be an Abelian category.

**Proposition 3**

*Let* **(I, F)** *be an object in* **pro(A)**. *The following statements are equivalent:*

(a) *the system $\mathbf{F}: \mathbf{I} \to \mathbf{A}$ satisfies the condition:*

(EM): *There is an index $j_0$ in* **I** *such that for any $\alpha: i \to j_0$ there is a $\beta: k \to i$ for which the natural map*

$$\operatorname{Ker}(\mathbf{F}(\beta)) \to \operatorname{Ker}(\mathbf{F}(\alpha\beta))$$

*is an isomorphism;*

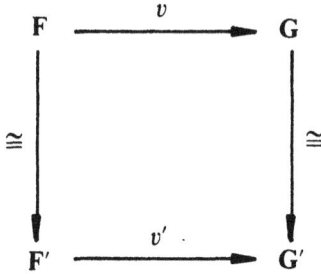

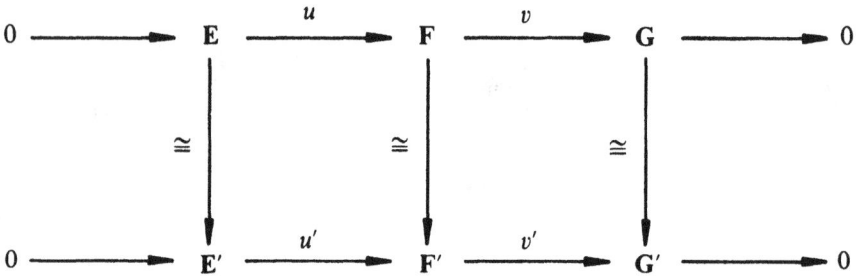

(*b*) $\mathbf{F}$ *is a subobject (in* **pro**$(\mathbf{A})$*) of some object of* $\mathbf{A}$;
(c) $(\mathbf{I}, \mathbf{F})$ *is isomorphic to a pro-object in* $\mathbf{A}$ *all of whose bonding maps are monomorphisms.*

*Proof*
Suppose $(\mathbf{I}, \mathbf{F})$ satisfies condition (EM) and consider the projection $p = pr_{j_0}: \mathbf{F} \to \mathbf{F}(j)$. We claim that this is a monomorphism in **pro**$(\mathbf{A})$. In fact using the Reindexing Lemma gives us a cofiltering category $\mathbf{H}_p$ and $\mathbf{Ker}\ p: \mathbf{H}_p \to \mathbf{A}$. As before, this indexing category has as objects the triples $(i, f, j)$ where $i$ is in $\mathbf{I}$, $j$ is in the indexing category $\mathbf{1}$, of the object $\mathbf{F}(j_0)$ and $f: \mathbf{F}(i) \to \mathbf{F}(j_0)$ represents $p$; of course this latter condition means that

$$\begin{array}{ccc} \mathbf{F} & \xrightarrow{pr_{j_0}} & \mathbf{F}(j_0) \\ {\scriptstyle pr_i}\big\downarrow & & \big\downarrow{\scriptstyle =} \\ \mathbf{F}(i) & \xrightarrow[f]{} & \mathbf{F}(j_0) \end{array}$$

commutes and as if $\alpha: i \to j_0$ in $\mathbf{I}$, $pr_{j_0} = \mathbf{F}(\alpha)pr_i$, $\mathbf{H}_p$ contains an initial subcategory isomorphic to $\mathbf{I}\downarrow j_0$, i.e. to the subcategory of objects smaller than $j_0$. Because of this, $\mathbf{Ker}\ p$ is isomorphic to the system $\mathbf{K}: \mathbf{I}\downarrow j_0 \to \mathbf{A}$ defined by $\mathbf{K}(\alpha: i \to j_0) = \mathrm{Ker}\ \mathbf{F}(\alpha)$. However, for any such index $\alpha$, there is a $\beta: k \to i$ such that the inclusion of $\mathrm{Ker}\ \mathbf{F}(\alpha\beta)$ into $\mathrm{Ker}\ \mathbf{F}(\beta)$ is an isomorphism (by the hypothesis that $\mathbf{F}$ satisfies (EM)). The kernel of $\mathbf{K}(\beta): \mathbf{K}(\alpha\beta) \to \mathbf{K}(\alpha)$ is the kernel of $\mathbf{F}(\beta)$ restricted to $\mathrm{Ker}(\mathbf{F}(\alpha\beta))$ and hence is $\mathrm{Ker}\ \mathbf{F}(\beta) \cap \mathrm{Ker}\ \mathbf{F}(\alpha\beta) = \mathrm{Ker}\ \mathbf{F}(\alpha\beta)$ so $\mathbf{K}(\beta)$ is the zero map and $\mathbf{K}$ is the zero pro-object. Thus $\mathbf{Ker}\ p$ is trivial, $p$ is a monomorphism as required, and $\mathbf{F}$ is a subobject of $\mathbf{F}(j_0)$.

Now suppose that $\mathbf{F}$ is a subobject of some object of $\mathbf{A}$, that is, there is a monomorphism

$$u: \mathbf{F} \to \mathbf{A}$$

in **pro**$(\mathbf{A})$. We have seen that $u$ can be replaced by a levelwise monomorphism

$$\bar{u}: \bar{\mathbf{F}} \to \bar{\mathbf{A}}$$

where, on inspection, we have as indexing system an initial subcategory of $\mathbf{I}$, $\bar{\mathbf{F}}(i) = \mathrm{Im}\ u(i)$, and $\bar{\mathbf{A}}$ is the corresponding constant functor. As each $\bar{u}(i)$ is then a monomorphism, each bonding map in $\bar{\mathbf{F}}$ must also be a monomorphism. As $\mathbf{F} \cong \bar{\mathbf{F}}$, this proves that (b) implies (c).

To complete the proof of the Proposition, we note that if two pro-objects $\mathbf{G}$ and $\mathbf{H}$ are isomorphic and $\mathbf{G}$ satisfies (EM), then so does $\mathbf{H}$. The proof is not hard and is left as an exercise. Now if $\mathbf{F}$ is isomorphic to a pro-object $\mathbf{F}'$ having monomorphic bonding maps, $\mathbf{F}'$ trivially satisfies (EM), hence $\mathbf{F}$ must satisfy (EM) as well.

**Definition**
An object of **pro**$(\mathbf{A})$ that satisfies any one of the equivalent conditions of Proposition 3 (above) will be said to be *essentially monomorphic*.

Thus the essentially monomorphic pro-objects are the subobjects of stable pro-objects.

Although quotients of constant pro-objects are pro-objects which are isomorphic to ones having epimorphic bonding morphisms, the converse is not necessarily true. A very simple example is that in which **A** is the category of finite dimensional real vector spaces. If we consider $\mathbb{N}$ as indexing category for the system **F** having $\mathbf{F}(n) = \mathbb{R}^n$ with the bonding from $\mathbf{F}(n+1)$ to $\mathbf{F}(n)$ being projection onto the first $n$ coordinates, then clearly there can be no vector space in **A** which projects onto all the $\mathbf{F}(i)$. Thus the result on essentially monomorphic systems cannot be expected to have a dual; however, the following is true.

**Proposition 4**

*Let* $(\mathbf{I}, \mathbf{F})$ *be a pro-object in* **A**; *then* **F** *is isomorphic to a pro-object all of whose bonds are epimorphisms if and only if* $(\mathbf{I}, \mathbf{F})$ *itself satisfies the Mittag–Leffler condition*:

(ML) *Given any* $i$ *in* **I** *there is a* $j_0$ *in* **I** *and* $\alpha: j_0 \to i$ *such that if* $\beta: j' \to j_0$, *then the natural morphism*

$$p: \operatorname{Im} \mathbf{F}(\alpha\beta) \to \operatorname{Im} \mathbf{F}(\alpha)$$

*is an isomorphism.*

**Remarks**

(i) In comparison with the condition (EM), it is often convenient to use the term 'essentially epimorphic' for a pro-object that satisfies the Mittag–Leffler condition. The term 'strict pro-object' is also used for pro-objects with epimorphic bonds.

(ii) Proposition 4 is, in fact, a special case of a more general result, which is true in those categories for which the idea of image makes sense and behaves well. Rather than search here for the most general possible statement of this result, we will give a proof of it that nowhere uses additivity. This proof will therefore generalise to handle the cases of essentially epimorphic prosets or progroups that we will need later.

*Proof of Proposition 4*

Suppose $(\mathbf{I}, \mathbf{F})$ satisfies the Mittag–Leffler condition and consider the category $\mathbf{I}^2$ of morphisms in **I**. This category $\mathbf{I}^2$ is cofiltering and there is an initial functor $\mathbf{\Delta}: \mathbf{I} \to \mathbf{I}^2$ given by $\mathbf{\Delta}(i) = \operatorname{Id}(i)$, the identity morphism on $i$.

There is a functor $\bar{\mathbf{F}}: \mathbf{I}^2 \to \mathbf{A}$ given by $\bar{\mathbf{F}}(u) = \operatorname{Im} \mathbf{F}(u)$ and if $(\bar{u}, \bar{u}')$ is a morphism from $u_1$ to $u_2$ in $\mathbf{I}^2$ (so that

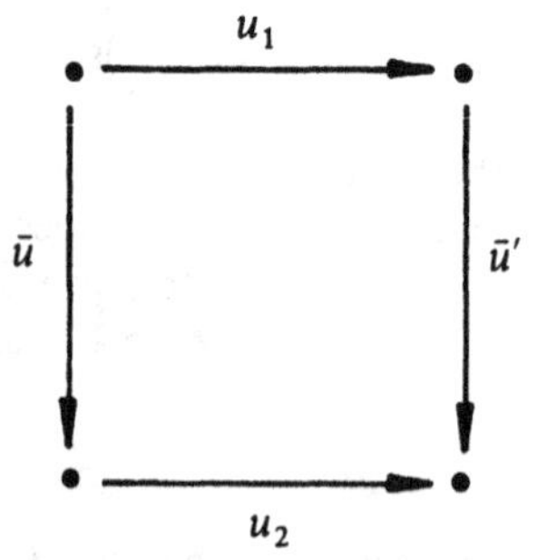


commutes in **I**) then $\bar{\mathbf{F}}(\bar{u}, \bar{u}') = \mathbf{F}(\bar{u}')|\operatorname{Im} \mathbf{F}(u)$.

We note that $\bar{\mathbf{F}}\Delta = \mathbf{F}$ so $(\mathbf{I}^2, \bar{\mathbf{F}}) \cong (\mathbf{I}, \mathbf{F})$.

Now let $\mathbf{J}$ be the full subcategory of $\mathbf{I}^2$ determined by:

if $i$ is in $\mathbf{I}$, $u: i' \to i$ is in $\mathbf{J}$ if and only if for all $u': i'' \to i'$ in $\mathbf{I}$, $\operatorname{Im} \mathbf{F}(u) = \operatorname{Im} \mathbf{F}(uu')$.

The Mittag–Leffler condition implies precisely that $\mathbf{J}$ is initial in $\mathbf{I}^2$, hence on defining $\mathbf{G}: \mathbf{J} \to \mathbf{A}$ to be $\bar{\mathbf{F}}\mathbf{J}$, i.e. $\mathbf{G}(u) = \operatorname{Im} \mathbf{F}(u)$, then $(\mathbf{J}, \mathbf{G}) \cong (\mathbf{I}, \mathbf{F})$ in **pro**(**A**). To complete one half of the proof it remains to check that for each morphism $v$ in $\mathbf{J}$, $\mathbf{G}(v)$ is an epimorphism. Let $v = (\bar{u}, \bar{u}'): u_1 \to u_2$ in $\mathbf{J}$ so $\bar{u}'u_1 = u_2\bar{u}$. Then $\operatorname{Im} \mathbf{F}(u_2) = \operatorname{Im} \mathbf{F}(u_2\bar{u}) = \operatorname{Im} \mathbf{F}(\bar{u}u_1) = \operatorname{Im}(\mathbf{F}(\bar{u})|\operatorname{Im} \mathbf{F}(u_1))$ but this is $\operatorname{Im} \mathbf{G}(v)$ by definition.

To prove the converse we first note that it is sufficient to prove the following lemma, which is stronger than we need for the proposition.

**Lemma**

*If* $\mathbf{F}$ *is a strict pro-object and* $u: \mathbf{F} \to \mathbf{G}$ *is an epimorphism in* **pro**(**A**) *then* $\mathbf{G}$ *satisfies the Mittag–Leffler condition.*

*Proof*

First we may assume that $u$ is a level morphism without loss of generality, since reindexing a strict pro-object yields a strict pro-object. We then note that $\mathbf{Im}\, u$ is a strict pro-object. We are thus left to prove that if

$$m: \mathbf{Im}\, u \rightarrowtail \mathbf{G}$$

is the 'inclusion' monomorphism (which is thus an isomorphism) and $\mathbf{Im}\, u$ is strict, then $\mathbf{G}$ satisfies Mittag–Leffler.

Since $m$ is an isomorphism, there is a map $\theta: |\mathbf{I}| \to |\mathbf{I}|$ and morphisms $\{s(i): \mathbf{G}(\theta(i)) \to \operatorname{Im} u(i): i \in I\}$ such that there is some $j_0$ with morphisms $\alpha: j_0 \to i$ and $\alpha': j_0 \to \theta(i)$ such that

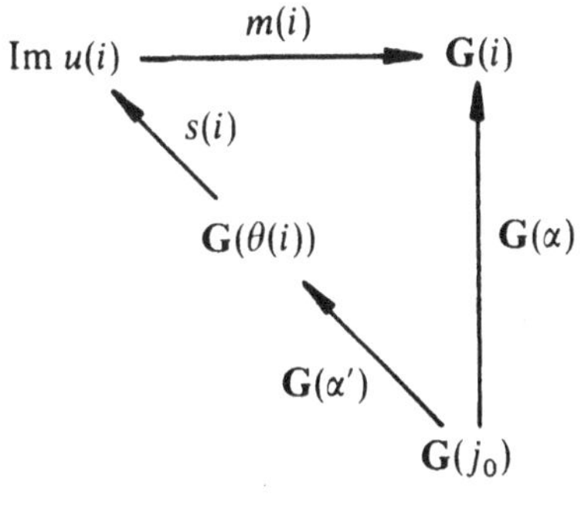


commutes. Hence $\operatorname{Im} \mathbf{G}(\alpha) \subset \operatorname{Im} u(i)$, but corresponding to $\alpha: j_0 \to i$ there is an epimorphism $(\mathbf{Im}\, u)(\alpha): \operatorname{Im} u(j_0) \to \operatorname{Im} u(i)$ such that $m(i)(\operatorname{Im} u)(\alpha) = \mathbf{G}(\alpha)m(j)$. Thus $\operatorname{Im} u(i) = \operatorname{Im}(\operatorname{Im} u(\alpha)) \subset \operatorname{Im} \mathbf{G}(\alpha)$. (We are abusing notation here and are pretending that $m(i)$ is an inclusion, not just a subobject.) Thus $\operatorname{Im} \mathbf{G}(\alpha) = \operatorname{Im} u(i)$.

Similarly if $\beta: j \to j_0$, then $\operatorname{Im} \mathbf{G}(\alpha\beta) = \operatorname{Im} u(i)$, hence $\operatorname{Im} \mathbf{G}(\alpha) = \operatorname{Im} \mathbf{G}(\alpha\beta)$ as required. This finishes the proof of both the lemma and the proposition.

Combining Propositions 3 and 4, we follow Verdier and Duskin in calling a pro-object which is both essentially monomorphic and essentially epimorphic *essentially constant*.

**Corollary**

*A pro-object* **F** *is essentially constant if and only if it is stable.*

*Proof*

If **F** is essentially monomorphic, we may replace it by a pro-object, **F**′, having monomorphic bonds. If, in addition, **F** satisfies (ML) then **F**′ will also satisfy (ML), as the last proposition and the lemma contained in its proof easily imply that being essentially epimorphic is (isomorphism) invariant. Hence we can assume **F** itself is monomorphic and satisfies (ML). We now feed it into the process used in the proof of Proposition 3, which replaces it by a strict pro-object, **G**, say. It is easily checked that at each stage of the process the bonds are restrictions of bonds of **F** and hence are monomorphisms. Thus the pro-object **G** that results from this process has all its bonds both epimorphic and monomorphic and so is an 'essentially isomorphic' system. Now it is a trivial matter to show that **G** is stable.

The converse is an easy consequence of the last two propositions and the lemma.

Thus we have with this corollary an internal characterisation of stability in pro-Abelian categories. The importance of this is two-fold. Firstly it enables us, up to a point, to use functors to Abelian categories as test functors for stability, but, and this is probably more important, it also allows us to gain insight into the ways in which stability may fail. The best route to this is via some more specific results to which we now turn.

We earlier pointed out that the Mittag–Leffler condition could not, in general, imply that the pro-object concerned was a quotient of a stable one. Our example was of a pro-finite-dimensional vector space. It is well known that taking limits, in general, destroys exactness but that limit functors on categories of systems of finite dimensional vector spaces are exact. This suggests that pro-finite-dimensional vector spaces are not far from being stable. Let us look at this in more detail.

Let $k$ be a field so that **Mod**-$k$ is the category of $k$-vector spaces. If $\mathbf{M}: \mathbf{I} \to \mathbf{Mod}\text{-}k$ is a pro-object in **Mod**-$k$, we will say that **M** is of *essentially bounded dimension* if there is an integer $n$ such that

$$\dim_k(\operatorname{Im} \mathbf{M}(\alpha)) < n$$

for all bonds $\alpha: i \to j$ in **I**.

**Proposition 5**

*If* (**I**, **M**) *is a provector space of essentially bounded dimension, then* **M** *is stable.*

*Proof*

We first check that the Mittag–Leffler condition is satisfied. Using the fact that each $\operatorname{Im} \mathbf{M}(\alpha)$ is finite dimensional, we may, if need be, replace **M** by an isomorphic provector space in which each $\mathbf{M}(i)$ is of finite dimension, hence we may as well assume this of **M** itself. For any index $i$ in $I$ the family $\{\operatorname{Im} \mathbf{M}(\alpha): \operatorname{codom}(\alpha) = i\}$ is a family of subspaces of $\mathbf{M}(i)$ and so has a minimal member which we denote $\operatorname{Im} \mathbf{M}(\alpha(i))$ where $\alpha(i): f(i) \to i$. (Here it is necessary to choose as in general many such $\operatorname{Im} \mathbf{M}(\alpha)$ will do.) If $\beta: j \to f(i)$, then $\operatorname{Im} \mathbf{M}(\alpha(i)\beta) \subseteq \operatorname{Im} \mathbf{M}(\alpha(i))$ but $\operatorname{Im} \mathbf{M}(\alpha(i))$ is minimal so $\operatorname{Im} \mathbf{M}(\alpha(i)\beta) = \operatorname{Im} \mathbf{M}(\alpha(i))$ as required and (ML) is satisfied. We next replace **M** by an isomorphic strict pro-object. (Following the method given above, this is possible without destroying the finite dimensional nature of the individual vector spaces in the pro-object.)

We now turn to (EM): let $i$ be fixed and consider the family of numbers $\dim_k(\text{Ker}\,\mathbf{M}(\alpha))$, $\text{codom}(\alpha) = i$. These are bounded above and non-decreasing (since each $\mathbf{M}(\alpha)$ is an epimorphism) and so there is some $\alpha: i' \to i$ such that for $\beta: j \to i'$

$$\dim_k(\text{Ker}\,\mathbf{M}(\alpha\beta)) = \dim_k(\text{Ker}\,\mathbf{M}(\alpha)).$$

This implies (EM) holds, but we can also use it to prove stability by a direct route. For such a $\beta$, we must have

$$\dim_k(\text{Ker}\,\mathbf{M}(\alpha\beta)) \geqslant \dim_k(\text{Ker}\,\mathbf{M}(\beta)) + \dim_k(\text{Ker}\,\mathbf{M}(\alpha))$$

since $\mathbf{M}(\beta)$ is onto. However, this implies that $\dim_k(\text{Ker}\,\mathbf{M}(\beta))$ is zero so $\mathbf{M}(\beta)$ is a monomorphism and hence an isomorphism. Thus the bonds in $\mathbf{M}$ are 'essentially isomorphic' and $\mathbf{M}$ is stable.

**Remark**

The above argument can easily be generalised to promodules over certain other rings. The most important class for which this is true is that of semi-simple Artinian rings. These rings look like sums of non-commutative fields and so concepts like essentially bounded dimension go across with little or no problem.

We have not the space here to explore in any great depth the nature of stability in specific module categories and it would divert our attention away from the categorical questions that are our prime target; however, one example of a non-stable essentially monomorphic pro-Abelian group deserves to be given. This is related to the $P$-adic solenoids of Chapter 3. The indexing category will be the natural numbers $\mathbb{N}$, for $n \in \mathbb{N}$, $\mathbf{M}(n) = \mathbb{Z}$, and for the unique bond corresponding to $\alpha: n+1 \to n$, we have $\mathbf{M}(\alpha)(i) = 2i$. Each bond is thus monomorphic, but not epimorphic and, in fact, the sequence of images within any given $\mathbf{M}(n)$ is, of course, an infinite strictly decreasing one. To see directly that $\mathbf{M}$ is not stable we note that $\text{Lim}\,\mathbf{M} = 0$, so if it were stable it would have to be the trivial system, and as its bonds are monomorphic, and $\mathbf{M(n)}$ is always non-zero, this is clearly impossible.

To finish this section we quote a result which, in part, explains the limitations that stability puts on pro-Abelian groups.

**Proposition**

*If* $\mathbf{M}$ *is in* **pro(Ab)**, $\mathbf{M}$ *is of bounded torsion free rank (i.e. the torsion free ranks of the* $\mathbf{M}(i)$*s are bounded) and for any* $\alpha$, Coker $\mathbf{M}(\alpha)$ *and* Ker $\mathbf{M}(\alpha)$ *are torsion free, then* $\mathbf{M}$ *is stable.*

A proof of this and various related results can be found in Porter [89–91].

## 6.5 THE DERIVED FUNCTORS OF Lim

If the Abelian category **A** has arbitrary products, then the 'constant pro-object' functor

$$\mathbf{c}: \mathbf{A} \to \mathbf{pro(A)}$$

has a right adjoint, namely the limit functor. As Lim is a right adjoint, it is left exact, but need not be right exact. The classic example of this is obtained by picking two coprime natural numbers $m$ and $n$ both greater than 1 and then forming an exact sequence in **pro(Ab)** as follows:

let $\mathbf{S}(\mathbb{Z}, m)$ be the inverse sequence indexed by the natural numbers, $\mathbb{N}$, and specified by:

(i) $\mathbf{S}(\mathbb{Z}, m)_i = \mathbb{Z}$ for all $i$, and
(ii) $p_i^{i+1}: \mathbf{S}(\mathbb{Z}, m)_{i+1} \to \mathbf{S}(\mathbb{Z}, m)_i$ is given by $p_i^{i+1}(1) = m$.

(The example we considered towards the end of the last section was an example of such a system, namely $\mathbf{S}(\mathbb{Z}, 2)$.)

Now consider the level monomorphism

$$u: \mathbf{S}(\mathbb{Z}, m) \to \mathbf{S}(\mathbb{Z}, m)$$

given by $u_i(1) = n$.

The cokernel of $u$ is isomorphic to $\mathbf{c}(\mathbb{Z}_n)$, the constant system on the integers modulo $n$. Now take limits: $\mathrm{Lim}\, \mathbf{S}(\mathbb{Z}, m) = 0$ so from the exact sequence

$$0 \to \mathbf{S}(\mathbb{Z}, m) \xrightarrow{u} \mathbf{S}(\mathbb{Z}, m) \to \mathbf{c}(\mathbb{Z}_n) \to 0$$

we obtain a sequence

$$0 \to 0 \to 0 \to \mathbb{Z}_n \to 0$$

which is certainly not exact.

This non-exactness complicates the use of limiting processes as we have already noted in our earlier discussion of Čech homology. As usual with non-exact functors on Abelian categories, we can try to measure the non-exactness of Lim using derived functor theory. The derived functors of Lim have been much studied elsewhere and as one can find in Jensen's lecture notes [61] a detailed, if now slightly old, description of how this theory is applied in module theory, we will limit ourselves to a brief introduction to the construction of these 'derived limits'. We will not, in this section, provide full proofs as these are easily available elsewhere (more often than not in Jensen's notes [61]). Perhaps it should be mentioned that those notes, [61], contain many examples of applications of the general approximation theoretic or shape theoretic method within algebra, and are thoroughly to be recommended.

Because we are not aiming for full generality in this section we will assume that $\mathbf{A}$ is a category of modules. We first consider the simplest case, namely that when the indexing category $\mathbf{I}$ for the chosen pro-object is initially countable (i.e. has a countable initial subcategory); in this case we may assume that the indexing category is $\mathbb{N}$. (This will, for instance, be the case when $\mathbf{I} = \mathrm{cov}(X)$, the directed set of open covers of a compact metric space $X$. It is thus of use when studying Čech homology and the higher Čech homotopy/shape groups of compact metric spaces.)

We know that if $\mathbf{M}: \mathbb{N} \to \mathbf{A}$ is an inverse sequence of objects of $\mathbf{A}$ (i.e. of modules) then Lim $\mathbf{M}$ can be described as a subobject of the product $\prod\{\mathbf{M}(i): i \in \mathbb{N}\}$ specified by the property that $\underline{m} \in \mathrm{Lim}\, M$ if and only if for each $i \in \mathbb{N}$, $p_i^{i+1}(m_{i+1}) = m_i$ where as usual we have written $p_i^{i+1}$ as being short for the bond, $\mathbf{M}(i+1 \to i)$.

Consider the complex

$$\prod_{i \in \mathbb{N}} \mathbf{M}(i) \xrightarrow{\partial_{\mathbf{M}}} \prod_{i \in \mathbb{N}} \mathbf{M}(i)$$

in dimensions 0 and 1, where $\partial_{\mathbf{M}}$ is defined by

$$\partial_M(x_1, \ldots, x_n, \ldots) = (x_1 - p_1^2 x_2, \ldots, x_n - p_n^{n+1} x_{n+1}, \ldots);$$

in other dimensions the complex is zero. It is clear that $\mathrm{Lim}\, \mathbf{M} = \mathrm{Ker}\, \partial_{\mathbf{M}}$.

If

$$0 \to \mathbf{L} \to \mathbf{M} \to \mathbf{N} \to 0$$

is an exact sequence in **Func**($\mathbb{N}$, **A**), then we obtain an exact sequence

$$\begin{array}{ccccccccc}
0 & \longrightarrow & \prod\mathbf{L} & \longrightarrow & \prod\mathbf{M} & \longrightarrow & \prod\mathbf{N} & \longrightarrow & 0 \\
 & & \Big\downarrow{\scriptstyle \partial_{\mathbf{L}}} & & \Big\downarrow{\scriptstyle \partial_{\mathbf{M}}} & & \Big\downarrow{\scriptstyle \partial_{\mathbf{N}}} & & \\
0 & \longrightarrow & \prod\mathbf{L} & \longrightarrow & \prod\mathbf{M} & \longrightarrow & \prod\mathbf{N} & \longrightarrow & 0
\end{array}$$

provided that products are exact in **A** as is the case in module categories. The well known 'snake lemma' of elementary homological algebra gives us a six-term exact sequence

$$0 \to \operatorname{Ker} \partial_{\mathbf{L}} \to \operatorname{Ker} \partial_{\mathbf{M}} \to \operatorname{Ker} \partial_{\mathbf{N}} \xrightarrow{\partial} \operatorname{Coker} \partial_{\mathbf{L}} \to \operatorname{Coker} \partial_{\mathbf{M}} \to \operatorname{Coker} \partial_{\mathbf{N}} \to 0$$

which thus yields

$$0 \to \operatorname{Lim} \mathbf{L} \to \operatorname{Lim} \mathbf{M} \to \operatorname{Lim} \mathbf{N} \to \operatorname{Lim}^{(1)} \mathbf{L} \to \operatorname{Lim}^{(1)} \mathbf{M} \to \operatorname{Lim}^{(1)} \mathbf{N} \to 0$$

where $\operatorname{Lim}^{(1)} \mathbf{M}$ is defined to be Coker $\partial_{\mathbf{M}}$ etc. These $\operatorname{Lim}^{(1)}$ terms thus measure the extent to which Lim has destroyed exactness. For instance, for our classical example

$$0 \to \mathbf{S}(\mathbb{Z}, m) \xrightarrow{u} \mathbf{S}(\mathbb{Z}, m) \to \mathbf{c}(\mathbb{Z}_n) \to 0$$

we see that $\operatorname{Lim}^{(1)} \mathbf{S}(\mathbb{Z}, m)$ contains a copy of $\mathbb{Z}_n$ for each $n \in \mathbb{N}$ since

$$0 \to \mathbb{Z}_n \to \operatorname{Lim}^{(1)} \mathbf{S}(\mathbb{Z}, m) \xrightarrow{\operatorname{Lim}^{(1)} u} \operatorname{Lim}^{(1)} \mathbf{S}(\mathbb{Z}, m) \to 0$$

is exact for each $n$. In fact it is much bigger since, as multiplication by $n$ is an epimorphism on $\operatorname{Lim}^{(1)} \mathbf{S}(\mathbb{Z}, m)$, this group is divisible. Further results on these types of groups can be found in Jensen [61].

If **I** is a general cofiltering index category and $\mathbf{M}: \mathbf{I} \to \mathbf{A}$ then the complex used is more complicated. First, for each $q = 0$ we form a product

$$\prod{}^q \mathbf{M} = \prod \left\{ \mathbf{M}(u) \,\middle|\, u = \left( i_q \xrightarrow{\alpha_q} \cdots \xrightarrow{\alpha_1} i_0 \right), \mathbf{M}(u) = \mathbf{M}(i_0) \right\};$$

thus the product is indexed by all composable $q$-tuples of morphisms in **I** and to each such $q$-tuple is assigned the value of **M** on its codomain. These will be the objects of $q$-cochains of the complex. The coboundary $\partial: \prod^q \mathbf{M} \to \prod^{q+1} \mathbf{M}$ is given by

$$p_v(\partial \underline{m}) = \mathbf{M}(\alpha_1) p_{d_0 v}(\underline{m}) + \sum_{k=1}^{q+1} (-1)^k p_{d_k v}(\underline{m})$$

where, if

$$\underline{v} = \left( i_{q+1} \xrightarrow{\alpha_{q+1}} \cdots \xrightarrow{\alpha_1} i_0 \right), \qquad d_0 \underline{v} = \left( i_{q+1} \longrightarrow \cdots \xrightarrow{\alpha_2} i_1 \right),$$

for $1 \leqslant k \leqslant q$,

$$d_k \underline{v} = \left( i_{q+1} \longrightarrow \cdots \xrightarrow{\alpha_k \alpha_{k+1}} \cdots \xrightarrow{\alpha_0} i_0 \right)$$

and

$$d_{q+1} \underline{v} = \left( i_q \xrightarrow{\alpha_q} \cdots \xrightarrow{\alpha_1} i_0 \right),$$

$p_u \colon \prod^q \mathbf{M} \to \mathbf{M}(u)$ being the $u$th projection of the product.

If

$$0 \to \mathbf{L} \to \mathbf{M} \to \mathbf{N} \to 0$$

is a short exact sequence in **Func(I, A)**, then as long as products exist and are exact in **A**, we obtain a short exact sequence of complexes in **A**

$$0 \to \prod^\bullet \mathbf{L} \to \prod^\bullet \mathbf{M} \to \prod^\bullet \mathbf{N} \to 0$$

and on defining $\mathrm{Lim}^{(i)}\, \mathbf{M} = H^i(\prod^\bullet \mathbf{M})$ the $i$th cohomology of $\prod^\bullet \mathbf{M}$, etc., we obtain a long exact sequence

$$0 \to \mathrm{Lim}^0\, \mathbf{L} \to \mathrm{Lim}^0\, \mathbf{M} \to \mathrm{Lim}^0\, \mathbf{N} \to \mathrm{Lim}^1\, \mathbf{L} \to \cdots$$
$$\cdots \to \mathrm{Lim}^{(i)}\, \mathbf{L} \to \mathrm{Lim}^{(i)}\, \mathbf{M} \to \mathrm{Lim}^{(i)}\, \mathbf{N} \to \mathrm{Lim}^{(i+1)}\, \mathbf{L} \to \cdots$$

and it is easily seen that Lim and $\mathrm{Lim}^0$ are isomorphic functors.

If $\phi \colon \mathbf{J} \to \mathbf{L}$ is an initial functor then the induced morphisms $\mathrm{Lim}^{(i)}\, \mathbf{M} \to \mathrm{Lim}^{(i)}\, \mathbf{M}\phi$ are all isomorphisms, so by using the Reindexing Lemma of section 6.3, we can easily extend these derived limits from the various functor categories **Func(I, A)** to **pro(A)**.

We next list some technical results for $\mathbf{A} = \mathbf{Mod}\text{-}\mathbf{A}$ without proof. (Proofs are mostly to be found in Jensen's notes [61].) These are given in an attempt to suggest how the non-vanishing of $\mathrm{Lim}^{(i)}\, \mathbf{M}$ should be interpreted.

(A) If **I** contains an initial subcategory of cardinality at most $\aleph_n$, then for all $\mathbf{M} \colon \mathbf{I} \to \mathbf{Mod}\text{-}\mathbf{A}$ and $q \geqslant n+2$, $\mathrm{Lim}^{(q)}\, \mathbf{M} = 0$.

(B) If $\mathbf{M} \colon \mathbf{I} \to \mathbf{Mod}\text{-}\mathbf{A}$, where **I** is a directed set and each module $\mathbf{M}(i)$ is Noetherian of Krull–Gabriel dimension $\leqslant n$, then $\mathrm{Lim}^{(q)}\, \mathbf{M} = 0$ for $q > n$. In particular all $\mathrm{Lim}^{(q)}\, \mathbf{M} = 0$ for $q > 0$ if all the $\mathbf{M}(i)$ are Artinian modules, e.g. if $A$ is a field and each $\mathbf{M}(i)$ is finite dimensional.

(C) If each $\mathbf{M}(i)$ is a compact topological module and the bonds are continuous then $\mathrm{Lim}^{(q)}\, \mathbf{M} = 0$ for all $q > 0$.

The study of the derived functors of Lim has many important and subtle interactions with other parts of algebra and logic.

**Remark**

A non-Abelian version of $\mathrm{Lim}^{(1)}$ has been defined by Bousfield and Kan [16]. We omit the details.

## 6.6 MOVABILITY AND MITTAG–LEFFLER

Since the $\mathrm{Lim}^{(q)}$ are functors and, clearly, $\mathrm{Lim}^{(q)}\, \mathbf{M} = 0$ if **M** is a constant system, it must be that these derived functors vanish on stable pro-objects and hence can be used as a test for non-stability. Of course one does not need stability to obtain vanishing derived limits, for instance pro-finite-dimensional vector spaces are not usually stable, but we have noted that for $q>0$ the $\mathrm{Lim}^{(q)}$ vanish on them. In this section we connect up the Mittag–Leffler condition with vanishing derived limits and discuss how one may try to extend this to non-additive contexts. (Here we will be needing to use the non-additive versions of our results on Mittag–Leffler and essentially epimorphic pro-objects as we mentioned in section 6.4.)

First we note how to extend the Mittag–Leffler condition to pro-objects in other than Abelian categories. The definition actually only needs a well behaved notion of image such as would be available in any category with an epi-monic factorisation, and we will briefly consider such a situation later, but for now we will restrict attention to prosets and progroups.

A proset $\mathbf{S}: \mathbf{I} \to \mathbf{Sets}$ (or progroup $\mathbf{S}: \mathbf{I} \to \mathbf{Groups}$) is said to satisfy the Mittag–Leffler condition if for each $i$ in $\mathbf{I}$, there is a morphism $u: i' \to i$ such that for every $u': i'' \to i'$

$$\mathrm{Im}(\mathbf{S}(u)) = \mathrm{Im}(\mathbf{S}(uu')).$$

**Lemma 1**

*Let* $\mathbf{S}: \mathbb{N} \to \mathbf{Sets}$ *be a countably indexed proset that satisfies the Mittag–Leffler condition. Writing* $p_n$: $\mathrm{Lim}\,\mathbf{S} \to \mathbf{S}(n)$ *for the* nth *projection, for any* $n \in \mathbb{N}$ *there is an* $m \geqslant n$ *such that*

$$\mathrm{Im}\ p_n = p_n^m(\mathbf{S}(m))$$

*where as usual for* $\mathbb{N}$*-indexed pro-objects* $p_n^m = \mathbf{S}(m \to n)$.

*Proof*

Using the Mittag–Leffler condition, we can construct a function $\theta: \mathbb{N} \to \mathbb{N}$ satisfying

(1) $\theta(n) \geqslant n$
(2) for each $n' \geqslant \theta(n)$, $\mathrm{Im}\ p_n^{n'} = \mathrm{Im}\ p_n^{\theta(n)}$.

Let $\mathbf{T}(n) = p_n^{\theta(n)}\mathbf{S}(\theta(n))$.

If $n \leqslant n'$, let $l \geqslant \max(\theta(n), \theta(n'))$, then one has $\mathbf{T}(n) = p_n^{\theta(n)}\mathbf{S}(\theta(n)) = p_n^l\mathbf{S}(l) = p_n^{n'}p_{n'}^l\mathbf{S}(l) = p_n^{n'}p_{n'}^{\theta(n')}\mathbf{S}(\theta(n')) = p_n^{n'}\mathbf{T}(n')$. It is thus clear that $(\mathbf{T}(n), p_n^{n'}|\mathbf{T}(n), \mathbb{N})$ defines a proset which has surjective bonds. (In fact the above derivation is a simpler version of part of the proof of Proposition 4 of section 6.4 in this simpler situation.) The projections $t_n$: $\mathrm{Lim}\ \mathbf{T}(n) \to \mathbf{T}(n)$ are also surjections and $T = \mathrm{Lim}\ \mathbf{T}(n) \subseteq \mathrm{Lim}\ \mathbf{S}(n)$.

Consequently $t_n(T) = \mathbf{T}(n) = p_n^{\theta(n)}\mathbf{S}(\theta(n)) \subseteq \mathrm{Im}\ p_n$ and as $\mathrm{Im}\ p_n = p_n^{\theta(n)}\ \mathrm{Im}\ p_{\theta(n)} \subset p_n^{\theta(n)}\mathbf{S}(\theta(n))$, we have $\mathrm{Im}\ p_n = p_n^{\theta(n)}\mathbf{S}(\theta(n))$ as required.

**Remark**

We note that in the non-Mittag–Leffler pro-object $\mathbf{S}(\mathbb{Z}, m)$ for $m > 1$, $\mathrm{Lim}\ \mathbf{S}(\mathbb{Z}, m) = 0$ but each $\mathrm{Im}\ p_n^m$ is infinite.

**Lemma 2**

*Let* $\mathbf{G}: \mathbb{N} \to \mathbf{Groups}$ *be a progroup satisfying the Mittag–Leffler condition. If* $\mathrm{Lim}\ \mathbf{G}(n) = 0$ *then* $\mathbf{G} \cong 0$ *in* **proGroups**.

*Proof*
For each $n \in \mathbb{N}$, there is a $\theta(n) \geq n$ such that

$$p_n(\text{Lim } \mathbf{G}) = p_n^{\theta(n)}\mathbf{G}(\theta(n)) = 0,$$

i.e. $p_n^{\theta(n)} = 0$, but this implies that $\mathbf{G} \cong 0$.

**Proposition 1**
*Let* $\{f_n: \mathbf{G}(n) \to \mathbf{H}(n) | n \in \mathbb{N}\}$ *be a level epimorphism of countably indexed progroups such that its kernel* (Ker $f_n$) satisfies (ML); *then* Lim $f$ *is an epimorphism.*

*Proof*
Let $(y_n)_{n \in \mathbb{N}}$ be an element of $H = \text{Lim } \mathbf{H}(n)$. As each $f_n$ is an epimorphism, there exist $x_n \in \mathbf{G}(n)$ such that $f_n(x_n) = y_n$; of course $(x_n)_{n \in \mathbb{N}}$ need not necessarily be an element of $G = \text{Lim } \mathbf{G}(n)$.

As $f_n p_n^{n+1}(x_{n+1}) = q_n^{n+1} f_n(x_{n+1}) = y_n = f_n(x_n)$, we have $p_n^{n+1}(x_{n+1}) - x_n \in \ker f_n$. By assumption the prokernel Ker $f$ satisfies ML; thus on reindexing we may assume its bonds are actually epimorphisms. Therefore there is some $z_{n+1} \in \text{Ker } f_{n+1}$ such that $p_n^{n+1}(z_{n+1}) = p_n^{n+1}(x_{n+1}) - x_n$. Set $x'_{n+1} = x_{n+1} - z_n$; we then have

(i) $f_{n+1}(x'_{n+1}) = f_{n+1}(x_{n+1}) = y_{n+1}$
(ii) $p_n^{n+1}(x'_{n+1}) = x_n$.

We can thus start with $n = 1$ and obtain by this method a coherent sequence $(x_n)_{n \in \mathbb{N}}$ belonging to $G = \text{Lim } \mathbf{G}(n)$ and such that $(\text{Lim } f)((x_n)_{n \in \mathbb{N}}) = (y_n)_{n \in \mathbb{N}}$ as required.

This proposition suggests the following:

**Proposition 2**
*If* $\mathbf{M}: \mathbb{N} \to$ **Mod-A** *satisfies Mittag–Leffler, then* $\text{Lim}^{(1)} \mathbf{M} = 0$.

*Proof*
As might be expected, the proof is similar to the preceding one. It consists in examining the coboundary $\partial_{\mathbf{M}}$ of the short complex for calculating $\text{Lim}^{(1)} \mathbf{M}$ when handling countably indexed systems. Clearly $\text{Lim}^{(1)} \mathbf{M}$ is zero if and only if $\partial_{\mathbf{M}}$ is onto.

Recall that $\partial_{\mathbf{M}}(x_1, \ldots, x_n, \ldots) = (x_1 - p_1^2 x_2, \ldots, x_n - p_n^{n+1} x_{n+1}, \ldots)$. As usual since $\mathbf{M}$ is Mittag–Leffler we may assume it to have surjective bonds.

Let $\underline{m} = (m_i) \in \prod \mathbf{M}(i)$; set $x_1 = 0$ and suppose $x_i$ is defined for $1 \leq i \leq n - 1$ such that $m_i + x_i = p_i^{i+1}(x_{i+1})$. Then since $p_n^{n+1}$ is surjective, there is some $x_{n+1} \in \mathbf{M}(n+1)$ such that $p_n^{n+1}(x_{n+1}) = x_n + m_m$. This defines an element $\underline{x} \in \prod \mathbf{M}(n)$ mapping down to $\underline{m}$.

**Remark**
If $\mathbf{M}: \mathbb{N} \to$ **Groups** satisfies Mittag–Leffler then the Bousfield–Kan $\text{Lim}^{(1)} M$ is trivial—see [16].

**Examples**
(1) Let $\mathbf{F}: \mathbf{I} \to$ **Sets** be an inverse system of finite sets; then $\mathbf{F}$ satisfies the Mittag–Leffler condition. In fact let $i$ be in $\mathbf{I}$ with card $\mathbf{F}(i) < \infty$; then there is an index $j$ in $\mathbf{I}$ and $u: j \to i$ such that

$$\text{card}(\text{Im } \mathbf{F}(u)) = \min\{\text{card}(\text{Im } \mathbf{F}(v)) | \text{codom}(v) = i\}.$$

If $u': j' \to j$ is a morphism in $\mathbf{I}$ then

$$\operatorname{Im} \mathbf{F}(uu') \subseteq \operatorname{Im} \mathbf{F}(u)$$

but $\operatorname{card}(\operatorname{Im} \mathbf{F}(uu')) \geqslant \operatorname{card}(\operatorname{Im} \mathbf{F}(u))$ so $\operatorname{Im} \mathbf{F}(u) = \operatorname{Im} \mathbf{F}(uu')$. This argument can be adapted to various other contexts. It immediately generalises to show that profinite groups satisfy the Mittag–Leffler condition.

(2) Let $K$ be a field, $\mathbf{F}: \mathbf{I} \to \mathbf{Mod}\text{-}K$ a pro 'finite dimensional $K$-vector space'. Then $\mathbf{F}$ satisfies the Mittag–Leffler condition. The proof is as in (1) but one replaces 'card' by 'rank'.

The argument in these two examples uses a type of well-ordering or induction. In more general examples we need to impose conditions that allow one to do 'downwards-induction'. (The following examples may be somewhat technical for some readers, but they are not used in the sequel.)

(3) Let $A$ be a ring, $\mathbf{F}: \mathbf{I} \to \mathbf{Mod}\text{-}A$ a pro 'Artinian right $A$-module'. So the lattice of submodules of each $\mathbf{F}(i)$ satisfies the descending chain condition and hence non-empty families of submodules of an $\mathbf{F}(i)$ have minimal elements. Pick an index $i$ in $\mathbf{I}$ and consider the family

$$\{\operatorname{Im} \mathbf{F}(u) \mid u \text{ in } \mathbf{I}, \operatorname{codom}(u) = i\}.$$

As $\mathbf{I}$ is cofiltering, minimal submodules in this family must be least elements since if $\operatorname{Im} \mathbf{F}(u)$ and $\operatorname{Im} \mathbf{F}(v)$ are minimal, there is a diagram

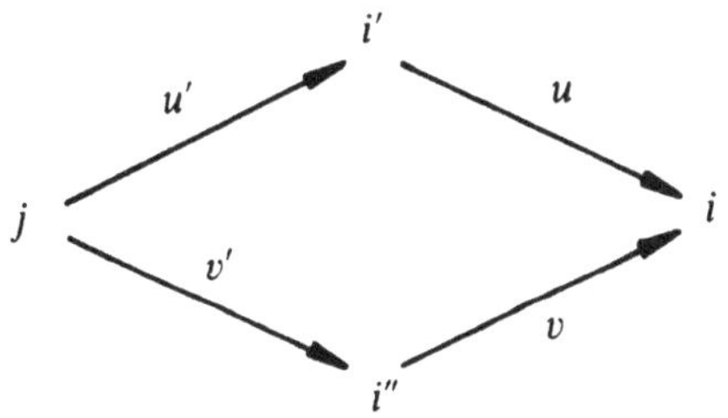


with $uu' = vv'$ and $\operatorname{Im} \mathbf{F}(uu') = \operatorname{Im} \mathbf{F}(u) \cap \operatorname{Im} \mathbf{F}(v)$. Hence $\operatorname{Im} \mathbf{F}(u) = \operatorname{Im} \mathbf{F}(v)$ by minimality. For each $i$ such a least image exists, but this clearly implies the Mittag–Leffler condition.

(4) Let $\mathbf{F}: \mathbf{I} \to \mathbf{TopGps}$ be a pro-object in the category of compact topological groups; then $\mathbf{F}$ satisfies the Mittag–Leffler condition. The proof is more or less the same.

Thus in each of these cases in which the higher derived limits are defined, i.e. in which module (or group) structure is present, the $\operatorname{Lim}^{(i)}$ are zero. Of course this makes sense only for $i = 1$ if the groups are not Abelian and this illustrates two closely related problems. The first is that of finding an analogue of the Mittag–Leffler condition that can be applied in categories where the notion of image does not exist, e.g. in the homotopy category. The second is that of studying the lack of stability of a pro-object is a non-Abelian category. As the higher derived limits are not available, how can one check the extent to which Lim has destroyed information?

We will concentrate on the first of these. The second one leads off in the direction of strong shape theory in its various forms and we have not the space to deal with that here.

Although the notion of movability introduced by Borsuk yields a suitable non-image-based version of the Mittag–Leffler condition, it would be incorrect and unhelpful to introduce it solely in that way. Movability and the related idea of strong movability may be viewed as versions of 'Mittag–Leffler' suitable for situations in which not only does 'image' not behave well, but also epimorphisms need not split. Thus in **Sets** the Axiom of Choice implies that 'movability' and 'Mittag–Leffler ' coincide, but even in **Groups**, this is no longer true. This suggests that movability might be a very subtle and useful condition. This impression is further strengthened when one observes that in the paper [11] in which Borsuk defines movable compacta, the idea that the definition of movable is something dreamt up to mimick (in the homotopy context) the algebraic Mittag–Leffler condition is nowhere in sight. Rather, Borsuk's vision of movability is that of a property satisfied by all ANRs (i.e. by the models in his setting) and yet not alone by them, as it is easy to give examples of spaces that are not ANRs, but are movable.

**Definition** (Borsuk [11])
Let $X$ be a compact metric space embedded in the Hilbert cube $I^{\infty}$. We say $X$ is *movable* if for each neighbourhood $U$ of $X$ in $I^{\infty}$, there is a neighbourhood $V$ of $X$, $V \subset U$, such that for any neighbourhood $W$ of $X$, $W \subset U$, there is a homotopy $H: V \times [0, 1] \to U$ such that

(1) $H \mid V \times \{0\} =$ inclusion of $V$ in $U$
(2) $H \mid V \times \{1\} \subset W$.

Borsuk showed in [11] that movability is an invariant for his shape theory and that each compact subspace of $\mathbb{R}^2$ is movable. The translation, via, say, the Morita form of shape theory, of this concept into the language of inverse systems of polyhedra and thence to the context of pro-objects in a category, **C**, is easily available in the literature (see [32] or [78]). The end result of this translation is the following:

**Definition**
Let $\mathbf{F}: \mathbf{I} \to \mathbf{C}$ be a pro-object in a category **C**. We say that $(\mathbf{I}, \mathbf{F})$ is *movable* if for each $i$ in **I**, there is a morphism $u: i' \to i$ in **I** such that for any $u': i'' \to i$ in **I**, there is a morphism $h^{i'}_{i''}: \mathbf{F}(i') \to \mathbf{F}(i'')$ such that the diagram

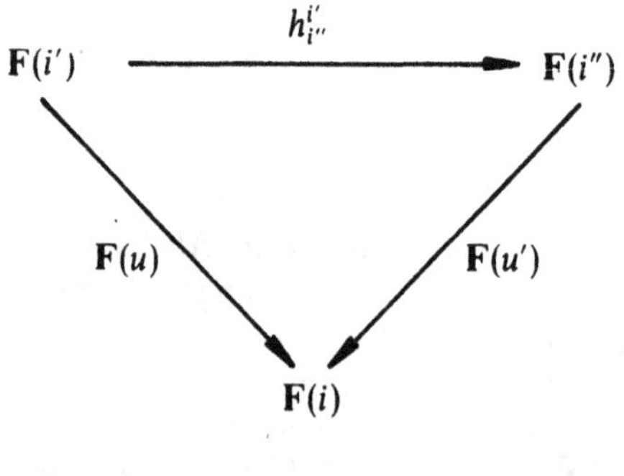


commutes.

**Example**

If $(\mathbf{I}, \mathbf{F})$ is a pro-set such that each $\mathbf{F}(u)$ is onto, then $\mathbf{F}$ is movable since if $u$ is a morphism, $u: i' \to i$, in $\mathbf{I}$ and $u': i'' \to i$ then for each $x' \in \mathbf{F}(i')$ choosing an $x'' \in \mathbf{F}(i'')$ such that $\mathbf{F}(u')(x'') = \mathbf{F}(u)(x')$, we can set $h_{i''}^{i'}(x') = x''$ to obtain the necessary structure. In particular if each $\mathbf{F}(i)$ is finite, we have seen that $\mathbf{F}$ satisfies Mittag–Leffler so is isomorphic to a strict pro-object and it remains to show that movability is preserved by isomorphism to conclude that profinite sets are movable. (We note the use of the Axiom of Choice in the above.)

**Theorem 1**

*Let* $(\mathbf{I}, \mathbf{F})$, $(\mathbf{J}, \mathbf{G})$ *be pro-objects in* $\mathbf{C}$ *such that* $(\mathbf{I}, \mathbf{F})$ *dominates* $(\mathbf{J}, \mathbf{G})$. *If* $(\mathbf{I}, \mathbf{F})$ *is movable, then so is* $(\mathbf{J}, \mathbf{G})$.

*Proof*

We first note that this is a non-additive version of the lemma in the proof of Proposition 4 of section 6.4. The proof is somewhat similar.

We have that $(\mathbf{I}, \mathbf{F})$ dominates $(\mathbf{J}, \mathbf{G})$, i.e. there are maps

$$\underline{f} = ((f_j)_{j \in |J|}, \theta): (\mathbf{I}, \mathbf{F}) \to (\mathbf{J}, \mathbf{G}) \quad \text{and} \quad \underline{g} = ((g_i)_{i \in |I|}, \gamma): (\mathbf{J}, \mathbf{G}) \to (\mathbf{I}, \mathbf{F})$$

such that $[\underline{f}][\underline{g}] = [\mathrm{Id}_{(\mathbf{J}, \mathbf{G})}]$ (in the notation introduced in section 2.3). Translating this, we have that for any $j$ in $\mathbf{J}$, there is some $\bar{j}$ in $\mathbf{J}$ and morphisms $\bar{v}: \bar{j} \to j, \bar{v}': \bar{j} \to \gamma\theta(j)$ such that the diagram

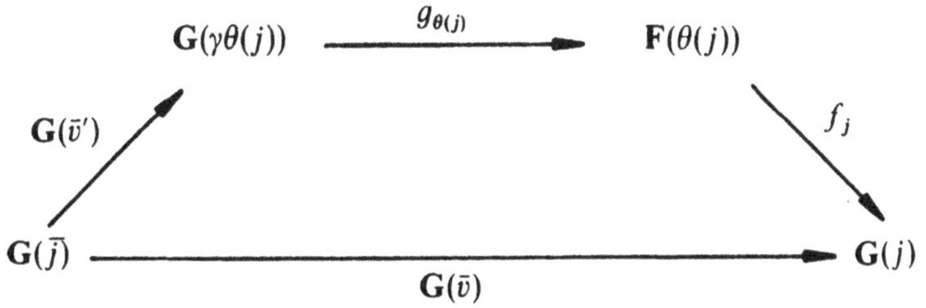


commutes.

As $(\mathbf{I}, \mathbf{F})$ is movable, there is some $\bar{\theta}(j)$ in $\mathbf{I}$ and $\bar{u}: \bar{\theta}(j) \to \theta(j)$ such that for every $u': i' \to \theta(j)$, there is some morphism $h_{i'}^{\bar{\theta}(j)}$ in $\mathbf{C}$ satisfying $\mathbf{F}(u') h_{i'}^{\bar{\theta}(j)} = \mathbf{F}(\bar{u})$.

As $\underline{g}$ is a morphism of systems, there exists a $k$ in $\mathbf{J}$ and $w: k \to \gamma\theta(j)$, $w': k \to \gamma\bar{\theta}(j)$ such that

$$g_{\theta(j)} \mathbf{G}(w) = \mathbf{F}(\bar{u}) g_{\bar{\theta}(j)} \mathbf{G}(w').$$

The category $\mathbf{J}$ is cofiltering so there is a $j'$ in $\mathbf{J}$ and morphisms $\bar{w}: j' \to \bar{j}$, $\bar{w}': j' \to k$ such that $\mathbf{G}(\bar{v}')\mathbf{G}(\bar{w}) = \mathbf{G}(w)\mathbf{G}(w')$.

Let $\bar{v} = \bar{v}\bar{w}: j' \to j$ and $v': j'' \to j$. As $\underline{f}$ is a morphism of systems, there is an $i'$ in $\mathbf{I}$ and $u': i' \to \theta(j)$, $u: i' \to \theta(j'')$ such that the following diagram commutes:

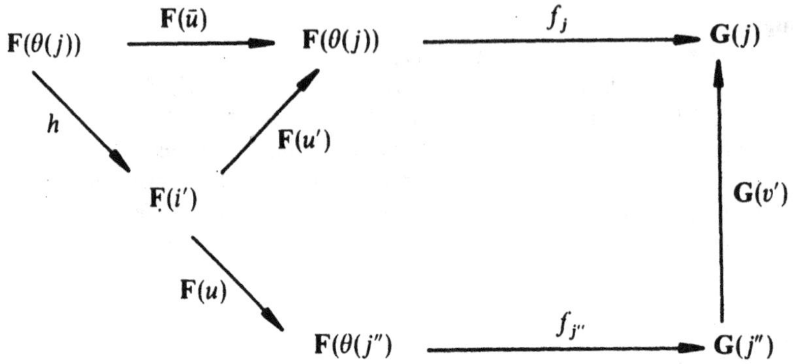


Now there is an $h_{i'}^{\bar{\theta}(j)}$ such that $\mathbf{F}(u')h_{i'}^{\bar{\theta}(j)} = \mathbf{F}(\bar{u})$ by the movability of $(\mathbf{I}, \mathbf{F})$. We set $g_{j''}^{j'}: \mathbf{G}(j') \to \mathbf{G}(j'')$ equal to $f_{j''}\mathbf{F}(u)h_{i'}^{\bar{\theta}(j)}g_{\bar{\theta}(j)}\mathbf{G}(w')\mathbf{G}(\bar{w}')$ and leave it to the reader to check that $\mathbf{G}(v')g_{j''}^{j'} = \mathbf{G}(v)$ as required.

Of course this implies:

**Corollary 1**
*If* $(\mathbf{I}, \mathbf{F})$ *and* $(\mathbf{J}, \mathbf{G})$ *are isomorphic as pro-objects then* $(\mathbf{I}, \mathbf{F})$ *is movable if and only if* $(\mathbf{J}, \mathbf{G})$ *is movable.*

**Corollary 2**
*Let* **K** *be the inclusion of a full proreflective subcategory* **A** *of* **B**, *i.e.* **K** *has a proadjoint. Then if B is a* **K**-*stable object of* **B**, *B is '***K**-*movable'.*

*Remark and Proof*
Corollary 1 shows that movability of the pro-object associated to $B$ in **B** is a **K**-shape invariant. In this case the term '**K**-movable' can clearly be applied to it. As any object of the form **K**$A$ is clearly **K**-movable, as its associated pro-object can be indexed by the one arrow category, **K**-stable objects are **K**-movable.

We next return to the context of Borsuk's shape theory.

Let $X$ be a compact metric space embedded in the Hilbert cube $I^\infty$. We have seen that there is a decreasing sequence $(X_n)_{n \in \mathbb{N}}$ of compact ANR neighbourhoods of $X$ in $I^\infty$ such that $X = \bigcap_{n \in \mathbb{N}} X_n$; $\underline{X} = (X_n, [i_n^{n+1}], \mathbb{N})$ is then an inverse system associated with $X$, $i_n^{n+1}$ being the inclusion of $X_{n+1}$ in $X_n$. In this context we will say that $X$ is movable if $\underline{X}$ is a movable pro-object. This double use of the term 'movable' leads to no confusion because of the following:

**Theorem 2**
*A compact metric space X is movable if and only if it is movable in the sense of Borsuk.*

*Proof*
Suppose $\underline{X}$ is movable and that $U$ is a neighbourhood of $X$ in $I^\infty$. There is some $n \in \mathbb{N}$ such that $X_n \subset U$ and as $\underline{X}$ is movable, there is $n' \geqslant n$ such that for all $n'' \geqslant n$, there is a continuous map $f_{n''}^{n'}: X_{n'} \to X_{n''}$ such that $i_n^{n''} f_{n''}^{n'} \simeq i_n^{n'}$ in $X_n$ and thus in $U$.

Set $V = X_{n'}$, then if $W$ is a neighbourhood of $X$ and $W \subset U$ there is some $n'' \geqslant n$ such that $X_{n''} \subset W$ and there is a homotopy

$$H: V \times [0, 1] \to X_n \hookrightarrow U$$

such that

$$H(x, 0) = x \quad \text{for all } x \in V$$

and

$$H(x, 1) = f^{n'}_{n''}(x) \in X_{n''} \subset W.$$

Conversely let $X$ be movable in the sense of Borsuk and let $n \in \mathbb{N}$; there is an $n' \geqslant n$ such that for all $n'' \geqslant n$, there is a homotopy

$$H: X_{n'} \times [0, 1] \to X_n$$

such that

$$H(x, 0) = x \quad \text{for all } x \in X_{n'},$$
$$H(x, 1) \in X_{n''}.$$

Let $f^{n'}_{n''}(x) = H(x, 1)$; $H$ is a homotopy from $i^{n'}_n$ to $i^{n''}_n f^{n'}_{n''}$ within $X_n$.

Earlier, in section 6.2, we gave some examples of non-stable compact metric spaces but deferred the proofs of non-stability until later. By Corollary 2 to Theorem 1, it in fact suffices to prove that these spaces are not movable. These results are therefore stronger than the corresponding results for stability.

**Examples**

1. The solenoids are not movable.

Let $P = (r_1, \ldots, r_n, \ldots)$ be a sequence of prime numbers, and $S^1_P$ the $P$-adic solenoid. Recall that this is defined as being Lim $\underline{X}$ where $\underline{X}_n = (X_n, p^{n+1}_n, \mathbb{N})$, each $X_n = S^1$ and $p^{n+1}_n(z) = z^{r_n}$, $z \in S^1 = \{z \in |\mathbb{C}| \, |z| = 1\}$.

We suppose $S^1_P$ is movable; then for $n \in \mathbb{N}$ there is an $n' \geqslant n$ such that for $n'' \geqslant n$, there exists a $f^{n'}_{n''}: X_{n'} \to X_{n''}$ such that $p^{n''}_n f^{n'}_{n''} \simeq p^{n'}_n$. Thus for the degrees of the composites, one has

$$\deg p^{n''}_n f^{n'}_{n''} = r_n \cdots r_{n''-1} \deg f^{n'}_{n''} = \deg p^{n'}_n = r_n \cdots r_{n'-1}.$$

If $n'' > n'$, we have $r_{n'} \ldots r_{n''-1} \deg f^{n'}_{n''} = 1$ which is clearly impossible; hence $S^1_P$ is not movable.

2. Let $X = \text{Lim}\, \underline{X}$ where $\underline{X} = (X_n, p^{n+1}_n, \mathbb{N})$, $X_n = S^1 \vee S^1$ and $p^{n+1}_n$ is a pointed continuous mapping, $p^{n+1}_n = p: (S' \vee S^1, 0) \to (S' \vee S^1, 0)$ where

$$p_*(a) = aba^{-1}b^{-1} \in \pi_1(S^1 \vee S^1)$$

$$p_*(b) = a^2b^2a^{-2}b^{-2} \in \pi_1(S^1 \vee S^1),$$

$a$ and $b$ being the two generators of $\pi_1(S^1 \vee S^1) \cong \mathbb{Z} * \mathbb{Z}$. Almost trivially we have $\check{\mathbf{H}}_q(X; G) = 0$ for any group $G$ and for any $q \geqslant 0$. Similarly $\check{\mathbf{H}}^q(X; G) = 0$, since the morphisms $p_*: \mathbf{H}^q(S \vee S^1; G) \to \mathbf{H}^q(S^1 \vee S^1; G)$ are all zero.

For each $c \in \pi_1(S^1 \vee S^1)$, we let $l(c)$ be the length of $c$, that is, the length of the minimal word in the symbols $a^j, b^j$, the powers of the generators—for instance:

$$l(p_*(a)) = l(p_*(b)) = 4.$$

We note that $l(c)=0$ if and only if $c$ is the empty word. By induction, one can calculate that $l(p_\#(c)) \geqslant 3l(c)$; thus if $c \neq 1$, we have, for $k \geqslant 0$,

$$l(p_\#^k(c)) \geqslant 3^k l(c) \quad \text{and} \quad p_\#^k(c) \neq 1.$$

If $\underline{X}$ is movable, then for each $n$, there is an $n' \geqslant n$ such that for all $n'' \geqslant n$, there is a $f_{n''}^{n'}: X_{n'} \to X_{n''}$ such that

$$p^{n''-n} f_{n''}^{n'} = p_n^{n''} f_{n''}^{n'} \simeq p_{n''}^{n'} = p^{n'-n}.$$

The maps induce morphisms on $\pi_1(S^1 \vee S^1)$ such that $(p_n^{n''} f_{n''}^{n'})_\# = (p_n^{n'})_\#$, so for the generator $a \in \pi_1(X_{n'})$, one has

$$l(p_\#^{n''-n}(f_{n''\#}^{n'}(a))) = l(p_\#^{n'-n}(a))$$

and $l(a) \neq 0$ implies $l(p_\#^{n'-n}(a)) \geqslant 1$.

If $(f_{n''}^{n'})_\#(a)=1$, $l(p_\#^{n'-n}(a)) = l(p_\#^{n''-n}(f_{n''}^{n'})_\#(a))=0$ which is impossible, thus one has $(f_{n''}^{n'})_\#(a) \neq 1$ and $l(p_\#^{n'-n}(a)) = l(p_\#^{n''-n}(f_{n''}^{n'})_\#(a)) \geqslant 3^{n''-n} l((f_{n''}^{n'})_\#(a)) \geqslant 3^{n''-n}$. The integers $n$ and $n'$ are given, but $l(p_\#^{n'-n}(a)) \geqslant 3^{n''-n}$ and $n''$ is arbitrary. As this is clearly impossible, $X$ is not $\mathbf{K}_0$-movable, i.e. movable in the sense of pointed shape theory. As we pointed out earlier, this space $X$ has the strange property that although it is not even movable, its suspension $\Sigma X$ has the shape of a point. We next turn to the latter result.

We know that if $Y$ is a pointed CW-complex, the reduced suspension $\Sigma^* Y$ has the same homotopy type as $\Sigma Y$. If we can show that $\Sigma p: \Sigma(S^1 \vee S^1) \to \Sigma(S^1 \vee S^1)$ is null homotopic, the proof that $\Sigma X$ has the shape of a point will be trivial. It is clear that it suffices to show that

$$\Sigma^* p: \Sigma^*(S^1 \vee S^1) \to \Sigma^*(S^1 \vee S^1)$$

is null homotopic.

We know $\pi_1(\Sigma^*(S^1 \vee S^1))=0$ and the restrictions of $p$ to the two cofactors of the coproduct, denoted $p_i: S^1 \to S^1 \vee S^1$ for $i=1,2$, induce zero homomorphisms

$$p_{i*}: \mathrm{H}_1(S^1; \mathbb{Z}) \to \mathrm{H}_1(S^1 \vee S^1; \mathbb{Z})$$

(this follows from the description of $p_\#(a)$ as $aba^{-1}b^{-1}$ and of $p_\#(b)$ as $a^2b^2a^{-2}b^{-2}$).

In the commutative diagram

$$\begin{array}{ccc}
\mathrm{H}_1(S^1; \mathbb{Z}) & \xrightarrow{\quad p_{i*} \quad} & \mathrm{H}_1(S^1 \vee S^1; \mathbb{Z}) \\
\cong \sigma(S^1) \Big\downarrow & & \Big\downarrow \cong \sigma(S^1 \vee S^1) \\
\mathrm{H}_2(\Sigma^*(S^1); \mathbb{Z}) & \xrightarrow{\quad (\Sigma p_i)_* \quad} & \mathrm{H}_2(\Sigma^*(S^1 \vee S^1); \mathbb{Z})
\end{array}$$

where the $\sigma$ are the suspension isomorphisms, $p_{i*}=0$ implies $(\Sigma p_{i*}=0)$ for $i=1,2$.

The naturality of the Hurewicz homomorphisms $\phi$ gives a commutative diagram

$$\begin{array}{ccc}
\pi_2(\Sigma^*S^1) & \xrightarrow{\;(\Sigma^*p_i)_\#\;} & \pi_2(\Sigma^*(S^1 \vee S^1)) \\
\Big\downarrow{\scriptstyle \phi(\Sigma^*S^1)} & & \Big\downarrow{\scriptstyle \phi(\Sigma^*(S^1 \vee S^1))} \\
\mathbf{H}_2(\Sigma^*S^1; \mathbb{Z}) & \xrightarrow{\;(\Sigma^*p_i)_*\;} & \mathbf{H}_2(\Sigma^*(S^1 \vee S^1); \mathbb{Z})
\end{array}$$

As $\phi(\Sigma^*(S^1 \vee S^1))$ is an isomorphism, we have $(\Sigma^*p_i)_\# = 0$ for $i = 1, 2$.

As $\pi_2(\Sigma^*S^1) = \mathbb{Z}$ and $(\Sigma^*p_i)_\#(c) = 0$ where $c$ is the generator of $\pi_2(\Sigma^*S^1)$, there exists for $i = 1, 2$ a pointed homotopy

$$h_i\colon \Sigma^*(S^1) \times [0, 1] \to \Sigma^*(S^1 \vee S^1)$$

such that $h_i|\Sigma^*(S^1) \times \{0\} = \Sigma^*p_i$ and $h_i|\Sigma^*(S^1) \times \{1\} = *$.

As $h_1$ and $h_2$ are constant on $* \times [0, 1]$, there is a pointed homotopy $h\colon (\Sigma^*(S^1) \vee \Sigma^*(S^1)) \times [0, 1] \to \Sigma^*(S^1 \vee S^1)$ such that $h$ is $h_1$ on the first cofactor and $h_2$ on the second. We note that $\Sigma^*(S^1) \vee \Sigma^*(S^1) \simeq \Sigma^*(S^1 \vee S^1)$ and that considering $h$ as being defined on $\Sigma^*(S^1 \vee S^1)$ we have $h|\Sigma^*(S^1 \vee S^1) \times \{0\} = \Sigma^*p$ and $h|\Sigma^*(S^1 \vee S^1) \times \{1\} = *$ as required.

Movability is, of course, preserved by functors. If we have a functor $\mathbf{F}\colon \mathbf{C} \to \mathbf{D}$ and a movable $X$ in **pro**(**C**) then its image **pro**(**F**)($X$) in **pro**(**D**) is movable.

Suppose, in the above, **D** is a suitable Abelian category such as the category of Abelian groups, then **pro**(**F**)($X$) will be a movable Abelian group. Our previous discussion suggests that **pro**(**F**)($X$) should therefore satisfy the Mittag–Leffler condition. This is handled by the following result.

**Proposition 3**
*Let* $\mathbf{F}\colon \mathbf{I} \to \mathbf{Sets}$ *be a movable proset. Then* **F** *satisfies the Mittag–Leffler condition.*

*Proof*
If $i$ is in **I**, as **F** is movable there is an $i'$ in **I** and $u\colon i' \to i$ such that for any $u'\colon i'' \to i$, there is an $h^{i'}_{i''}$ satisfying $\mathbf{F}(u')h^{i'}_{i''} = \mathbf{F}(u)$. Now suppose $u'\colon i'' \to i'$; there is then a second morphism $h^{i'}_{i''}$ such that $\mathbf{F}(uu')h^{i'}_{i''} = \mathbf{F}(u)$ and hence $\operatorname{Im} \mathbf{F}(u) \subseteq \operatorname{Im} \mathbf{F}(uu')$. As the other inclusion is clear, $\operatorname{Im} \mathbf{F}(u) = \operatorname{Im} \mathbf{F}(uu')$ as required.

**Remark**
Nowhere in this proof are elements used. It is another example of a result true wherever a good notion of image is available.

Returning to our movable $X$ and its image **pro**(**F**)($X$), we now know this latter object satisfies the Mittag–Leffler condition. A particularly nice and useful example of this occurs in the topological setting. We recall from earlier in his chapter that if $(X, A)$ was a stable pair of spaces then its Čech homology long sequence was exact, a similar result holding for stable pointed pairs on considering Čech homotopy groups (Propositions 4 and 5 of section 6.2). If $(X, A)$ is a pair (or a pointed pair) of spaces so that its associated propair is movable in the category of pairs of ANRs (respectively of pointed pairs of ANRs), then we will simply say that $(X, A)$ is a movable pair (resp. a movable pointed pair) of spaces.

The construction of the long homology sequence of a pair of ANRs yields a functor **H** from $\mathbf{W}^2$ to the category **LES** of long exact sequences. Thus $\mathbf{H}(X, A)$ will be a movable pro-long exact sequence. Therefore, if, in addition, $(X, A)$ is a compact metric pair, $\mathbf{H}(X, A)$ will be movable and countably indexed. Splitting up $\mathbf{H}(X, A)$ into a series of short exact sequences in the usual way and applying Lim to each of these yields a long exact sequence (after reassembly) which is precisely the long Čech homology sequence of $(X, A)$. (At each stage of this, one uses (ML) and the fact that $(X, A)$ is countably indexed to allow application of Proposition 1 (of this section).)

We therefore have:

**Proposition 4**

*Let $(X, A)$ be a movable pair of compact metric spaces. Then for any coefficients, the long Čech homology sequence of $(X, A)$ is exact.*

**Remarks**

(i) A similar result is true for the Čech homotopy sequence of a movable pointed compact metric pair $(X, A, x_0)$.

(ii) Various other results such as a form of the Hurewicz isomorphism theorem hold for movable compact metric spaces. These shape versions of properties of polyhedral spaces indicate to some extent how close movable compact metric spaces are to being ANRs.

(iii) Various choices of coefficients for the homology groups in the long Čech homology sequence of a pair $(X, A)$ allow one to avoid movability. Classically it was known that coefficients in a finite dimensional vector space or in a compact Abelian group or module guaranteed exactness in the limit (see for instance Eilenberg and Steenrod [38]). More recently, Garavaglia [45] has given a description of those modules which, when taken as coefficients for Čech homology, yield long exact sequences for all compact metric spaces, $(X, A)$. This should be compared with our earlier comment on examples of movability in profinite sets, provector spaces, etc. It indicates how various of the usual test functors of algebraic topology can create movability.

(v) The non-exactness of the long Čech homology sequence can be avoided by using Steenrod–Sitnikov homology. The main difference between Čech and Steenrod–Sitnikov homology is that whereas in Čech homology one first applies the homology to the prospace and then takes a limit, in the Steenrod–Sitnikov construction one uses the prospace to construct a prochain complex, and then one takes its limit and finally homology. The prochain complex one uses is related to that which we used when constructing the derived functors of Lim in section 6.5. This is related to a form of algebraic homotopy limit (see Cordier [18, 19]). This is one of the first indications we have met in this monograph that topological shape theory has some weaknesses in its construction. By taking the homotopies between the approximating maps more into account one obtains strong shape theory. This provides a setting that allows a much better behaved collection of invariants to be defined; unfortunately the descriptions of strong shape theory so far available are technically difficult so this theory cannot be adequately summarised in this book.

## 6.7 STRONG MOVABILITY

There are two questions raised by the notion of movability that merit close attention.

(i) In the application of movability to the Čech homology sequence, it was necessary to assume the movable pair was compact metric. This was so that the indexing category for the associated propair would have a countable initial subcategory and hence that the Mittag–Leffler condition could be used to ensure exactness in the limit. Is there a stronger form of movability that allows one to do away with the countability assumption?

(ii) Movability can be thought of as a weak form of stability. Is there a condition equivalent to stability and yet phrased in similar terms to movability? In the Abelian case we did have the internal description of stability given by the 'essentially constant' condition; for what categories does a similar result hold?

Borsuk [12] noted that stable spaces satisfied a stronger condition to movability. The details can be found in his book ([13], p. 263). The general condition for pro-objects was given by Mardešić [73] (see also Mardešić and Segal [78], p. 225). We use a slightly modified version given by Seymour [95].

Suppose $(\mathbf{I}, \mathbf{F})$ is a pro-object in a category $\mathbf{C}$. We say $(\mathbf{I}, \mathbf{F})$ is *strongly movable* if, for each object $i$ in $\mathbf{I}$, there is a morphism $\alpha: j \to i$ such that for each $\beta: k \to j$ there is an $\gamma: l \to k$ and a morphism $h$ in $\mathbf{C}$ such that

$$
\begin{array}{ccc}
\mathbf{F}(l) & \xrightarrow{\mathbf{F}(\gamma)} & \mathbf{F}(k) \\
\downarrow{\scriptstyle \mathbf{F}(\beta\gamma)} & {\scriptstyle h} \nearrow & \downarrow{\scriptstyle \mathbf{F}(\alpha\beta)} \\
\mathbf{F}(j) & \xrightarrow[\mathbf{F}(\alpha)]{} & \mathbf{F}(i)
\end{array}
$$

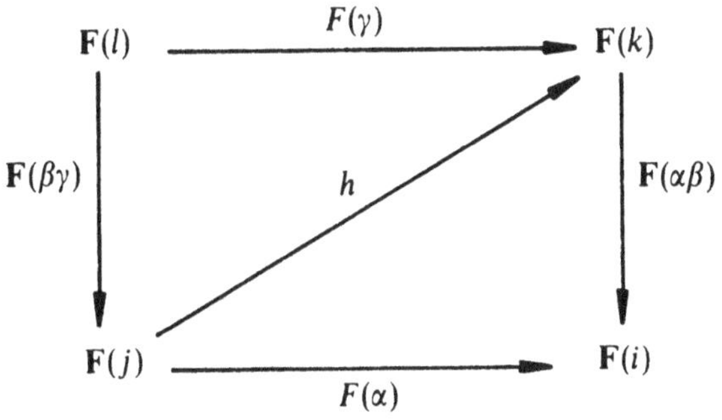

commutes.

### Remarks

(i) It is clear that any object $X$ of $\mathbf{C}$, considered as a pro-object, will be strongly movable. To show that stable pro-objects are strongly movable we will shortly prove that strong movability is an invariant of pro-isomorphism.

(ii) The term 'strong movability' would be a bad one unless 'strong movability' implied 'movability'. Because of this we start by checking that this is the case.

### Proposition 1

*If $(\mathbf{I}, \mathbf{F})$ is a strongly movable pro-object then it is movable.*

### Proof

Suppose given $i$ in $\mathbf{I}$, we claim that a special 'movability morphism' $u: i' \to i$ for $(\mathbf{I}, \mathbf{F})$ (as in the definition of movability) can be chosen to be the 'strong movability morphism' $\alpha: j \to i$ of the definition just given. To see this, suppose $u': i'' \to i$ is given. As $I$ is cofiltering, there is a $\beta: k \to j$, $v: k \to i''$ such that

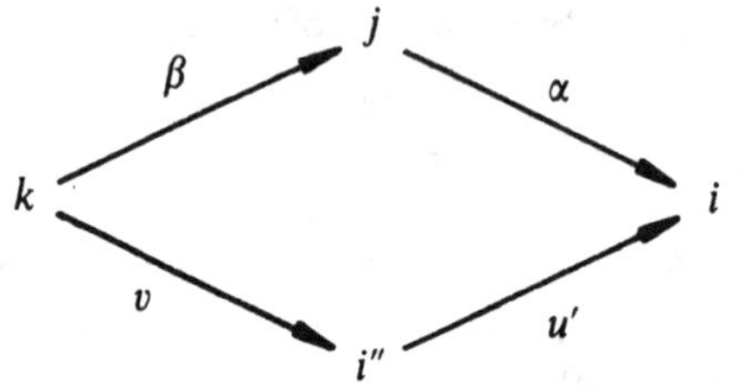


commutes, i.e. $u'v = \alpha\beta$. We feed this $\beta: k \to j$ into our definition of strong movability to obtain an $h: \mathbf{F}(j) \to \mathbf{F}(k)$ and a $\gamma: l \to k$ such that

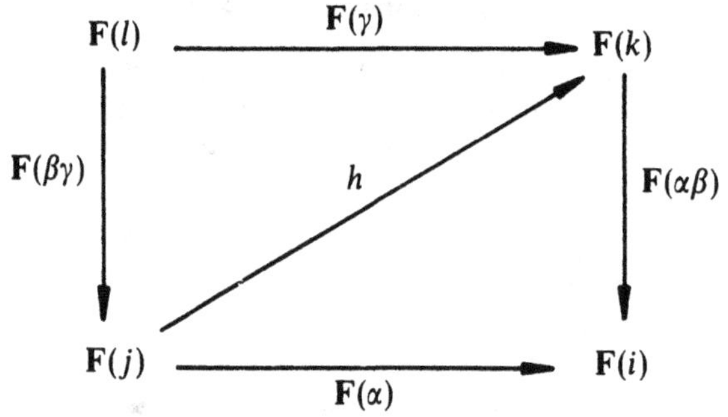


commutes, but then in particular the bottom right triangle commutes and $\mathbf{F}(\alpha) = \mathbf{F}(\alpha\beta)h = \mathbf{F}(u')\mathbf{F}(v)h$. Clearly a suitable $h': \mathbf{F}(i') \to \mathbf{F}(i'')$ can be constructed by taking $h' = \mathbf{F}(v)h$.

This proof shows how strong movability involves a certain amount of 'duality'. We only used the bottom right triangle in order to obtain movability. This suggests that, since 'movability' implies 'Mittag–Leffler' which in turn implies 'essentially epimorphic' in well behaved contexts, perhaps 'strong movability' might, in some categories **C**, imply 'essentially monomorphic' as well and hence 'essentially constant', i.e. 'stable'. We will explore this possibility shortly. First, however, we will state the analogue for strong movability of Theorem 1 of section 6.6.

**Theorem 1**
*Let* $(\mathbf{I}, \mathbf{F})$, $(\mathbf{J}, \mathbf{G})$ *be pro-objects in* **C** *such that* $(\mathbf{I}, \mathbf{F})$ *dominates* $(\mathbf{J}, \mathbf{G})$. *If* $(\mathbf{I}, \mathbf{F})$ *is strongly movable, then so is* $(\mathbf{J}, \mathbf{G})$.

The proof is, for nearly all its length, just a rerun of that of Theorem 1 of section 6.6. The definition of the $h$ requires a bit of care, needing one or two more applications of cofiltering and the definition of maps of systems, but the necessary adaptation can be safely left to the reader. (It is very useful to draw the corresponding diagram.)

'Domination' within an Abelian category corresponds to 'direct summand', so we see that if **A** is an Abelian category and $(\mathbf{I}, \mathbf{F})$ is strongly movable in **pro(A)**, then any direct summand of $(\mathbf{I}, \mathbf{F})$ is strongly movable. Of course, any direct summand of a constant pro-object in **pro(A)** is both a subobject and a quotient object of that object, hence must be essentially constant, i.e. stable.

**Corollary**

*If* (**I**, **F**) *and* (**J**, **G**) *are isomorphic in* **pro**(**A**) *and* (**I**, **F**) *is strongly movable, then* (**J**, **G**) *is strongly movable.*

**Corollary**

*Any stable object in* **pro**(**C**) *is strongly movable.*

In the first section of this chapter we introduced the term '**K**-equal'. Recall that $B$ and $B'$ were **K**-equal relative to the shape theory ($B \underset{\mathbf{K}}{\equiv} B'$) if $B \underset{\mathbf{K}}{\leqslant} B'$ and $B' \underset{\mathbf{K}}{\leqslant} B$, i.e. $B$ **K**-dominates and is **K**-dominated by $B'$. In the context of procategories, we will drop the reference to **K** (which in this generic situation is just the inclusion, 'constant pro-objects', of the category **C** into **pro**(**C**)) and merely write $(\mathbf{I}, \mathbf{F}) \equiv (\mathbf{J}, \mathbf{G})$.

**Corollary**

*If* $(\mathbf{I}, \mathbf{F}) \equiv (\mathbf{J}, \mathbf{G})$ *and* (**I**, **F**) *is strongly movable then so is* (**J**, **G**).

Earlier on in this section we noted how strong movability seemed to have a certain amount of duality built into it. This is clear if one looks at the Abelian case. We include the following result even though it is an immediate consequence of another result that we will prove shortly.

**Lemma 1**

*If* **A** *is an Abelian category, and* (**I**, **F**) *is a strongly movable pro-object in* **A**, *then* (**I**, **F**) *satisfies* (*EM*).

*Proof*

Given any $i$ in **I**, take the $j_0$ in condition (EM) to be the domain of $\alpha: j \to i$ in the definition of strong movability. Now given any $\beta: k \to j_0$, there is a $\gamma: l \to k$ and a morphism $h$ in **A** such that the diagram

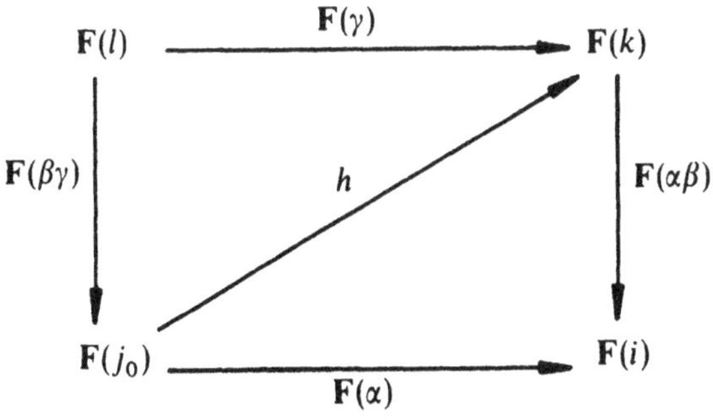


commutes. This time we ignore the bottom right triangle and note only that $\mathbf{F}(\gamma) = h\mathbf{F}(\beta\gamma)$; hence

$$\text{Ker } \mathbf{F}(\gamma) \supseteq \text{Ker } \mathbf{F}(\beta\gamma).$$

Since the opposite inclusion is obvious, this proves that the natural inclusion of Ker $\mathbf{F}(\gamma)$ into Ker $\mathbf{F}(\beta\gamma)$ is an isomorphism, i.e. that (EM) is satisfied.

Combining this with earlier results we obtain:

**Theorem 2**
*If* (**I**, **F**) *is strongly movable in* **pro**(**A**) *for* **A** *Abelian, then* (**I**, **F**) *is essentially constant and hence stable.*

Earlier, in our discussion of the non-additive cases of the Mittag–Leffler condition, we mentioned that many of the results would remain true if one had a category **C** in which there is a well behaved notion of image. We next turn to this in more detail; it will involve a brief look at (E, M) factorisations in categories.

We suppose that we have a category **C**. For two morphisms $\varphi: A \to B$, $\mu: X \to Y$ in **C**, we say $\varphi$ has the unique left lifting property (ULLP) for $\mu$, or, equivalently, $\mu$ has the unique right lifting property (URLP) for $\varphi$, if for each commutative diagram

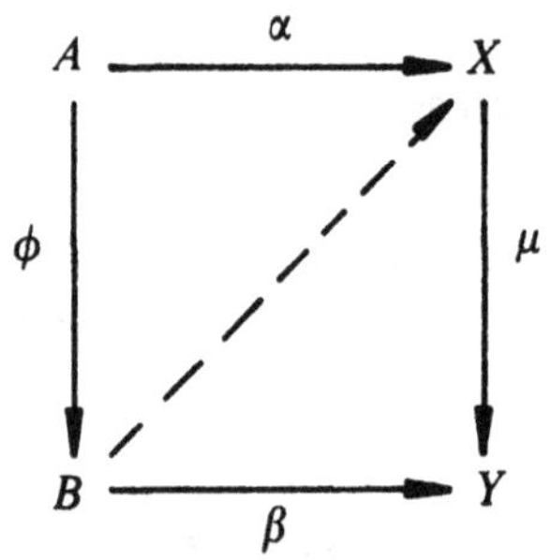


in **C**, there exists a unique map $\gamma$ such that $\gamma\varphi = \alpha$ and $\mu\gamma = \beta$.

For a class $S$ of morphisms in **C**, we set

$$\mathscr{E}(S) = \{\varphi \mid \varphi \text{ has the ULLP for each } \mu \in S\}$$

$$\mathscr{M}(S) = \{\mu \mid \mu \text{ has the URLP for each } \varphi \in S\}.$$

**Definition**
A factorisation system $(E, M)$ in **C** consists of classes of morphisms $E$ and $M$ such that

(i) $E = \mathscr{E}(M)$ and $M = \mathscr{M}(E)$.
(ii) For every morphism $f$ in **C**, there exists $f_m \in M$, $f_e \in E$ such that $f = f_m f_e$.

**Remark**
The factorisation in (ii) is clearly unique up to natural isomorphism and is natural, i.e. given a commutative diagram

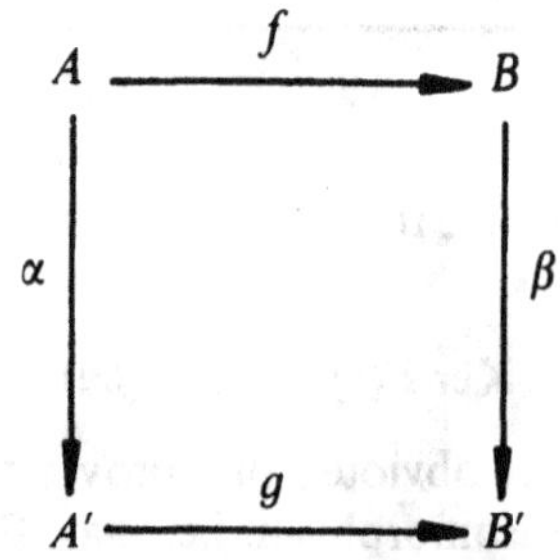

there is a commutative diagram

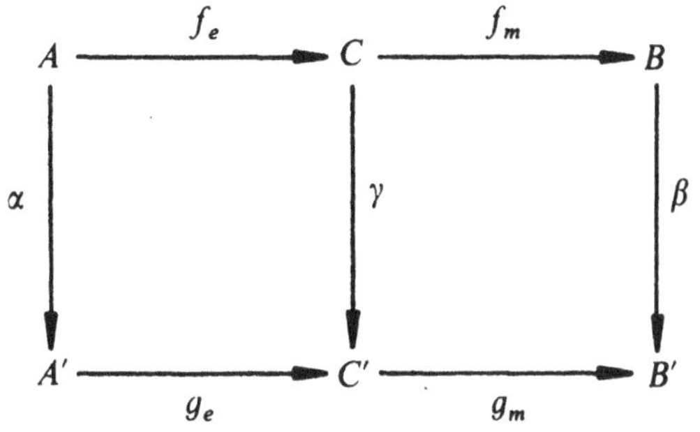


in which $\gamma$ is unique with this property.

**Proposition**

*A pair $(E, M)$ of classes of morphisms in $\mathbf{C}$ forms a factorisation system if and only if the following properties hold.*

(*a*) *Every isomorphism is in both $E$ and $M$.*
(*b*) *Both $E$ and $M$ are closed under composition.*
(*c*) *If $\varphi \in E$ and $\mu \in M$, then $\varphi$ has the* ULLP *for $\mu$.*
(*d*) *For every morphism $f$ in $\mathbf{C}$, there exists $f_m \in M$ and $f_e \in E$ such that $f = f_m f_e$.*

*Proof*

We limit ourselves to proving that any $(E, M)$ satisfying (a) to (d) is a factorisation system. The converse is easy.

So we will assume $(E, M)$ satisfies (a) to (d). We start by proving $\mathscr{E}(M) \subseteq E$. If $f \in \mathscr{E}(M)$, then we have some factorisation $f = f_m f_e$ with $f_m \in M$, $f_e \in E$. We consider the square

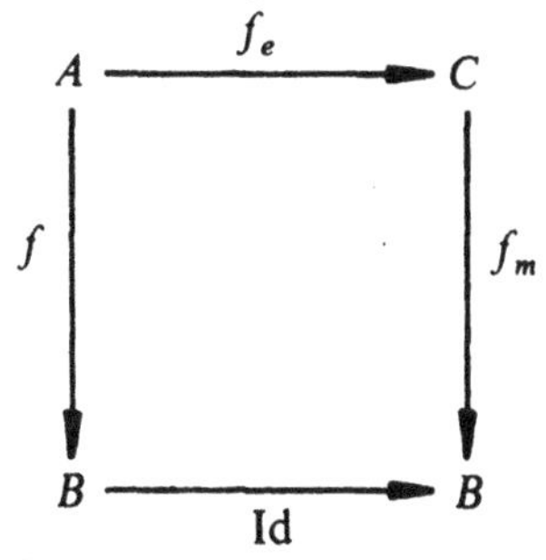


which, since $f \in \mathscr{E}(M)$, gives us a unique $u\colon B \to C$ satisfying $uf = f_e$, $f_m u = \mathrm{Id}_B$. Now $f_e$ has the ULLP for $f_m$, so since the diagram

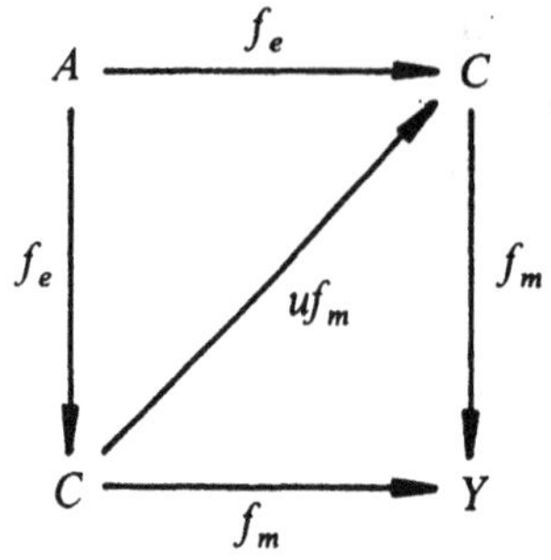


commutes, 'uniqueness' implies $uf_m = \mathrm{Id}_C$; thus $f_m$ is a isomorphism and hence $f \in E$. Condition (c) implies $E \subseteq \mathscr{E}(M)$ so we have equality. As $M \subseteq \mathscr{M}(\mathscr{E}(M))$, this also shows that $M \subseteq \mathscr{M}(E)$, i.e that the dual of (c) is satisfied.

We then note that a dual proof shows $\mathscr{M}(E) \subseteq M$, and hence $\mathscr{M}(E) = M$, which completes the proof.

**Examples**

In **Sets**, **Groups** and in Abelian categories, one can take $E =$ epimorphisms, $M =$ monomorphisms.

In any category, one can take $E =$ isomorphisms, $M =$ all maps or vice versa.

In any case $E \cap M$ consists exactly of the isomorphisms.

**Remark**

The theory of factorisation systems can be found in some textbooks on category theory. The basic aspects of this theory can be found in Kelly's article [62]. Bousfield in [15] gives methods of constructing factorisation systems in various contexts and explores their connection with localisations and completions. If **C** has a terminal object $t$, there is a functor $\mathbf{T}: \mathbf{C} \to \mathbf{C}$ and a transformation $\eta: \mathbf{1} \to \mathbf{T}$ where $X \xrightarrow{\eta(X)} \mathbf{T}X \to t$ is the factorisation (with respect to the given $(E, M)$) of the unique morphism $X \to t$. The pair $(\mathbf{T}, \eta)$ gives an idempotent monad on **C**. Letting **LocC** denote the full subcategory of **C** specified by those objects $X$ for which $X \to t$ is in $M$, **T** provides a left adjoint for the inclusion of **LocC** into **C**.

For example, taking $\mathbf{C} = \mathbf{Top}$, $E$ to be the class of continuous maps $f: X \to Y$ inducing a bijection $\mathrm{Top}(Y, I) \cong \mathrm{Top}(X, I)$ where $I$ is the closed unit interval, then $(E, \mathscr{M}(E))$ is a factorisation system on Top and the corresponding localisation, $T$, is the Stone–Čech compactification.

Returning from these general considerations to a more special class of factorisation systems, we suppose that all morphisms in $E$ are epimorphisms and all morphisms in $M$ are monomorphisms, i.e the factorisation system is proper in the sense of Freyd and Kelly [44]. For such a system and for any $f: A \to B$ the factorisation $f_e, f_m$ determines a subobject of $B$, namely the class of $f_m: C \to B$; this merits the name of the image of $f$.

It should now be clear how to adapt the results, proved earlier, on the Mittag–Leffler condition to the context of a category **C** with a proper factorisation system.

**Proposition 2**
*Suppose* **C** *is a category having a proper factorisation system* $(E, M)$.

(i) *If* $(\mathbf{I}, \mathbf{F})$ *satisfies the Mittag–Leffler condition (relative to* $(E, M)$*) then* $(\mathbf{I}, \mathbf{F})$ *is isomorphic to a pro-object all of whose bonds are in* $E$.
(ii) *If* $(\mathbf{I}, \mathbf{F})$ *is a movable pro-object then* $(\mathbf{I}, \mathbf{F})$ *satisfies the Mittag–Leffler condition (relative to* $(E, M)$*).*
(iii) *If* $(\mathbf{I}, \mathbf{F})$ *is strongly movable, it is isomorphic to a strongly movable pro-object all of whose bonding morphisms are in* $E$.

The proofs are essentially the same as in our previous discussions.

We next need a lemma. The point of the lemma is that we do not have kernels, but can still use the sort of argument that we used to prove that strongly movable pro-objects are essentially monomorphic in the Abelian case. The result is given in Seymour [95].

**Lemma 2**
*Let* $(\mathbf{I}, \mathbf{F})$ *be a strongly movable pro-object in* **C** *all of whose bonds are epimorphisms. Then given* $i$, *there is a morphism* $\alpha: j \to i$ *such that for any* $\beta: k \to j$, $\mathbf{F}(\beta)$ *is an isomorphism.*

*Proof*
We choose $\alpha: j \to i$ as in the definition of strong movability. We thus obtain for any $\beta: k \to j$ a morphism $\gamma: l \to k$ and a diagram

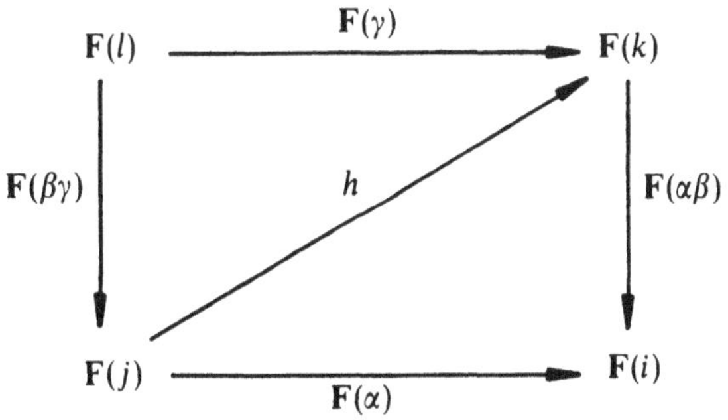


We have $h\mathbf{F}(\beta)\mathbf{F}(\gamma) = \mathbf{F}(\gamma)$ and as $\mathbf{F}(\gamma)$ is an epimorphism, $h\mathbf{F}(\beta)$ is the identity on $\mathbf{F}(k)$.

Now consider $\mathbf{F}(\beta)h: \mathbf{F}(j) \to \mathbf{F}(j)$; we have $\mathbf{F}(\beta)h\mathbf{F}(\beta\gamma) = \mathbf{F}(\beta\gamma)$ so again using that $\mathbf{F}(\beta\gamma)$ is an epimorphism yields that $\mathbf{F}(\beta)h$ is the identity and $\mathbf{F}(\beta)$ is an isomorphism.

Putting this lemma together with Proposition 2, we obtain:

**Theorem 3**
*Let* **C** *be a category with a proper factorisation system. A pro-object* $(\mathbf{I}, \mathbf{F})$ *is strongly movable if and only if it is stable.*

*Proof*
We have already seen the 'if' part of this. For the 'only if' we note the lemma gives an initial subcategory on which **F** has isomorphisms as bonds, hence is stable.

Thus if the idea of an 'image of a morphism' is available within **C**, strong movability and stability are the same. Of course, within the various homotopy categories of spaces, polyhedra, etc., 'images' are not available. The machinery necessary for a full treatment of strong movability in those situations would, if properly done, involve us in a discussion of homotopy limits, strong shape theory and the non-splitting of homotopy idempotents. Although this would be of great value, it already exists to some extent in the book by Mardešić and Segal [78] and as it would lengthen this monograph by quite a few pages, we have decided not to include it. (References to articles on topological stability are given in the Notes on Sources that follows.)

Our aim in the last half of this final chapter has been to search for a condition that was equivalent to stability but which was internal to the pro-object to which it referred, i.e. it did not make reference to some other object of **pro**(**C**). For Abelian categories, we had the conjunction of the two conditions (ML) and (EM), and in more generality, if **C** has a proper factorisation system, strong movability fitted the bill. In practice it is usual to check the image of a pro-object under test functors with values in, say, the category of Abelian groups.

This raises the interesting possibility of analysing, for a given non-stable pro-object, the test functors that detect that non-stability, e.g. for the pro-Abelian group $\mathbf{S}(\mathbb{Z}, p)$, $p$ prime, taking $\mathbb{Z}_{(q)}$ to be integers localised at a prime $q$, the system $\mathbf{S}(\mathbb{Z}, p) \otimes_{\mathbb{Z}} \mathbb{Z}_{(q)}$ is stable if $p = q$ and is non-stable otherwise. Thus important information on a non-stable system can be obtained by using different test functors. These techniques have as yet not been exploited to any great extent.

## NOTES ON SOURCES

The idea of studying the interpretation of categorical notions within a shape category first appears in Porter [87] (see also [20] and [21]). The concept of stability was introduced within a shape theory context in the thesis of the second author. Of course within pro-Abelian categories, conditions on pro-objects which implied 'stability' were known to Verdier [100]. (We have here used notes of Duskin [30].) Verdier's work was aimed at applications in the theory of derived functors, in which the theory of pro and ind categories takes an important role (see Verdier [101], Illusie [60] and Deligne [28]). Results on weak forms of stability of promodules can be found in [90]–[92].

The Reindexing Lemmas seem to have originated in the work of Grothendieck, Verdier *et al.* [50] and [51]. The usual reference is Artin and Mazur [4], but they only provide a sketch proof. Complete proofs and the generalisations mentioned in the text can be found in Meyer [79].

The literature on the derived functors of Lim is extensive. The best basic source for the general theory still seems to be Jensen [61] with the non-Abelian theory found in Bousfield and Kan [16]. Control over the vanishing of higher derived limits is difficult to achieve; however, some results are known—see Gruson and Jensen [52], [53] and Porter [92]. The connection between homotopy limits and the derived functors of Lim has been fully explored in an article [18] by the first author. (This provides a good example of many of the techniques introduced here including procategories, and stability, but proved to be too long to summarise in these notes.)

The notion of movability was introduced by Borsuk in [11]. The ANR-systems version was given by Mardešić and Segal [75]. The exactness of the Čech homology sequence for movable pairs was first noted by Overton [82]. Segal has introduced, in [93], the notion of movability relative to a functor; we have not given this explicitly, but his idea is clearly behind the way in which we have presented the material. Many other results on movability can be found in Segal's notes [94], in Dydak and Segal [32] and in Mardešić and Segal [78].

As mentioned in the text, strong movability was introduced by Borsuk in [12], and is developed by Mardešić in [73]. Mardešić and Segal have a section on it in [78]. Seymour has a very interesting, but apparently unpublished, article [95], which presents various different and new aspects of the idea. Some of our proofs are adapted from his paper. The idea of using proper factorisation systems is basically contained in [95]. This idea clearly leads on to various intriguing questions relating to the corresponding theory relative to an arbitrary (E, M) factorisation system.

The original work on factorisation systems is in Kelly [62]. Some useful ideas are to be found in Bousfield's paper [15]. Much of the use of these systems has been in a slightly different form in categorical topology. Another link with procategories is to be found in Tozzi's paper [99] and the references therein.

Results on stability of prospaces and hence on stability within shape theory can be found in Mardešić and Segal [78]. The original articles include Dydak [31], Edwards and Geoghegan [33–35], Geoghegan [47] and Porter [88].

# Appendix

# Categorical Shape Theory and Pattern Recognition, a possible link

The basic problems of Pattern Recognition are (i) to decide perhaps by automatic means whether two images are the same (this could be called the Sorting or Classification Problem) and (ii) give some image, to decide if it is possible to associate to it some definite pattern, name, label or, as we shall say, archetype or model (i.e. a Recognition Problem).

Both images and models can be transformed. If the objects (images and models) are of geometrical nature then typically one has a large class of continuous deformations, rotations, etc. defined as well as ways of comparing disparate images (such as projections and inclusions). Allowing for composition of these transformations one arrives at two categories **B**, a category of objects of interest and **A**, a category of archetypes or models. We must have some way of comparing images with models, and this we suppose is done via a functor $\mathbf{K}: \mathbf{A} \to \mathbf{B}$.

Thus our basic data is $\mathbf{K}: \mathbf{A} \to \mathbf{B}$, which can be thought of as a basic abstract recognition system. To start with, one may know very little about **K** or for that matter **A** or **B**, so the aim of modelling pattern recognition and perception theoretic questions in terms of an arbitrary **K** is that of seeking to explore:

(i) those properties of recognition systems (e.g. their associated shape theories) which are independent of any special properties **K** may have, and
(ii) by testing the consequences of imposing structural conditions on **K**, **A** and **B**, to try to reveal more of the structures of these constructs in normal and aberrant situations (e.g. malfunctioning of the system, optical illusion, etc.).

The theory based on categorical shape theory is, of course, still fairly primitive, as it does not take into account the stochastic nature of many of the phenomena involved in the 'real-life' situation which it hopes to model. However, this very simplicity means that it should prove more accessible to workers in these areas who may then be able to develop more valid models for recognition systems, possibly involving an enriched probabilistic version of this theory.

Thus far we have discussed the interpretation of the categories **A** and **B** and the functor **K**. We next turn to Chapter 2 and the categories $B\downarrow\mathbf{K}$. Here $B$ is an object of **B**. An object of $B\downarrow\mathbf{K}$ consists of a transformation (morphism) $f\colon B\to\mathbf{K}A$ from $B$ to an object of the form $\mathbf{K}A$, together with the object $A$. Such an object should be thought of as being an approximation to $B$ by $\mathbf{K}A$. As a particular **K** may send distinct objects of **A** to the same object in **B** (i.e. $A_1\neq A_2$ but $\mathbf{K}A_1=\mathbf{K}A_2$) it is best to represent such an approximation as a pair $(f, A)$ as we have done in Chapter 2.

The definition of a morphism $k\colon(f, A)\to(f', A')$ corresponds to saying that if $B\xrightarrow{f'}\mathbf{K}A'$ can be written as a composite

$$B\xrightarrow{f}\mathbf{K}A\xrightarrow{\mathbf{K}k}\mathbf{K}A'$$

then $(f, A)$ already contains the information encoded in $(f', A')$; thus in some way $(f, A)$ is a 'finer' approximation to $B$ than is $(f', A')$. Of course it can happen that there are morphisms from $(f, A)$ to $(f', A')$ and vice versa, so 'finer' is not really a correct way of describing the relationship between $(f, A)$ and $(f', A')$.

The study of conditions on $B\downarrow\mathbf{K}$ is the main subject of Chapter 2. If **K** has an adjoint then the structure is simple; each $B\downarrow\mathbf{K}$ has an initial object, i.e. a best approximation to $B$. This latter condition can occur for specific $B$ even when **K** has no particular special properties and corresponds to the notion of stability that is studied in Chapter 6. In less technical terms, if $B$ is stable then one can solve the recognition problem for it within the theoretical limits set by the recognition system. That this is not always as useful as it may at first appear is suggested by the examples in section 2.2.

If $B\downarrow\mathbf{K}$ encodes the only information obtainable on $B$ using approximating objects from **A** and the recognition system **K**, then clearly it is comparison of $B\downarrow\mathbf{K}$ and $B'\downarrow\mathbf{K}$ which will measure to what extent the abstract recognition system can tell the difference between $B$ and $B'$. This idea is behind the interpretation of the Holsztyński model for shape. A shape morphism from $B$ to $B'$ compares the information encoded in their corresponding categories of approximations $B\downarrow\mathbf{K}$ and $B'\downarrow\mathbf{K}$.

Initiality conditions on the various $B\downarrow\mathbf{K}$ correspond to conditions on **K** itself and to solutions of embedding questions of the shape category into more easily studied categories such as **próA**. The advantage of such conditions is that they imply that each $B\downarrow\mathbf{K}$ can be handled as if it were a directed set. Thus these conditions are potentially of great theoretical significance: for instance they imply that given any two approximations, one can find a mutual refinement of them. It should be said, however, that other initiality conditions may suggest themselves when the implications of this theory for descriptions of actual phenomena have been further explored.

The other chapters are of varying applicability to this area. Chapter 3 relates mostly to the central geometric shape theoretic case study. Chapter 4 gives a set of powerful categorical tools which are generally available for any **K**. Thus the material of this chapter will be of use, in particular, if it is necessary to handle recognition systems with initiality conditions other than those considered in Chapter 2. In Chapter 5, the question of changing the models is studied. This is once again of some significance for pattern recognition, since if one uses information from previous searches to augment the stock of archetypal models, it is necessary to be able to study how the performance of the recognition system changes. This might be measured in the success of solving the recognition problem, i.e. by the size of the class of stable objects. However, the signficance of the result in Chapter 5 is that if one only adds stable objects into the models then no increase in the performance (as measured above) can be expected. (Of course, in practical computing situations, it may be that increasing the class of archetypes may change the speed of the decision process, but this cannot be studied via our model here.)

Chapter 6 explores stability and movability. The non-algebraic side of this, essentially the first two sections, would seem to be of most significance for the applications considered briefly here.

We expect that the basic categorical shape theory disussed in this monograph is able to provide part of the foundations for a mathematical metatheory of pattern recognition and perhaps also of perception. However, this expectation is partially provisional on the development of a richer theory incorporating probabilistic and stochastic considerations. Category theory in its applications is a language and as such provides a vehicle for thought processes and a means for encoding logical and philosophical ideas. It does not seem to be of the nature of categorical language to provide direct quantitative results on what it describes, but it can and frequently does provide an excellent qualitative and descriptive theory in which by the addition of more analytic ideas, one can formulate and test hypotheses which do have quantitative implications. It is this function that we envisage for the theory described in this monograph not only within the limits of applications to pattern recognition and perception, but also with regard to applications elsewhere within pure mathematics.

# References


[1] Alexandroff, P. (1928) Mengen beliebiger Dimension, *Ann. of Maths.* (2) **30**, 101–197.

[2] Alo, R. A., and Shapiro, H. (1974) Normal topological spaces, Cambridge Tracts in Maths. No. 65, Cambridge University Press.

[3] Applegate, H., and Tierney, M. (1969) Categories with models, Lecture Notes in Maths. 80, Springer-Verlag, pp. 156–244.

[4] Artin, M., and Mazur, B. (1968) Etale homotopy, Lecture Notes in Maths. 100, Springer-Verlag.

[5] Bacon, P. (1975) Continuous functors on paracompacta, *Gen. Top. and its Appl.* **5**, 321–331.

[6] Bacon, P. (1975) Axiomatic Shape Theory, *Proc. Amer. Math. Soc.* **53**, 489–496.

[7] Bénabou, J. (1967) Introduction to bicategories. Lecture Notes in Maths. 47, Springer-Verlag, pp. 1–77.

[8] Bénabou, J. (1973) Les distributeurs, Rapport 33, Inst. Math. Pure Appl., Univ. Louvain-la-Neuve.

[9] Borsuk, K. (1967) Theory of Retracts, Monografie Mat. 44, Warsaw.

[10] Borsuk, K. (1968) Concerning homotopy properties of compacta, *Fund. Math.* **62**, 223–254.

[11] Borsuk, K. (1969) On movable compacta, *Fund. Math.* **66**, 137–146.

[12] Borsuk, K. (1970) A note on the theory of shape of compacta, *Fund. Math.* **67**, 265–278.

[13] Borsuk, K. (1975) Theory of Shape, Monografie Mat. 59, Warsaw.

[14] Bourn, D., and Cordier, J.-M. (1980) Distributeurs et Théorie de la Forme, *Cahiers Top. et Géom. Diff.* **21**, 161–189.

[15] Bousfield, A. K. (1977) Constructions of factorisation systems, *J. Pure Appl. Algebra* **9**, 207–220.

[16] Bousfield, A. K., and Kan, D. M. (1972) Homotopy limits, Completions and Localisations, Lecture Notes in Maths. 304, Springer-Verlag.

[17] Christie, D. E. (1944) Net homotopy for compacta, *Trans. Amer. Math. Soc.* **56**, 275–308.

[18] Cordier, J.-M. (1984) The algebraic homotopy limit functor, U.C.N.W., Pure Maths Preprint 84.12.

[19] Cordier, J.-M. (1987) L'homologie de Steenrod–Sitnikov et limites homotopiques algébriques, *Manu. Math.* **59**, 35–52.

[20] Corider, J.-M., and Porter, T. (1978) Introduction à la théorie de la forme, *Esquisses Math.* **30**, Amiens.

[21] Cordier, J.-M., and Porter T. (1983) Functors between shape theories, *J. Pure Appl. Algebra* **27**, 1–13.

[22] Deleanu, A., and Hilton, P. J. (1968) Some remarks on general cohomology theories, *Math. Scand.* **22**, 227–240.

[23] Deleau, A., and Hilton, P. J. (1970) On the generalized Čech construction of cohomology theories, Battelle Institute Report 28, Geneva, 1969, also; *Symposia Math. Rome* **4**, 193–218.

[24] Deleanu, A., and Hilton, P. J. (1970) Remark on Čech extensions of cohomology functors, Proc. Adv. St. Inst., Aarhus, pp. 44–66.

[25] Deleanu, A., and Hilton, P. J. (1971) On Kan extensions of cohomology theories and Serre classes of groups, Battelle Institute Report 34, Geneva, 1970; also: *Fund. Math.* **73**, 143–165.

[26] Deleanu, A., and Hilton, P. J. (1976) Borsuk shape and Grothendieck categories of pro-objects, *Math. Proc. Camb. Phil. Soc.* **79**, 473–482.

[27] Deleanu, A., and Hilton, P. J. (1977) On the categorical shape of a functor, *Fund. Math.* **97**, 157–176.

[28] Deligne, P. (1973) Cohomologie à supports propres, S.G.A.4, exp. 17, in Lecture Notes in Maths. 305, Springer-Verlag.

[29] Dold, A. (1972) Lectures on Algebraic Topology, Grundlehren der Math. Wiss., Band 200, Springer-Verlag.

[30] Duskin, J. (1966) Pro-objects (after Verdier), Sém. Heidelberg–Strasbourg, Exposé 6, I.R.M.A., Strasbourg.

[31] Dydak, J. (1979) The Whitehead and Smale theorems in shape theory, *Dissertationes Math.* **156**, 1–55.

[32] Dydak, J., and Segal, J. (1978) Shape Theory, an Introduction, Lecture Notes in Maths. 688, Springer-Verlag.

[33] Edwards, D. A., and Geoghegan, R. (1975) The stability problem in shape and a Whitehead theorem in prohomotopy, *Trans. Amer. Math. Soc.* **214**, 261–277.

[34] Edwards, D. A., and Geoghegan, R. (1976) Shapes of complexes, ends of manifolds, homotopy limits and the Wall obstruction, *Ann. Math.* **101**, 521–535; Erratum, *Ann. Math.* **104**, 389.

[35] Edwards, D. A., and Geoghegan, R. (1976) Stability theorems in shape and prohomotopy, *Trans. Amer. Math. Soc.* **222**, 389–403.

[36] Edwards, D. A., and Hastings, H. M. (1976) Čech and Steenrod homotopy theories with applications to geometric topology, Lecture Notes in Maths. 542, Springer-Verlag.

[37] Eilenberg, S., and Moore J. (1962) Limits and spectral sequences, *Topology* **1**, 1–23.

[38] Eilenberg, S., and Steenrod, N. (1952) *Foundations of Algebraic Topology*, Princeton University Press.

[39] Faux, I. D., and Pratt, M. J. (1979) *Computational Geometry for Design and Manufacture*, Ellis Horwood, Chichester.

[40] Fox, R. H. (1972) On shape, *Fund. Math.* **74**, 47–71.

[41] Fox, R. H. (1974) Shape theory and covering spaces, Lecture Notes in Maths. 375, Springer-Verlag, pp. 71–90.

[42] Frei, A. (1974) On completion and shape, *Boletim Soc. Brasil Mat.* **5**, 147–159.

[43] Frei, A. (1976) On categorical shape theory, *Cahiers Top. et Géom. Diff.* **17**, 261–294.

[44] Freyd, P. J., and Kelly, M. (1972) Categories of continuous functors I, *J. Pure Appl. Algebra* **2**, 169–191; Erratum, *J. Pure Appl. Algebra* **4**, 121 (1974).

[45] Garavaglia, S. (1978) Homology with equationally compact coefficients, *Fund. Math.* **100**, 89–95.

[46] Gasson, P. C. (1983) *Geometry of Spatial Forms*, Ellis Horwood, Chichester.

[47] Geoghegan, R. (1978) Elementary proofs of stability theorems in pro-homotopy and shape,*Gen. Top. and its Appl.* **8**, 265–281.

[48] Giuli, E. (1981) Relations between reflective subcategories and shape theories, *Glasnik Mat.* **16**, (36), 205–210.

[49] Gouzou, M.-F., and Grunig, R. (1973) Caractérisation de Dist., *Comtes Rendus Acad. Sci., Paris* **276**, 519–521.

[50] Grothendieck, A. (1959) Techniques de descente et théorèmes d'existence en géométrie algébrique, II, Sém. Bourbaki, 12 année, Exp. 195, pp. 1–22.

[51] Grothendieck, A., and Verdier, J.-L. (1972) Théorie des Topos et Cohomologie Etale des Schémas, Exposé I, Sém Géom. Alg. 4, Tome I, Lecture Notes in Maths. 269, Springer-Verlag.

[52] Gruson, L., and Jensen, C. U. (1973) Modules algébriquement compacts et foncteurs $\lim^{(i)}$, *Comptes Rendus Acad. Sci., Paris* **276**, 1651–1653.

[53] Gruson, L., and Jensen, C. U. (1981) Dimensions cohomologiques reliées aux foncteurs $\lim^{(i)}$, in P. Dubreil and M.-P. Malliavin, Algebra Seminar, 33rd year (Paris 1980) Lecture Notes in Maths. 867, 234–294.

[54] Guitart, R. (1980) Relations et Carrés Exacts, *Ann. Sci. Math. Québec* **4**, 103–125.

[55] Harting, R. (1977) Distributoren und Kan-Erweiterung, *Archiv Math.* **21**, 398–405.

[56] Hilton, P. J. (1971) General cohomology theory and K-theory, London Math. Soc. Lecture Notes 1, Cambridge University Press.

[57] Hofmann, K. H. (1976) Categorical theoretical methods in topological algebra, in: Categorical Topology (eds E. Binz and H. Herrlich), Lecture Notes in Maths. 540, Springer-Verlag, pp. 343–403.

[58] Holsztyński, W. (1971) An extension and axiomatic characterisation of Borsuk's theory of shape, *Fund. Math.* **70**, 157–168.

[59] Holsztyński, W. (1971) Continuity of Borsuk's shape functor, *Bull. Acad. Polon. Sci. sér. Sci. Math. Astron. Phys.* **19**, 1105–1108.

[60] Illusie, L. (1972) Complexe cotangent et déformations II, Lecture Notes in Maths. 283, Springer-Verlag.

[61] Jensen, C. U. (1972) Les foncteurs dérivés de lim et leurs applications en théorie des modules, Lecture Notes in Maths. 254, Springer-Verlag.

[62] Kelly, M. (1969) Monomorphisms, epimorphsms and pullbacks, *J. Austr. Math. Soc.* **9**, 124–142.

[63] Kuratowsi, K. (1948) Topologie I, Monografie Mat., Warsaw.

[64] Le Van, J. H. (1973) Shape Theory, Thesis, University of Kentucky.

[65] Lee, C. N., and Raymond, F. (1968) Čech extension of contravariant functors, *Trans. Amer. Math. Soc.* **133**, 415—434.

[66] Lefschetz, S. (1942) *Algebraic Topology*, American Math. Society Publication 27.

[67] Linton, F. E. J. (1966) Some aspects of equational categories, Proc. Conf. Categorical Algebra, La Jolla, 1965, Springer-Verlag, pp. 84–95.

[68] Linton, F. E. J. (1969) An outline of functorial semantics, Lecture Notes in Maths. 80, Springer-Verlag, pp. 7–59.

[69] Lord, E. A., and Wilson, C. B. (1984) *The Mathematical Description of Shape and Form*, Ellis Horwood, Chichester.

[70] MacDonald, J. L. (1976) Natural factorisations and the Kan extension of chomology, *Cahiers Top. et Géom. Diff.* **18**, 69–93.

[71] MacLane, S. (1971) Categories for the working mathematician, Grad. Texts in Maths. 5, Springer-Verlag.

[72] Mardešić, S. (1973) Shapes for topological spaces, *Gen. Top. and its Appl.* **3**, 265–282.

[73] Mardešić, S. (1973) Strongly movable compacta and shape retracts, Proc. Intern. Sym. on Top. and its Appl. (Budva 1972), Savez Društava Mat. Fiz. Astron., Beograd, pp. 163–166.

[74] Mardešić, S. (1978) Foundations of Shape Theory, University of Kentucky, mimeographed.

[75] Mardešić, S., and Segal, J. (1970) Movable compacta and the ANR-system approach to shapes, *Bull. Acad. Polon. Sci. sér. Sci., Math., Astronom. Phys.* **18**, 649–654.

[76] Mardešić, S., and Segal, J. (1971) Shapes of Compacta and ANR-systems, *Fund. Math.* **72**, 41–59.

[77] Mardešić, S., and Segal, J. (1971) Equivalence of the Borsuk and the ANR-system approach to shapes, *Fund. Math.* **72**, 61–68.

[78] Mardešić, S., and Segal, J. (1982) Shape theory, the inverse systems approach, North-Holland Mathematical Library vol 26, North-Holland, Amsterdam.

[79] Meyer, C. V. (1980) Approximation filtrante de diagrammes finis par pro-C, *Ann. Sci. Math. Québec* **IV**, 35–57.

[80] Morita, K. (1975) On shapes of topological spaces, *Fund. Math.* **86**, 251–259.

[81] Morita, K. (1975) Čech cohomology and covering dimension for topological spaces, *Fund. Math.* **87**, 31–52.
[82] Overton, R. H. (1973) Čech homology for movable compacta, *Fund. Math.* **77**, 241–251.
[83] Pavel, M. (1983) 'Shape theory' and Pattern Recognition, *Pattern Recognition* **16**, 349–356.
[84] Popescu, N. (1973) Abelian categories with applications to rings and modules, London Math. Soc. Monographs 3, Academic Press.
[85] Porter, T. (1973) Borsuk's theory of Shape and Čech Homotopy, *Math. Scandinav.* **33**, 83–89.
[86] Porter, T. (1974) On the $T_0$-shape of the suspension of a space, *Bull. Acad. Polon. Sci. sér. Sci., Math., Astronom. Phys.* **22**, 847–850.
[87] Porter, T. (1974) Generalised shape theory, *Proc. Royal Irish Acad., Sec. A.* **74**, 33–48.
[88] Porter, T. (1974) Stability results for topological spaces, *Math. Zeit.* **140**, 1–21.
[89] Porter, T. (1976) Stability results for algebraic inverse systems, *Proc. Royal Irish Acad.* **76**, 79–83.
[90] Porter, T. (1976) Torsion theoretic results in procategories I, II, and III, *Proc. Royal Irish Acad.* **76**, 145–154, 155–164, and 165–172.
[91] Porter, T. (1978) Stability of algebraic inverse systems, I: Stability, weak stability and the weakly-stable socle, *Fund. Math.* **100**, 17–33.
[92] Porter, T. (1979) Essential properties of proobjects in Grothendieck categories, *Cahiers Top. et Géom. Diff.* **20**, 3–57.
[93] Segal, J. (1974) Movable shapes, in Lecture Notes in Maths. 375, Springer-Verlag.
[94] Segal, J. (1976) Lecture Notes on Shape Theory, University of Washington, Seattle, mimeographed.
[95] Seymour, R. (1985) Stability in pro-homotopy theory, Preprint, University College London.
[96] Spanier, E. H. (1966) *Algebraic Topology*, McGraw-Hill.
[97] Stramaccia, L. (1982) Reflective subcategories and dense subcategories, *Rend. Sem. Mat. Univ. Padova* **67**, 191–198.
[98] Thiébaud, M. (1971) Self-dual structure-semantics and algebraic categories, Thesis, Dalhousie University, Halifax, Nova Scotia.
[99] Tozzi, A. (1984) On the characterisation of (E, M)-structures in procategories, Proc. Convegno di Topologia, L'Aquila, 28–30 March 1983, supplement to *Rendi. Circ. mat. di Palermo* **4**, 133–147.
[100] Verdier, J.-L. (1965) Equivalence essentielle des systèmes projectifs, *Comptes Rendus Acad. Sci., Paris* **261**, 4950–4953.
[101] Verdier, J.-L. (1977) *Catégories dérivées, état 0, in S.G.A.* $4\frac{1}{2}$, (Lecture Notes in Maths, 569), Springer-Verlag.
[102] Weber, C. (1973) La forme d'un espace topologique est une completion, *Comptes Rendus Acad. Sci., Paris* **277**, 7–9

# Index

www.ingramcontent.com/pod-product-compliance
Lightning Source LLC
Chambersburg PA
CBHW060847280326
41934CB00007B/951
* 9 7 8 0 4 8 6 4 6 6 2 3 1 *